ANALYSIS
OF
LINEAR SYSTEMS

THIS BOOK IS IN
THE ADDISON-WESLEY SERIES
IN ELECTRICAL ENGINEERING

WILLIAM H. HUGGINS AND WILLIAM A. LYNCH
Consulting Editors

ANALYSIS
OF
LINEAR SYSTEMS

by

D A V I D   K.   C H E N G

Department of Electrical Engineering
Syracuse University

A D D I S O N - W E S L E Y   P U B L I S H I N G   C O M P A N Y ,   I N C.

READING, MASSACHUSETTS, U.S.A.

LONDON, ENGLAND

To

my parents

PREFACE

Among the myriad and ever-increasing subjects which an engineering student is required to master, few are more important than the techniques for analyzing linear systems. The study of linear systems is important for several reasons. First, a great majority of engineering situations are linear, at least within specified ranges. Second, exact solutions of the behavior of linear systems can usually be found by standard techniques. Third, the techniques remain the same irrespective of whether the problem at hand is one on electrical circuits, mechanical vibration, heat conduction, motion of elastic beams, or diffusion of liquids. Except for a very few special cases, there are no exact methods for analyzing nonlinear systems. Practical ways of solving nonlinear problems involve graphical or experimental approaches. Approximations are often necessary, and each situation usually requires special handling. Two essential steps are involved in the analysis of a physical system, namely, the formulation of mathematical equations that describe the system in accordance with physical laws, and the solution of these equations subject to appropriate initial or boundary conditions. This book attempts to furnish a thorough exposition of the techniques that are important in executing both of these steps for linear systems.

In the formulation of the equations that describe a physical system, emphasis is oriented toward electrical circuits. To deal with systems other than electrical, a chapter (Chapter 4) on analogous systems is included which treats in detail methods for drawing electrical circuits analogous to linear mechanical and electromechanical systems. This approach is advantageous because electrical engineers have developed a set of convenient symbols for circuit elements, so that a complex system can be set down with conventional symbols in the form of a circuit diagram from which the behavior of the system can be analyzed. Once the circuit diagram of the analogous electrical system is determined, it is possible to visualize and often predict important system behaviors by inspection. Moreover, electrical circuit-theory techniques, such as the use of the impedance concept and the various network theorems, can be applied in the actual analysis of the system.

One of the primary purposes of this book is to introduce the Laplace transform method of solving linear differential and integrodifferential equations. Fourier series and Fourier integral are first reviewed, and a discussion of Fourier transforms leads logically and directly to Laplace transforms. This method of introducing Laplace transformation is preferable to the unsatisfying approach of pulling the defining formulas out of thin air and applying them in a mechanical manner.

Although the Laplace transform method of solving linear differential and integrodifferential equations is, in many circumstances, simpler and more convenient to use than classical methods, I do not wish to minimize the importance of understanding the classical methods. I do not feel that the student should be led to believe that the Laplace transform method is superior to all other methods under all circumstances. First of all, there are definite limitations to the applicability of transform methods. For example, the Laplace transform method cannot conveniently be used to solve linear differential equations with variable coefficients even of the first order, while the classical approach can yield solutions to many such equations of practical importance. Second, when the known conditions of a problem are specified at values of the independent variable other than zero, the Laplace transform method becomes cumbersome to use even when the physical situation can be described by differential equations with constant coefficients. On the other hand, the application of classical methods is not modified by the way in which the known conditions are specified. Third, the separation of the general solution to a differential equation into a complementary function and a particular integral in the classical approach helps the understanding of the general nature of system response. It is not difficult to cite situations for which the complementary-function and particular-integral parts of the solutions can be written from the equations by inspection, while all steps in the formal procedure will have to be carried out in the transform method. Systems with constant or sinusoidal excitations are typical examples of such situations. A separate chapter (Chapter 2) is devoted to the discussion of classical methods for solving linear differential equations.

In applying the Laplace transform method, over-reliance on tables of transforms is discouraged. I strongly feel that a few fundamental transform pairs together with several important theorems should be remembered. It is realized that one cannot remember everything, but a good engineer or scientist should not be hopelessly ineffective without his tables or handbooks. The memory work involved is really very little. The tables of transforms in Appendix B are for reference purposes only, and students should not have to refer to them when they are learning the subject.

The complex Laplace inversion integral is derived in Chapter 6 from the Fourier integral, but evaluation of inverse Laplace transforms by contour integration in the complex plane is not attempted. This book does not include a chapter on the theory of functions of a complex variable. I am convinced that a superficial knowledge of the theory of functions of a complex variable serves no useful purpose in a book like this one. The inclusion of some introductory material on function theory may make the

level of the book appear more advanced, but it would be a rather unnecessary and unrewarding digression. Experience has indicated that in
order for a student to be able to evaluate inverse transforms of irrational
functions (functions with branch points) a much better background on
the theory of functions of a complex variable than could be offered in one
or two short chapters is necessary. The use of function theory and Cauchy's
residue theorem in connection with functions having pole singularities
results in little advantage since Heaviside's expansion theorem can be
applied with ease in these cases.   I have chosen to discuss the inverse
Laplace transforms of irrational functions in a direct manner (Sections
8–6 and 11–4) and carry the development far enough so that typical important systems with distributed parameters can be analyzed completely.
A rigorous discussion of the intricacies of the inversion integral from the
point of view of function theory is left to more advanced treatises.

The book begins with a chapter which explains in detail the characteristics of linear systems from both a physical and a mathematical viewpoint.
General properties of linear differential equations are discussed. Chapter
2 presents the essentials of classical methods for solving linear differential
equations.

Electrical circuit theory and methods of analyzing lumped-element
electrical systems are carefully presented in Chapter 3, which should be
well within the grasp of students in all branches of engineering, physics,
and applied mathematics. Chapter 4 deals with analogous systems and
discusses in detail methods for drawing electrical circuits analogous to
linear mechanical and electromechanical systems.

Chapter 5 reviews Fourier series and Fourier integral. A discussion of
Fourier transforms leads to Laplace transforms, which are introduced in
Chapter 6.

Chapter 7 illustrates the applications of Laplace transformation. Impulse response, step response, convolution and superposition integrals,
and other system concepts are discussed in Chapter 8. Inverse Laplace
transforms of some irrational functions are also derived there.

Chapter 9 treats systems with feedback where both block-diagram
and signal flow graph representations are used. System stability requirements are developed in detail.

Chapter 10 deals with sampled-data systems. z transformation is introduced and stability requirements for sampled-data systems are examined. Also included are the method of solving difference equations by
z transformation and a modified z transformation for determining the
response between sampling instants.

Chapter 11 discusses systems with distributed parameters. Two ap-
pendixes, one on numerical solution of algebraic equations and the other
containing transform tables, complete the book.

This book may be used by either advanced undergraduates or beginning graduate students. A few of the possible combinations of chapters that may serve various groups of students are suggested below:

1. Electrical engineering students with no prior knowledge of differential equations: Chapters 1, 2, 3, 5, 6, 7, 8.
2. Electrical engineering students who have had a course in ordinary differential equations: Chapters 1, 3, 4, 5, 6, 7, 8.
3. Electrical engineering graduate students: Chapters 1, 4, 6, 7, 8, 9, 10, 11.
4. Graduate students in mechanical engineering, physics, or applied mathematics: Chapters 1, 3, 4, 6, 7, 8, 9, 11.

Various other selections of material are of course possible, depending upon the nature of the course in the curriculum. The book provides enough material to prepare a student to go on to more advanced work in network theory, control systems, and vibrations.

This book was originally to be a joint project with Professor Norman Balabanian. Due to other commitments Professor Balabanian had to withdraw from this venture. I wish to thank him for a number of ideas which he contributed in the planning stage. To Professors William H. Huggins and William A. Lynch I wish to express my sincere appreciation for their many constructive suggestions. Thanks are also due Professor Richard A. Johnson, who reviewed parts of the manuscript and suggested improvements. The assistance of Mr. Mark Ma, who carefully read the galley proofs, is much appreciated. My special thanks go to my wife Enid, whose patience and understanding made the tedious book-writing task much easier to endure.

March, 1959   D. K. C.

CONTENTS

CHAPTER 1

CHARACTERISTICS OF A LINEAR SYSTEM

1-1 Introduction. The study of linear systems is important for two reasons: (1) a great majority of engineering situations are linear, at least within specified ranges; and (2) exact solutions of the behavior of linear systems can usually be found by standard techniques. Except for a very few special types, there are no standard methods for analyzing nonlinear systems. The practical ways of solving nonlinear problems involve graphical or experimental approaches. Approximations are often necessary, and each situation usually requires special handling. The present state of the art is such that there is neither a standard technique which can be used to solve nonlinear problems exactly, nor is there any assurance that a good solution can be obtained at all for a given nonlinear system. Hence, we are indeed fortunate that a great majority of engineering problems are linear and can be solved. However, we must realize that not all physical systems are linear without restrictions.

We are all familiar with the Ohm's law that governs the relation between the voltage across and the current through a resistor. It is a *linear* relationship because the voltage across a resistor is (linearly) proportional to the current through it. But even for this simple situation, the linear relationship does not apply under all conditions. For instance, as the current in a resistor is greatly increased, the value of its resistance will increase due to heat developed in the resistor, the amount of increase being dependent upon the magnitude of the current; and it is no longer correct to say that the voltage across the resistor bears a linear relationship to the current through it. The same can be said about Hooke's law, which states that the stress is (linearly) proportional to the strain of a spring. But this linear relationship breaks down when the stress on the spring is too great. When the stress exceeds the elastic limit of the material of which the spring is made, stress and strain are no longer linearly related. The actual relationship is much more complicated than the Hooke's law situation. We are therefore forewarned that restrictions always exist for linear physical situations; saturation, breakdown, or material changes will ultimately set in and destroy linearity. Under ordinary circumstances, however, physical conditions in many engineering problems stay well within the restrictions, and the linear relationship holds.

Ohm's law and Hooke's law describe only special linear systems. There exist systems that are much more complicated and so cannot be conveniently described by simple voltage-current or stress-strain relationships.

1

Other more universal criteria are necessary to establish that a system is linear. Linear systems are characterized by certain definite properties which make them simpler to describe physically and easier to solve mathematically. In the following sections, we shall examine the characteristics of a linear system from both a physical and a mathematical viewpoint.

1–2 Linear system from a physical viewpoint. An engineer's interest in a physical situation is very frequently the determination of the response of a system to a given excitation. Both the excitation and the response may be any physically measurable quantity, depending upon the particular problem. Figure 1–1 depicts such a situation. Suppose that an excitation function $e_1(t)$, which varies with time in a specified manner, produces a response function $w_1(t)$, and that a second excitation function $e_2(t)$ produces a second response function $w_2(t)$.

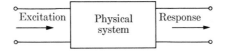

Fig. 1–1.   A physical situation.

Symbolically, we may write

$$e_1(t) \rightarrow w_1(t), \tag{1–1}$$

$$e_2(t) \rightarrow w_2(t). \tag{1–2}$$

Then, for a linear system,

$$e_1(t) + e_2(t) \rightarrow w_1(t) + w_2(t). \tag{1–3}$$

Relation (1–3), in conjunction with (1–1) and (1–2), states that a superposition of excitation functions results in a response which is the superposition of the individual response functions. Hence, from a physical point of view, we may say that *a necessary condition for a system to be linear is that the principle of superposition applies.* We note in passing that the different excitations do not have to be applied on the same part of the system.

The validity of the principle of superposition means that the presence of one excitation does not affect the responses due to other excitations; there are no interactions among responses of different excitations within a linear system. To analyze the combined effect of a number of excitations on a linear system, we can start with the analysis of the effect of each individual excitation as if the other excitations were not present, and then combine (add, or superpose) the results.

If there are n identical excitations applied to the same part of the system, that is, if

$$e_1(t) = e_2(t) = \cdots = e_n(t),\qquad(1\text{--}4)$$

then, for a linear system,

$$\sum_{k=1}^{n} e_k(t) = n e_1(t) \rightarrow \sum_{k=1}^{n} w_k(t) = n w_1(t).\qquad(1\text{--}5)$$

Comparing relation (1–5) with (1–1), we see that n appears as a scale factor (a magnitude change). Hence, *a characteristic of linear systems is that the magnitude scale factor is preserved.* This characteristic is sometimes referred to as the property of *homogeneity.*

At this point the reader must be warned that although the "derivation" of (1–5) from (1–3) seemed flawless, there are situations in which we cannot automatically assume the property of homogeneity (1–5) when the principle of superposition (1–3) holds. This may be illustrated by the following example. Let Fig. 1–2 represent a *nonlinear* system in which the filters 1 and 2 separate the input signal or excitation into two nonoverlapping spectral bands. Then if the spectrum of $e_1(t)$ falls entirely inside the passband of filter 1 and that of $e_2(t)$ falls entirely inside the passband of filter 2, relation (1–3) would be satisfied and yet the system remains nonlinear. Here, then, we have a situation where relation (1–3) does not imply relation (1–5). It is for this reason that the properties of superposition and homogeneity should be regarded as two *separate* requirements for a linear system. *A system is linear if and only if both* (1–3) *and* (1–5) *are satisfied.*

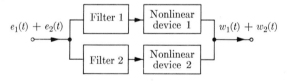

FIG. 1–2. A nonlinear system.

There is another physical aspect that characterizes a linear system with constant parameters. If the excitation function $e(t)$ applied to such a system is an alternating function of time with frequency f, then the steady-state response $w(t)$, after the initial transient has died out, appearing in any part of the system will also be alternating with the frequency f. We were aware of this fact when we solved a-c circuit problems. When a 60-cycle source is applied to a network of *fixed, linear* elements R, L, and C, the voltages and currents in all parts of the network will also be of

60-cycle frequency; no frequencies other than that of the source can exist in the network after transients have died out. In other words, *stationary (non-time-varying) linear systems create no new frequencies.* The qualification of *stationarity* implies that if

$$e(t) \rightarrow w(t) \tag{1-6}$$

then

$$e(t - \tau) \rightarrow w(t - \tau), \tag{1-7}$$

where τ is an arbitrary time delay. This qualification is to exclude situations with variable system parameters. A familiar example for such a situation is the carbon microphone circuit, in which a sinusoidal variation of the resistance in an R-L circuit will produce currents of harmonic frequencies. Another example is a linear radar system in which a moving target will cause a so-called Doppler frequency shift.

We occasionally hear the use of the terms "linear oscillators," "linear modulators," and "linear detectors." These are unfortunate choices of words. Oscillators are generators of definite frequencies, in which the only sources are d-c (zero frequency). Linear systems with constant parameters cannot do this. It is also evident that some sort of nonlinear process is there to limit the oscillation amplitude. Modulators inherently involve conversion of frequencies and are not linear devices. The term "linear detectors" is rather commonly used for large-signal detectors where the detected output follows the envelope of the modulated carrier at the input. But large-signal detectors operate under class C conditions and are basically nonlinear. They are sometimes called "linear detectors" perhaps to emphasize their difference from small-signal or square-law detectors.

The physical viewpoints that have been discussed in this section will become clearer and can all be proved after we have examined the characteristics of a linear system from a mathematical point of view. This will be done in the next section.

1-3 Linear system from a mathematical viewpoint. In mathematical language we can define linear systems as systems whose behavior is governed by linear equations, whether linear algebraic equations, linear difference equations, or linear differential equations. Let us be more specific with a typical linear differential equation, since we shall be dealing with differential equations throughout this book:

$$\frac{d^2w}{dt^2} + a_1 \frac{dw}{dt} + a_0 w = e(t). \tag{1-8}$$

In Eq. (1–8), t is used as the independent variable,* e is the excitation function, and w is the response function. Coefficients a_1 and a_0 are system parameters determined entirely by the number, type, and arrangement of the elements in the system; they may or may not be functions of the independent variable t. Since there are no partial derivatives (there is only one independent variable) in Eq. (1–8), and the highest order of the derivative is 2, Eq. (1–8) is an ordinary differential equation of the second order.† Equation (1–8) is a *linear* ordinary differential equation of the second order because neither the dependent variable w nor any of its derivatives is raised to a power greater than one and because none of its terms contains a product of two or more derivatives of the dependent variable or a product of the dependent variable and one of its derivatives.

The validity of the principle of superposition here can be verified as follows. We assume that the excitations $e_1(t)$ and $e_2(t)$ give rise to responses $w_1(t)$ and $w_2(t)$ respectively, as before. Hence

$$\frac{d^2w_1}{dt^2} + a_1 \frac{dw_1}{dt} + a_0 w_1 = e_1, \tag{1–9}$$

$$\frac{d^2w_2}{dt^2} + a_1 \frac{dw_2}{dt} + a_0 w_2 = e_2. \tag{1–10}$$

Adding Eqs. (1–9) and (1–10), we have directly

$$\frac{d^2}{dt^2}(w_1 + w_2) + a_1 \frac{d}{dt}(w_1 + w_2) + a_0(w_1 + w_2) = (e_1 + e_2). \tag{1–11}$$

---

* Although the symbol t is used here, the independent variable does *not* have to be time. It is just a mathematical symbol. What it is in a physical system depends upon the problem; it may be time, distance, angle, or some other physical quantity.

† The *degree* of a differential equation is the same as the power of the highest derivative in the equation. Hence Eq. (1–8) is of the first degree; an equation like

$$\left(\frac{d^2w}{dt^2}\right)^3 - 3\frac{d^2w}{dt^2}\frac{dw}{dt} + \left(\frac{dw}{dt}\right)^4 = 0$$

is of the third degree; and an equation like

$$w + t\frac{dw}{dt} = \sqrt{\frac{dw}{dt}},$$

which can be reduced to

$$\left(w + t\frac{dw}{dt}\right)^2 = \frac{dw}{dt}, \quad \text{or} \quad t^2\left(\frac{dw}{dt}\right)^2 + (2wt - 1)\frac{dw}{dt} + w^2 = 0,$$

is of the second degree.

Equation (1–11) states that the response of the system to an excitation $e_1(t) + e_2(t)$ is equal to the sum of the responses to the individual excitations, $w_1(t) + w_2(t)$. Note that *the principle of superposition applies and the system is linear even when the coefficients a_1 and a_0 are functions of the independent variable t.* The property of homogeneity (preservation of the magnitude scale factor) can also be easily verified.

The reader can satisfy himself in proving that the principle of superposition applies to *none* of the following equations:

$$3\frac{d^2y}{dx^2} + y\frac{dy}{dx} + 2y = 5x^2, \tag{1–12}$$

$$\frac{du}{d\theta} + u + u^2 = \sin^3\theta, \tag{1–13}$$

$$t\left(\frac{d^2v}{dt^2}\right)^2 + 5\frac{dv}{dt} + t^2v = \epsilon^{-t}. \tag{1–14}$$

Equation (1–12) is nonlinear because the second term, $y(dy/dx)$ is a product of the dependent variable and its derivative; Eq. (1–13) is nonlinear because the third term, u^2, is a second power of the dependent variable; and Eq. (1–14) is nonlinear because the first term $t(d^2v/dt^2)^2$ contains a second power of a derivative of the dependent variable. *The existence of powers or other nonlinear functions of the independent variable does not make an equation nonlinear.*

1–4 General properties of linear differential equations. There exist certain properties which are characteristic of all linear differential equations. We shall discuss these general properties here without referring to any particular physical situation but with a view toward understanding the nature of linear systems better. We shall make no attempt to solve the equations in this section.

An ordinary linear differential equation of an arbitrary order n may be written as

$$a_n(t)\frac{d^nw}{dt^n} + a_{n-1}(t)\frac{d^{n-1}w}{dt^{n-1}} + \cdots + a_1(t)\frac{dw}{dt} + a_0(t)w = e(t), \tag{1–15}$$

where the coefficients $a_n(t), a_{n-1}(t), \ldots, a_1(t), a_0(t)$ and the right member of the equation, $e(t)$, are given functions of the independent variable t, determined by the system and the excitation function respectively. The equation is said to be *homogeneous* if $e(t) = 0$, and *nonhomogeneous* if $e(t) \neq 0$.

It is convenient to employ an abbreviated symbol for the long left side of the equation. Thus, if $e(t) = 0$, we represent the homogeneous linear differential equation as follows:

$$\left[a_n(t) \frac{d^n}{dt^n} + a_{n-1}(t) \frac{d^{n-1}}{dt^{n-1}} + \cdots + a_1(t) \frac{d}{dt} + a_0(t) \right] w(t) = 0.$$

Using the abbreviation

$$L = a_n(t) \frac{d^n}{dt^n} + a_{n-1}(t) \frac{d^{n-1}}{dt^{n-1}} + \cdots + a_1(t) \frac{d}{dt} + a_0(t), \quad (1\text{-}16)$$

we write

$$L[w] = 0, \quad (1\text{-}17)$$

where L can be regarded as an operator, operating on the dependent variable w.

(A) Since multiplying the dependent variable w by a constant multiplies each term in the equation by the same constant, we have

$$L[cw] = cL[w] \quad (1\text{-}18)$$

and

$$L[cw] = 0, \quad \text{if} \quad L[w] = 0. \quad (1\text{-}19)$$

Relations (1–18) and (1–19) state that if $w(t)$ is a solution of the homogeneous equation $L[w] = 0$, then so also is $cw(t)$.

(B) Since replacing w by $w_1 + w_2$ replaces each term by the sum of two similar terms, one in w_1 and one in w_2, we have

$$L[w_1 + w_2] = L[w_1] + L[w_2] \quad (1\text{-}20)$$

and

$$L[w_1 + w_2] = 0, \quad \text{if} \quad L[w_1] = 0 \quad \text{and} \quad L[w_2] = 0. \quad (1\text{-}21)$$

Relations (1–20) and (1–21) state that if $w_1(t)$ and $w_2(t)$ are solutions of the homogeneous equation $L[w] = 0$, then so also is $w_1(t) + w_2(t)$.

By combining the results in (A) and (B), we see that *if $w_1(t)$, $w_2(t)$, ..., $w_n(t)$ are solutions of the homogeneous linear differential equation $L[w] = 0$, then so also is a linear combination of them: $c_1w_1(t) + c_2w_2(t) + \cdots + c_nw_n(t)$, where the c's are arbitrary constants.*

(C) The solution $w_c(t) = c_1w_1(t) + c_2w_2(t) + \cdots + c_nw_n(t)$ with n (the order of the original differential equation) arbitrary constants is a general solution of the homogeneous equation (1–17) provided the n individual solutions $w_1(t)$, $w_2(t)$, ..., $w_n(t)$ are linearly independent. The solutions are linearly independent if none of them can be expressed as a linear combination of the others.* This general solution of the homogeneous equation is called the *complementary function* of the nonhomogeneous equation (1–15).

(D) If $w_p(t)$ is any particular solution of the nonhomogeneous equation such that

$$L[w_p] = e(t), \tag{1-23}$$

then the sum of this particular solution (called a *particular integral*) and the complementary function

$$w(t) = w_c(t) + w_p(t)$$
$$= c_1 w_1(t) + c_2 w_2(t) + \cdots + c_n w_n(t) + w_p(t) \tag{1-24}$$

is the *general* or *complete* solution of the nonhomogeneous equation (1–15). In other words, any solution whatsoever of Eq. (1–15) can be written as a combination of the complementary function and a particular integral as

---

* The n solutions $w_1, w_2, \ldots, w_n$ are *linearly dependent* if constants $b_1, b_2, \ldots,$ b_n (which are not all zero) can be found such that

$$b_1 w_1 + b_2 w_2 + \cdots + b_n w_n = 0. \tag{1-22}$$

Hence $w_1 = \epsilon^{-(1+j2)t}$, $w_2 = \epsilon^{-(1-j2)t}$, and $w_3 = \epsilon^{-t} \sin(2t - \pi/4)$ are linearly dependent because

$$w_3 = \epsilon^{-t} (\cos \pi/4 \sin 2t - \sin \pi/4 \cos 2t) = \frac{\epsilon^{-t}}{\sqrt{2}} (\sin 2t - \cos 2t)$$

$$= \frac{\epsilon^{-t}}{\sqrt{2}} \left[\frac{1}{2j} (\epsilon^{j2t} - \epsilon^{-j2t}) - \frac{1}{2} (\epsilon^{j2t} + \epsilon^{-j2t}) \right]$$

$$= -\frac{\epsilon^{-t}}{2\sqrt{2}} [(1+j)\epsilon^{j2t} + (1-j)\epsilon^{-j2t}]$$

$$= -\frac{1}{2\sqrt{2}} [(1+j)w_2 + (1-j)w_1]$$

or

$$(1-j)w_1 + (1+j)w_2 + 2\sqrt{2}\, w_3 = 0.$$

Compared with Eq. (1–22), we have $b_1 = (1-j)$, $b_2 = (1+j)$, and $b_3 = 2\sqrt{2}$.

There is an elegant method for testing whether a set of n solutions are linearly independent. They are linearly independent if their *Wronskian* does not vanish. The Wronskian of n solutions $w_1(t), w_2(t), \ldots, w_n(t)$ is the determinant formed by these functions and their first $n - 1$ derivatives. A detailed discussion of this method is beyond the scope of this book. Interested readers are referred to E. L. Ince, *Ordinary Differential Equations*, Sec. 5.2, Dover Publications, 1944.

Eq. (1–24). For, if w is any solution of Eq. (1–15) and w_p is a particular integral, then, from Eq. (1–20), we can write

$$L[w - w_p] = L[w] - L[w_p]$$
$$= e(t) - e(t) = 0.$$

Hence $w - w_p = w_c$ is a solution of the homogeneous equation (1–17), which, by property (C), must be expressible as

$$w - w_p = c_1 w_1(t) + c_2 w_2(t) + \cdots + c_n w_n(t).$$

Transferring w_p to the right side, we obtain the solution in the form of (1–24).

(E) The n arbitrary constants $c_1, c_2, \ldots, c_n$ in the complete solution (1–24) are determined by n known values* of the response function or its derivatives for specific values of the independent variable.

Remarks (A) through (E) above apply to general linear differential equations of an arbitrary order. If all the coefficients $a_n, a_{n-1}, \ldots, a_1$, and a_0 are constants, we have a linear differential equation with constant coefficients. Linear differential equations with constant coefficients are of extreme importance because they characterize a large number of physical and engineering situations. They are of such a nature that transformation methods can be applied with advantage. They will receive our prime attention throughout this book.

1–5 Illustrative examples. A few examples are given below to illustrate the properties of linear differential equations and their solutions.

EXAMPLE 1–1. Verify that the function

$$y = c_1 \sin x + c_2 \cos x - \tfrac{1}{2} x \cos x$$

is a general solution of the linear differential equation

$$\frac{d^2 y}{dx^2} + y = \sin x. \tag{1–25}$$

Solution. Let us examine the complementary function and the particular integral separately: $y = y_c + y_p$.

---

* These are commonly referred to as *initial conditions*, but this term is inappropriate when the independent variable is not time. Even when the independent variable is time, this term does not always apply because final conditions or conditions given at any t are just as useful as initial conditions in determining the arbitrary constants.

Complementary function:

$$y_c = c_1 \sin x + c_2 \cos x. \qquad (1\text{–}26)$$

Differentiating twice, we obtain

$$\frac{d^2 y_c}{dx^2} = -c_1 \sin x - c_2 \cos x. \qquad (1\text{–}27)$$

It is obvious that

$$\frac{d^2 y_c}{dx^2} + y_c = 0, \qquad (1\text{–}28)$$

which is the homogeneous equation of Eq. (1–25). Since $\sin x$ and $\cos x$ are linearly independent of each other, it follows that y_c in (1–26) with two arbitrary constants is the complementary function of Eq. (1–25).

Particular integral:*

$$y_p = -\tfrac{1}{2}x \cos x. \qquad (1\text{–}29)$$

Differentiating twice, we obtain

$$\frac{dy_p}{dx} = -\tfrac{1}{2}(\cos x - x \sin x), \qquad (1\text{–}30)$$

$$\frac{d^2 y_p}{dx^2} = \tfrac{1}{2}(x \cos x + 2 \sin x). \qquad (1\text{–}31)$$

Adding (1–29) and (1–31), we see that y_p satisfies Eq. (1–25) and is a particular integral. Therefore

$$y = y_c + y_p = c_1 \sin x + c_2 \cos x - \tfrac{1}{2}x \cos x \qquad (1\text{–}32)$$

is a general solution of the linear differential equation (1–25).

   EXAMPLE 1–2.   Find the differential equation which has a general solution of the form

$$w = c_1 \epsilon^{-t} + c_2 \epsilon^{-2t} + 3. \qquad (1\text{–}33)$$

---

   * The particular-integral part of the general solution to a differential equation is generally considered as the solution with all arbitrary constants set to equal zero. This agreement will prevent ambiguity. For instance,

$$y = 2 \sin x + 3 \cos x - \tfrac{1}{2}x \cos x$$

is also a particular solution of Eq. (1–25), as can be proved readily by direct substitution. But when we refer to the particular integral, we agree that it is

$$y_p = y|_{c_1=0,\ c_2=0} = -\tfrac{1}{2}x \cos x$$

Solution. Since there are two arbitrary constants, we expect it to be the solution of a second-order differential equation. Differentiating it with respect to t twice, we have

$$\frac{dw}{dt} = -c_1\epsilon^{-t} - 2c_2\epsilon^{-2t} \tag{1–34}$$

and

$$\frac{d^2w}{dt^2} = c_1\epsilon^{-t} + 4c_2\epsilon^{-2t}. \tag{1–35}$$

The two constants c_1 and c_2 can be eliminated from the three expressions above by substitution. We can proceed as follows:

$$(1\text{–}34) + (1\text{–}33): \quad \frac{dw}{dt} + w = -c_2\epsilon^{-2t} + 3, \tag{1–36}$$

$$(1\text{–}35) - (1\text{–}33): \quad \frac{d^2w}{dt^2} - w = 3c_2\epsilon^{-2t} - 3, \tag{1–37}$$

$$3(1\text{–}36) + (1\text{–}37): \quad \frac{d^2w}{dt^2} + 3\frac{dw}{dt} + 2w = 6, \tag{1–38}$$

which is the required differential equation.

EXAMPLE 1–3. Assume in the above example that the values of w and its first derivative are known at $t = 0$.

$$\text{At } t = 0: \quad w = 0 \quad \text{and} \quad \frac{dw}{dt} = 5.$$

Determine the arbitrary constants c_1 and c_2 in the general solution.

Solution. Substitute the known values into Eqs. (1–33) and (1–34):

$$t = 0, \quad 0 = c_1 + c_2 + 3,$$
$$t = 0, \quad 5 = -c_1 - 2c_2.$$

Solving for c_1 and c_2, we have

$$c_1 = -1, \quad c_2 = -2.$$

Hence the complete solution of Eq. (1–38) under the given conditions is

$$w = -\epsilon^{-t} - 2\epsilon^{-2t} + 3.$$

PROBLEMS

1-1. Prove by mathematical reasoning the statement in Section 1-2 that stationary linear systems create no new frequencies.

1-2. Determine which of the following differential equations are linear and which are nonlinear:

(a) $4\dfrac{d^2w}{dx^2} = w\dfrac{dw}{dx}$

(b) $x^2\dfrac{d^3y}{dx^3} - \epsilon^{-x}\dfrac{dy}{dx} + 2y = \sin x$

(c) $\dfrac{1}{r}\dfrac{d^2r}{dt^2} + \dfrac{1}{r}\dfrac{dr}{dt} + 1 = 0$

(d) $\dfrac{du}{d\theta} = \sqrt{\theta}$

(e) $\dfrac{d^2w}{dt^2} + \dfrac{1}{w}\dfrac{dw}{dt} - 3 = 0$

(f) $\dfrac{d^2e}{dt^2} + \dfrac{1}{t}\dfrac{de}{dt} + \left(1 - \dfrac{n^2}{t^2}\right)e = 0$, n is a constant

(g) $(1 - x^2)\dfrac{d^2y}{dx^2} - 2x\dfrac{dy}{dx} + n(n + 1)y = 0$, n is a constant

1-3. Verify that $i = c_1 \sinh 2t + c_2 \cosh 2t$ is a general solution of the equation $(d^2i/dt^2) - 4i = 0$.

1-4. Suppose it is found that the functions $3 \cos 2t$, $4 \cos (2t + 1)$, and $5 \sin 2(t - \frac{1}{3})$ all satisfy the following homogeneous equation:

$$\frac{d^2w}{dt^2} + a_1 \frac{dw}{dt} + a_0 w = 0.$$

(a) Determine the general solution of the given equation. (b) Determine the coefficients a_1 and a_0.

1-5. Verify that $y = c_1x + c_2x^2 + \frac{1}{2}x^3$ is a general solution of the equation

$$x^2 \frac{d^2y}{dx^2} - 2x\frac{dy}{dx} + 2y = x^3.$$

1-6. Find the differential equation which has a general solution of the form $e = E \cos (377t + \theta)$, where E and θ are arbitrary constants.

1-7. (a) Find the differential equation which has a general solution of the form

$$y = c_1\epsilon^{-x} + c_2x\epsilon^{-x} + 2 \sin x.$$

(b) If at $x = 0$, $y = \frac{1}{2}$ and $dy/dx = 2$, determine the constants c_1 and c_2.

1-8. Determine which of the following sets of functions are linearly independent for all x, and which are linearly dependent. Give your reasons.

(a) ϵ^x, ϵ^{-x}

(b) ϵ^x, ϵ^{3x}, $\epsilon^{2x} \cosh x$

(c) $3x$, $1 - \dfrac{x}{2}$, $2(1 + x)$

(d) $3x$, $1 + x$, x^2

(e) $\sin x$, $\cos x$, $\sin 2x$

(f) $\epsilon^{(-\alpha+j\beta)x}$, $\epsilon^{(-\alpha-j\beta)x}$, $\epsilon^{-\alpha x} \cos (\beta x + \Phi)$

1–9. (a) Show that the particular integral of the following linear differential equation can only be of the form $K\epsilon^{kt}$:

$$\frac{d^2w}{dt^2} + A\frac{dw}{dt} + Bw = C\epsilon^{kt},$$

where A, B, C, and k are given constants, from which the constant K is determined.

(b) From the result of part (a), show further that the particular integral of the following linear differential equation can only be of the form $K_1 \sin \omega t + K_2 \cos \omega t$:

$$\frac{d^2w}{dt^2} + A\frac{dw}{dt} + Bw = C \sin \omega t.$$

CHAPTER 2

CLASSICAL SOLUTIONS OF LINEAR DIFFERENTIAL EQUATIONS

2–1 Introduction. In this chapter we shall review some of the classical methods of solving linear ordinary differential equations. These methods are referred to as classical methods because they do not involve the transformation of functions and operations, the use of which has become popular only relatively recently. Although one of the primary purposes of this book is to introduce the Laplace transform method of solving linear differential and integro-differential equations and to show how, in many circumstances, it is simpler and more convenient to use than classical methods, we must not minimize the importance of understanding the classical methods. First of all, there are definite limitations to the applicability of transform methods. For example, the Laplace transform method cannot conveniently be used to solve linear differential equations with variable coefficients even of the first order, while the classical approach can yield solutions to many such equations of practical importance. If the known conditions of a problem are specified at values of the independent variable other than zero, the Laplace transform method becomes cumbersome to use even when the physical situation can be described by differential equations with constant coefficients. On the other hand, the application of classical methods is not modified by the way in which the known conditions are specified. Secondly, the separation of the general solution to a differential equation into a complementary function and a particular integral in the classical approach helps the understanding of the general nature of system response. As we shall see more clearly later on, the complementary function, being the solution of the homogeneous equation (with no excitation), represents the transient response of the system and depends entirely on the type, size, and arrangement of system elements, whereas the particular integral represents the steady-state response of the system and depends not only on the system itself but also on the excitation. For many types of excitation which we encounter in engineering both the complementary-function and the particular-integral parts of the response can be obtained quite readily without having to go through the formal, and sometimes rather tedious, transformation procedures.

In this chapter we shall first discuss classical methods of solving general linear differential equations of the first order. Since no general formulas are available for the solution of equations with variable coefficients of

14

order greater than one, we shall devote the rest of the chapter to solutions of higher-order and simultaneous linear differential equations with constant coefficients. We need not be unduly dismayed by these restrictions because, fortunately, most engineering systems can be described or approximated within useful ranges by equations with constant coefficients. We shall not attempt to present an exhaustive account of all classical methods. Only a few important techniques will be discussed.

2–2 Linear equations of the first order. A general linear differential equation of the first order can be written in the following form:*

$$\frac{dw}{dt} + a(t)w = e(t). \tag{2–1}$$

Here we have used t as the independent variable. What it represents in a physical system is immaterial to the mathematical problem at hand. To solve this equation, we note that the presence of dw/dt and w in the two terms on the left side of Eq. (2–1) strongly suggests that it may be possible to arrange these two terms as the derivative of a product containing w as a factor, such as (wu), where u is as yet an unknown function of t. We know that if we could express the left side of Eq. (2–1) as an exact derivative of (wu), then w could be found by integrating the right side of the equation and dividing the result by u. But first we have to determine the function u. Now

$$\frac{d}{dt}(wu) = u\frac{dw}{dt} + \frac{du}{dt}w. \tag{2–2}$$

Comparing the right side of Eq. (2–2) with the two terms on the left side of Eq. (2–1), we are led to try multiplying the entire equation by the unknown function, u, of t, which yields

$$u\frac{dw}{dt} + ua(t)w = ue(t). \tag{2–3}$$

---

* The fact that the coefficient of dw/dt is unity in Eq. (2–1) represents no loss in generality, for if we have

$$a_1(t)\frac{dw}{dt} + a_0(t)w = e_1(t)$$

we can always divide the entire equation by the coefficient $a_1(t)$ and write

$$\frac{dw}{dt} + \frac{a_0(t)}{a_1(t)}w = \frac{e_1(t)}{a_1(t)}.$$

Equation (2–1) is obtained if we let

$$\frac{a_0(t)}{a_1(t)} = a(t) \qquad \text{and} \qquad \frac{e_1(t)}{a_1(t)} = e(t).$$

Hence for the left side of Eq. (2–3) to be an exact derivative of the product (wu), we require

$$\frac{du}{dt} = ua(t)$$

or, separating the variables,

$$\frac{du}{u} = a(t)\, dt.$$

Integrating, we obtain

$$\ln u = \int a(t)\, dt$$

or

$$u = \epsilon^{\int a(t)\, dt}. \tag{2–4}*$$

Since the coefficient $a(t)$ is a given function, u can be found from Eq. (2–4) as a function of t. Substituting back into Eq. (2–3), we write

$$\left[\frac{dw}{dt} + a(t)w\right] \epsilon^{\int a(t)\, dt} = e(t)\epsilon^{\int a(t)\, dt},$$

$$\frac{d}{dt}[w\epsilon^{\int a(t)\, dt}] = e(t)\epsilon^{\int a(t)\, dt}.$$

Integrating both sides and rearranging terms, we have finally

$$w = \epsilon^{-\int a(t)\, dt}\left[\int e(t)\epsilon^{\int a(t)\, dt}\, dt + c\right], \tag{2–5}$$

which is the required solution for Eq. (2–1). The function u as given in Eq. (2–4), introduced to make the left side of the original equation an exact derivative (so that it can be integrated), is called an *integrating factor*.

Recalling the general properties of the solutions of a linear differential equation from Section 1–4, we note that the solution in Eq. (2–5) can actually be separated into two parts:

$$w = w_c + w_p, \tag{2–6}$$

where

$$w_c = \frac{c}{u}, \qquad u = \epsilon^{\int a(t)\, dt} \tag{2–7}$$

---

* The reader need not be concerned with the integration constant which accompanies the indefinite integral of $a(t)$ for, as will be apparent from Eq. (2–5), the integration constant will be cancelled, and has no effect on the final solution of the differential equation.

is the complementary function which satisfies the homogeneous equation with $e(t) = 0$, and

$$w_p = \frac{1}{u} \int u e(t)\, dt, \qquad u = \epsilon^{\int a(t)\, dt} \tag{2-8}$$

is the particular integral which depends upon the excitation function $e(t)$.

EXAMPLE 2–1. Solve the following linear first-order differential equation:

$$\frac{dw}{dt} - w = \epsilon^t.$$

Solution. Here the integrating factor is

$$u = \epsilon^{\int(-1)\, dt} = \epsilon^{-t}.$$

Complementary function:

$$w_c = \frac{c}{u} = c\epsilon^t.$$

Particular integral:

$$w_p = \frac{1}{u} \int u e(t)\, dt = \epsilon^t \int \epsilon^{-t} \cdot \epsilon^t\, dt = t\epsilon^t.$$

Hence the general solution is

$$w = w_c + w_p = \epsilon^t(c + t).$$

EXAMPLE 2–2. Solve the following linear first-order differential equation:

$$(\cos x)\frac{dy}{dx} + (\sin x)\, y = 1.$$

Solution. Dividing the given equation by $\cos x$, we have

$$\frac{dy}{dx} + (\tan x)\, y = \sec x.$$

Integrating factor:

$$u = \epsilon^{\int \tan x\, dx} = \epsilon^{\ln \sec x} = \sec x.^*$$

Complementary function:

$$y_c = \frac{c}{u} = c \cos x$$

Particular integral:

$$y_p = \frac{1}{u} \int u\, (\sec x)\, dx = \cos x \int \sec^2 x\, dx = \sin x.$$

Hence the general solution is

$$y = y_c + y_p = c \cos x + \sin x.$$

---

* The relation $\epsilon^{\ln z} = z$ holds for any z. This can be seen quite easily by taking the natural logarithm of both sides.

2–3 Higher-order linear equations with constant coefficients. As we have indicated in Section 2–1, no general formulas are available for the solution of linear differential equations with variable coefficients of order greater than one. Hence we are unable to extend the general method of the preceding section to equations of a higher order. Fortunately, a great many engineering systems can be described by equations with constant coefficients, and general methods of solving higher-order linear differential equations with constant coefficients do exist. In this section we shall present a method which makes use of an operator notation and which can be used to solve almost all problems of practical importance.

We first introduce a symbol D to denote the differentiating operation with respect to the independent variable. If t is used as the independent variable, then symbolically we have

$$D \equiv \frac{d}{dt}. \tag{2–9}$$

Note that D is not simply a function; it is an *operator*. When D is placed to the left of a function, it implies that this function is to be differentiated once with respect to the independent variable t. *D must have something to operate on.* Hence

$$Dw = \frac{dw}{dt} \tag{2–10}$$

and

$$D^n w = \frac{d^n w}{dt^n}. \tag{2–11}$$

But *wD and wD^n mean nothing at all!* Although D is not an algebraic quantity, an operation which contains more than one D can be rearranged or regrouped in accordance with the fundamental laws of algebra so long as the relative positions of the operators and the functions to be operated on are not interchanged. Thus

(A) For addition:

    The commutative law: $(D^m + D^n)w = D^m w + D^n w.$ (2–12)

    The associative law: $[D^l + (D^m + D^n)]w = [(D^l + D^m) + D^n]w.$

$$\tag{2–13}$$

(B) For multiplication:

    The commutative law: $(D^m \cdot D^n)w = (D^n \cdot D^m)w = D^{m+n}w.$ (2–14)

    The associative law: $D^l(D^m \cdot D^n)w = (D^l \cdot D^m)D^n w.$ (2–15)

    The distributive law: $D^l(D^m + D^n)w = (D^{l+m} + D^{l+n})w.$ (2–16)

Before we apply the operator notation to an equation of an arbitrary order, let us apply it to a first-order equation with constant coefficients:

$$\frac{dw}{dt} + aw = e(t). \tag{2-17}$$

Equation (2–17) is the same as Eq. (2–1) except that now the coefficient is a constant (independent of t). With operator notation (2–10), we write Eq. (2–17) as

$$(D + a)w = e(t). \tag{2-18}$$

We must always remember that *equations containing D, such as Eq. (2–18), are not algebraic equations.* Now, what is the general solution of Eq. (2–18)? We have the answer to this question because we have just obtained the solution in the preceding section. We know that the appropriate integrating factor is

$$u = \epsilon^{\int a\,dt} = \epsilon^{at} \tag{2-19}$$

and that the general solution consists of two parts:

(1) *Complementary function* [which is the general solution of the homogeneous equation $(D + a)w_c = 0$]:

$$w_c = \frac{c}{u} = c\epsilon^{-at}. \tag{2-20}$$

(2) *Particular integral* [which is a particular solution of the nonhomogeneous equation $(D + a)w_p = e(t)$]:

$$w_p = \frac{1}{D + a}\, e(t). \tag{2-21}$$

By virtue of Eq. (2–8),

$$w_p = \frac{1}{u} \int ue(t)\, dt = \epsilon^{-at} \int \epsilon^{at} e(t)\, dt. \tag{2-22}$$

Thus, the operation of $1/(D + a)$ on a function $e(t)$ in Eq. (2–21) can be considered as a shorthand way of writing the right side of Eq. (2–22).

The general solution of Eq. (2–18) is then

$$w = w_c + w_p = \epsilon^{-at} \left[\int \epsilon^{at} e(t)\, dt + c \right]. \tag{2-23}$$

Next, consider an equation of the second order:

$$a_2 \frac{d^2w}{dt^2} + a_1 \frac{dw}{dt} + a_0 w = e_1(t), \tag{2-24}$$

where a_2, a_1, a_0 are constants and $e_1(t)$ is an arbitrary excitation function.

Dividing Eq. (2–24) by a_2 and writing it in operator notation, we have

$$(D^2 + b_1 D + b_0)w = \frac{1}{a_2} e_1(t) = e(t), \qquad (2\text{–}25)$$

where b_1 is written for a_1/a_2 and b_0 for a_0/a_2. To find the general solution of Eq. (2–25), we consider the complementary function and the particular integral separately.

(1) *Complementary function, w_c.* The homogeneous equation to be solved is

$$(D^2 + b_1 D + b_0)w_c = 0, \qquad (2\text{–}26)$$

which can be factored to give

$$(D - s_1)(D - s_2)w_c = 0, \qquad (2\text{–}27)$$

where*

$$s_1 = -\tfrac{1}{2}(b_1 + \sqrt{b_1^2 - 4b_0}), \qquad (2\text{–}28)$$

$$s_2 = -\tfrac{1}{2}(b_1 - \sqrt{b_1^2 - 4b_0}). \qquad (2\text{–}29)$$

The solution of either the equation

$$(D - s_1)w_{c_1} = 0 \qquad (2\text{–}30)$$

or the equation

$$(D - s_2)w_{c_2} = 0 \qquad (2\text{–}31)$$

will then be a solution of Eq. (2–27). Equations (2–30) and (2–31) are both first-order homogeneous linear equations and their solutions (similar to Eq. 2–20) are, respectively,

$$w_{c_1} = c_1 \epsilon^{s_1 t} \qquad (2\text{–}32)$$

and

$$w_{c_2} = c_2 \epsilon^{s_2 t}. \qquad (2\text{–}33)$$

Hence the complementary function of Eq. (2–25) is

$$w_c = w_{c_1} + w_{c_2} = c_1 \epsilon^{s_1 t} + c_2 \epsilon^{s_2 t}, \qquad (2\text{–}34)$$

which correctly contains two arbitrary constants c_1 and c_2.

Another way of arriving at Eq. (2–34) is to start with a trial solution of the type ϵ^{st} for the homogeneous equation (2–26). The reason for choosing an exponential solution of this type is that all derivatives of the exponential function contain the function itself, which can then be factored out,

---

* We assume here that $s_1 \neq s_2$. The case of multiple roots will be discussed in Section 2–5.

leaving a purely algebraic equation. Thus, substituting ϵ^{st} for w_c in Eq. (2–26), we have

$$(s^2 + b_1 s + b_0)\epsilon^{st} = 0. \tag{2–35}$$

Inasmuch as ϵ^{st} does not vanish, Eq. (2–35) can be satisfied only if

$$s^2 + b_1 s + b_0 = 0. \tag{2–36}$$

Equation (2–36) is called the *characteristic* or *auxiliary equation* of either Eq. (2–25) or Eq. (2–26). It is an algebraic equation obtained by replacing D by s in the terms within the parentheses on the left side of Eq. (2–25) or Eq. (2–26). Equation (2–36) has two roots, namely,

$$s = s_1 \quad \text{and} \quad s = s_2$$

if we use the notations in Eqs. (2–28) and (2–29). This tells us that there exist two independent solutions, $\epsilon^{s_1 t}$ and $\epsilon^{s_2 t}$ for the homogeneous equation (2–26). By property (C) stated in Section 1–4, the complementary function is then of the form shown in Eq. (2–34).

(2) *Particular integral, w_p.* The nonhomogeneous equation to be solved is

$$(D^2 + b_1 D + b_0)w_p = e(t). \tag{2–37}$$

We factor the operators as we did in Eq. (2–27) to give

$$(D - s_1)(D - s_2)w_p = e(t), \tag{2–38}$$

where s_1 and s_2 take the values given in Eqs. (2–28) and (2–29) respectively. The solution of Eq. (2–38) can be written in operator form as follows:

$$w_p = \frac{1}{(D - s_1)(D - s_2)} e(t). \tag{2–39}$$

We emphasize here again that Eq. (2–39) is not an algebraic equation and its right side does not represent a simple division. We can determine w_p from Eq. (2–39) by the *method of successive integrations.*

We write the two factors that appear on the right side of Eq. (2–39) separately and carry out the operations successively, one at a time. Thus,

$$w_p = \frac{1}{D - s_1}\left[\frac{1}{D - s_2} e(t)\right]. \tag{2–40}$$

From Eqs. (2–21) and (2–22) we can write the result of the operation contained in the brackets of Eq. (2–40) immediately:

$$\frac{1}{D - s_2} e(t) = \epsilon^{s_2 t}\int \epsilon^{-s_2 t} e(t)\, dt.$$

Substituting this back into Eq. (2–40) and performing the remaining operation, we have

$$
\begin{aligned}
w_p &= \frac{1}{D - s_1} \left[\epsilon^{s_2 t} \int \epsilon^{-s_2 t} e(t) \, dt \right] \\
&= \epsilon^{s_1 t} \int \epsilon^{-s_1 t} \left[\epsilon^{s_2 t} \int \epsilon^{-s_2 t} e(t) \, dt \right] dt \\
&= \epsilon^{s_1 t} \int \epsilon^{(s_2 - s_1) t} \int \epsilon^{-s_2 t} e(t) \, (dt)^2.
\end{aligned}
\tag{2–41}
$$

The general solution of the original second-order equation (2–24) is then the sum of the complementary function as given in Eq. (2–34) and the particular integral as given in Eq. (2–41).

In general, if the roots of the characteristic or auxiliary equation of an nth-order differential equation are $s_1, s_2, s_3, \ldots, s_n$, then

$$
\begin{aligned}
w_p &= \frac{1}{(D - s_1)(D - s_2)(D - s_3) \cdots (D - s_n)} e(t) \\
&= \epsilon^{s_1 t} \int \epsilon^{(s_2 - s_1) t} \int \epsilon^{(s_3 - s_2) t} \int \cdots \int \epsilon^{(s_n - s_{n-1}) t} \int \epsilon^{-s_n t} e(t) \, (dt)^n.
\end{aligned}
\tag{2–42}
$$

EXAMPLE 2–3. Find the solution for the following second-order linear differential equation:

$$
2 \frac{d^2 w}{dt^2} + 3 \frac{dw}{dt} + w = 10 \epsilon^{-3t},
\tag{2–43}
$$

for which $w = 1$ and $dw/dt = 0$ at $t = 0$.

Solution. First let us divide the given equation by 2 and rewrite it in operator notation as

$$
(D^2 + \tfrac{3}{2} D + \tfrac{1}{2}) w = 5 \epsilon^{-3t}.
\tag{2–44}
$$

Factoring the left side, we have

$$
(D + 1)(D + \tfrac{1}{2}) w = 5 \epsilon^{-3t}.
\tag{2–45}
$$

(1) *Complementary function.* Since $s_1 = -1$, $s_2 = -\tfrac{1}{2}$, we can write the complementary function directly from Eq. (2–34):

$$
w_c = c_1 \epsilon^{-t} + c_2 \epsilon^{-t/2}.
\tag{2–46}
$$

(2) *Particular integral.* Direct substitution of the values of s_1, s_2, and $e(t)$ into Eq. (2–41) yields

$$
w_p = \epsilon^{-t} \int \epsilon^{t/2} \int \epsilon^{t/2} (5 \epsilon^{-3t}) \, (dt)^2 = \epsilon^{-3t}.
\tag{2–47}
$$

Note that the exponential form of the particular integral is the same as that of the excitation function. This could have been predicted from Problem 1–9. The correctness of the particular integral can always be checked by substituting it back into the given differential equation. The general solution of the given equation is

$$w = w_c + w_p = c_1 \epsilon^{-t} + c_2 \epsilon^{-t/2} + \epsilon^{-3t}. \tag{2-48}$$

Now we apply the two given initial conditions to determine the constants c_1 and c_2:

$$\text{At } t = 0: \quad w = c_1 + c_2 + 1 = 1,$$

$$\frac{dw}{dt} = -c_1 - \tfrac{1}{2}c_2 - 3 = 0,$$

from which we find

$$c_1 = -6 \quad \text{and} \quad c_2 = 6.$$

Hence the desired solution is

$$w = -6\epsilon^{-t} + 6\epsilon^{-t/2} + \epsilon^{-3t}. \tag{2-49}$$

EXAMPLE 2–4. Find the general solution for the following third-order linear differential equation:

$$\frac{d^3y}{dx^3} + 4\frac{d^2y}{dx^2} + 5\frac{dy}{dx} = x. \tag{2-50}$$

Solution. Writing Eq. (2–50) in operator notation, we have

$$(D^3 + 4D^2 + 5D)y = x. \tag{2-51}$$

The characteristic or auxiliary equation is obtained by substituting s for D in the operator part of the homogeneous equation:

$$s^3 + 4s^2 + 5s = 0. \tag{2-52}$$

This third-degree algebraic equation has three roots and can be factored as

$$(s - s_1)(s - s_2)(s - s_3) = 0, \tag{2-53}$$

where

$$s_1 = 0, \quad s_2 = -(2 + j), \quad s_3 = -(2 - j) = s_2^*. \tag{2-54}$$

Note that the roots of Eq. (2–52) or Eq. (2–53) are 0, s_2, and s_3, and that s_2 and s_3 are complex conjugates of each other. *It is a property of algebraic equations with real coefficients that complex roots always occur in conjugate pairs.*

(1) *Complementary function.* With the roots of the auxiliary equation known, we can write the complementary function in the standard exponential form

$$y_c = c_1 \epsilon^{s_1 x} + c_2 \epsilon^{s_2 x} + c_3 \epsilon^{s_3 x}$$
$$= c_1 + c_2 \epsilon^{-(2+j)x} + c_3 \epsilon^{-(2--j)x}. \qquad (2\text{--}55)$$

But Eq. (2–55) can be put in a better form by combining the terms with conjugate imaginary exponents:

$$y_c = c_1 + \epsilon^{-2x}(c_2 \epsilon^{-jx} + c_3 \epsilon^{jx}). \qquad (2\text{--}56)$$

Recalling the relationship

$$\epsilon^{\pm jx} = \cos x \pm j \sin x, \qquad (2\text{--}57)$$

we write Eq. (2–56) as

$$y_c = c_1 + \epsilon^{-2x}[(c_2 + c_3) \cos x + j(c_3 - c_2) \sin x]. \qquad (2\text{--}58)$$

For the sake of simplicity, we call

$$A = c_2 + c_3, \qquad (2\text{--}59)$$

$$B = j(c_3 - c_2). \qquad (2\text{--}60)$$

Hence an equivalent form of the complementary function is

$$y_c = c_1 + \epsilon^{-2x}(A \cos x + B \sin x), \qquad (2\text{--}61)$$

which contains the three arbitrary constants c_1, A, and B.

At this point we can make a generalization for the form of the complementary function $y_c(x)$: *each distinct* real root s_k (including zero) of the characteristic or auxiliary equation gives rise to a $c_k \epsilon^{s_k x}$ term, and each distinct pair of complex conjugate roots $-\alpha \pm j\beta$ (including purely imaginary conjugate roots when $\alpha = 0$) yields two terms which can be written as* $\epsilon^{-\alpha x}(A \cos \beta x + B \sin \beta x)$.

(2) *Particular integral.*

$$y_p = \frac{1}{D(D + 2 + j)(D + 2 - j)} x. \qquad (2\text{--}62)$$

---

* A root of an equation is said to be *distinct* when it is not equal to any other root of the equation. Special treatment is needed when the characteristic or auxiliary equation of a differential equation has multiple or repeated roots (see Section 2–5).

Applying Eq. (2–42), we have

$$y_p = \int \epsilon^{-(2+j)x} \int \epsilon^{j2x} \int \epsilon^{(2-j)x} x \, (dx)^3 = \frac{x}{5} \left(\frac{x}{2} - \frac{4}{5} \right). \qquad (2\text{–}63)$$

That this particular integral is a solution of the given equation (2–50) or (2–51) can be readily checked by direct substitution. The general solution is the sum of the complementary function y_c, as found in Eq. (2–61), and this particular integral y_p:

$$y = y_c + y_p = c_1 + \epsilon^{-2x} (A \cos x + B \sin x) + \frac{x}{5} \left(\frac{x}{2} - \frac{4}{5} \right). \qquad (2\text{–}64)$$

The preceding discussions pertaining to second-order and third-order equations can be readily extended to linear differential equations with *constant* coefficients of an arbitrary order n:

$$a_n \frac{d^n w}{dt^n} + a_{n-1} \frac{d^{n-1} w}{dt^{n-1}} + \cdots + a_1 \frac{dw}{dt} + a_0 w = e_1(t), \qquad (2\text{–}65)$$

where $a_n \neq 0$. Dividing the entire equation by a_n and writing it in operator notation, we have

$$(D^n + b_{n-1}D^{n-1} + \cdots + b_1 D + b_0)w = \frac{1}{a_n} e_1(t) = e(t), \qquad (2\text{–}66)$$

where b_{n-1} is written for $a_{n-1}/a_n \ldots$, b_1 for a_1/a_n, and b_0 for a_0/a_n.

The general solution of Eq. (2–65) or (2–66) can be found systematically in the following manner:

(1) We form the characteristic or auxiliary equation by replacing D's by s's in the operator part of the given differential equation and setting it equal to zero:

$$s^n + b_{n-1}s^{n-1} + \cdots + b_1 s + b_0 = 0. \qquad (2\text{–}67)$$

(2) We then determine the n roots of the nth-degree algebraic equation (2–67),* and call these roots $s_1, s_2, s_3, \ldots, s_n$.

(3) If the n roots are all *distinct* (unequal), then the complementary function of Eq. (2–65) or (2–66) is

$$w_c = c_1 \epsilon^{s_1 t} + c_2 \epsilon^{s_2 t} + \cdots + c_n \epsilon^{s_n t} = \sum_{k=1}^{n} c_k \epsilon^{s_k t}. \qquad (2\text{–}68)$$

---

* When n is large, it may be difficult to determine the roots of Eq. (2–67). No formulas are available for finding the exact roots of algebraic equations of a degree higher than the fourth. In such cases, we must resort to graphical or numerical methods for determining the roots approximately. Appendix A discusses numerical solution of algebraic equations.

(4) The particular integral can be found by the method of successive integrations from an integral of the nth order:

$$w_p = \frac{1}{(D - s_1)(D - s_2) \cdots (D - s_n)} e(t)$$

$$= \epsilon^{s_1 t} \int \epsilon^{(s_2 - s_1)t} \int \epsilon^{(s_3 - s_2)t} \int \cdots \int \epsilon^{(s_n - s_{n-1})t} \int \epsilon^{-s_n t} e(t) \, (dt)^n. \quad (2\text{--}69)$$

(5) The general solution of the given equation is the sum of the complementary function w_c in (2–68) and the particular integral w_p in (2–69).

The expression for the complementary function, Eq. (2–68), must be modified in accordance with Section 2–5 when the characteristic or auxiliary equation has multiple or repeated roots. The above procedure also assumes that no terms in w_p duplicate any of the exponential terms in w_c. If this assumption does not hold, the general solution should be modified as though multiple roots existed (see Example 2–9).

2–4 Method of undetermined coefficients. In the preceding section we have described the method of successive integrations for determining a particular integral. Other methods are available for determining particular integrals, among which the method of undetermined coefficients is particularly simple and easy to apply. This method has an advantage over others in that it involves only differentiations and no integrations are needed. As we well know, differentiations are always easier than integrations. The method of undetermined coefficients does not apply to all types of the excitation function $e(t)$, but it is useful when $e(t)$ is composed of functions of the following types:

A constant, K.
A power of the independent variable, t^k (k a positive integer).
An exponential function, $\epsilon^{\gamma t}$.
A cosine function, $\cos \gamma t$
A sine function, $\sin \gamma t$.

It so happens that a majority of excitation functions we encounter in practice can be represented in terms of the functions listed above. Let us consider these functions separately. For simplicity we shall describe this method with a second-order equation:

$$\frac{d^2 w}{dt^2} + b_1 \frac{dw}{dt} + b_0 w = e(t). \quad (2\text{--}70)$$

However, the following development will hold for linear differential equations with constant coefficients of any order.

I. $e(t) = K$ (*a constant*). The differential equation (2–70) becomes

$$(D^2 + b_1 D + b_0)w = K. \qquad (2\text{–}71)$$

It is obvious that the particular integral is

$$w_p = A \text{ (a constant)}. \qquad (2\text{–}72)$$

Since $D^2 A = 0$ and $DA = 0$, we can determine A immediately by substituting it into Eq. (2–71):

$$b_0 A = K, \qquad w_p = A = \frac{K}{b_0}. \qquad (2\text{–}73)$$

II. $e(t) = t^k$ (*k a positive integer*).*

$$(D^2 + b_1 D + b_0)w_p = t^k. \qquad (2\text{–}74)$$

Here it is reasonable to assume that w_p will be a polynomial in t, because a positive integral power of t such as t^k can be obtained as the derivative of only another power of t. Moreover, if $b_0 \neq 0$, there can be no terms in w_p that are of a power higher than k. Let us then assume that

$$w_p = A_0 + A_1 t + A_2 t^2 + \cdots + A_k t^k. \qquad (2\text{–}75)$$

Hence

$$b_0 w_p = b_0 A_0 + b_0 A_1 t + b_0 A_2 t^2 + \cdots + b_0 A_k t^k, \qquad (2\text{–}76)$$

$$b_1 D w_p = b_1 A_1 + 2b_1 A_2 t + 3b_1 A_3 t^2 + \cdots + kb_1 A_k t^{k-1}, \qquad (2\text{–}77)$$

$$D^2 w_p = 2A_2 + 6A_3 t + \cdots + k(k-1)A_k t^{k-2}. \qquad (2\text{–}78)$$

Adding Eqs. (2–76), (2–77), and (2–78), and equating the result to t^k, we have

$$
\begin{aligned}
(D^2 + b_1 D + b_0)w_p = {} & (b_0 A_0 + b_1 A_1 + 2A_2) \\
& + (b_0 A_1 + 2b_1 A_2 + 6A_3)t + \cdots \\
& + (b_0 A_{k-1} + kb_1 A_k)t^{k-1} + b_0 A_k t^k = t^k.
\end{aligned}
$$

Equating the coefficients of the like powers in t, we obtain $(k + 1)$ equations:

$$
\begin{aligned}
b_0 A_0 \;&+\; b_1 A_1 + 2A_2 = 0, \\
b_0 A_1 \;&+\; 2b_1 A_2 + 6A_3 = 0, \\
&\;\vdots \\
b_0 A_{k-1} \;&+\; kb_1 A_k = 0, \\
b_0 A_k &= 1,
\end{aligned}
$$

---

* We note that case I, where $e(t)$ equals a constant, is a special case with $k = 0$.

from which the $(k + 1)$ coefficients, A_0, A_1, A_2, $\ldots$, A_{k-1}, and A_k in the assumed solution (2–75) can be determined.

EXAMPLE 2–5. Find the general solution of the equation

$$\frac{d^2w}{dt^2} - w = 2t^2. \tag{2–79}$$

Solution. Here we have

$$(D^2 - 1)w = 2t^2,$$

$$(D + 1)(D - 1)w = 2t^2.$$

Complementary function:

$$w_c = c_1\epsilon^{-t} + c_2\epsilon^t. \tag{2–80}$$

Particular integral: Since the excitation function has a second power in t, we assume w_p to be a polynomial in t of the second degree:

$$w_p = A_0 + A_1t + A_2t^2, \tag{2–81}$$

$$D^2w_p = 2A_2. \tag{2–82}$$

Subtracting Eq. (2–81) from Eq. (2–82) and equating the difference to $2t^2$, we have

$$(D^2 - 1)w_p = (2A_2 - A_0) - A_1t - A_2t^2 = 2t^2.$$

We can now write three simultaneous algebraic equations relating the coefficients:

$$2A_2 - A_0 = 0, \qquad -A_1 = 0, \qquad -A_2 = 2. \tag{2–83}$$

Hence $A_0 = -4$, $A_1 = 0$, $A_2 = -2$, and

$$w_p = -2(2 + t^2). \tag{2–84}$$

The general solution is obtained by adding the results in Eqs. (2–80) and (2–84):

$$w = w_c + w_p = c_1\epsilon^{-t} + c_2\epsilon^t - 2(2 + t^2). \tag{2–85}$$

Note that we did not need to evaluate any integrals in obtaining the solution.

EXAMPLE 2–6. Find the general solution of the equation

$$\frac{d^3w}{dt^3} - \frac{dw}{dt} = 2t^2. \tag{2–86}$$

Solution. Writing the given equation in operator notation, we have

$$D(D^2 - 1)w = 2t^2. \tag{2-87}$$

The characteristic equation has three distinct roots:

$$s_1 = 0, \qquad s_2 = +1, \qquad s_3 = -1.$$

Hence the complementary function is

$$w_c = c_1 + c_2\epsilon^t + c_3\epsilon^{-t}. \tag{2-88}$$

We note the similarity between the given equation and Eq. (2-79) in Example 2-5. Although the excitation function now is $2t^2$, as before, an assumed w_p in accordance with Eq. (2-81),

$$w_p = A_0 + A_1 t + A_2 t^2,$$

will not work, because $D^3 w_p = 0$ and

$$D(D^2 - 1)w_p = -(A_1 + 2A_2 t) = 2t^2$$

cannot be satisfied for any constant A_0, A_1, and A_2.

We can look at the given equation (2-86) or (2-87) as

$$(D^2 - 1)w = \int 2t^2 \, dt = \tfrac{2}{3}t^3 + K. \tag{2-89}$$

The right side of Eq. (2-89) is a polynomial in t of the third degree ($k = 3$). For a w_p to satisfy Eq. (2-89) we should assume

$$w_p = A_0 + A_1 t + A_2 t^2 + A_3 t^3. \tag{2-90}$$

Substituting the above w_p in the given equation, we have

$$D(D^2 - 1)w_p = 6A_3 - (A_1 + 2A_2 t + 3A_3 t^2) = 2t^2.$$

Three simultaneous equations are obtained by equating the corresponding coefficients:

$$6A_3 - A_1 = 0, \qquad -2A_2 = 0, \qquad -3A_3 = 2,$$

from which we find

$$A_1 = -4, \qquad A_2 = 0, \qquad A_3 = -\tfrac{2}{3}.$$

The value of A_0 is immaterial [since it will eventually be absorbed in the arbitrary constant c_1 in the complementary function w_c in **Eq. (2-88)**].

We set $A_0 = 0$. We have finally

$$w_p = -4t - \tfrac{2}{3}t^3, \tag{2-91}$$

and the complete solution is

$$w = w_c + w_p = c_1 + c_2\epsilon^t + c_3\epsilon^{-t} - 2(2 + \tfrac{1}{3}t^2)t. \tag{2-92}$$

We see that the particular integral in Eq. (2–91) is the integral of that in Eq. (2–84) found in Example 2–5. This enables us to make the following general statement: *when the characteristic equation of a differential equation has a zero root, the assumed form of the particular integral should be the integral of that which would normally be used in the method of undetermined coefficients.* This statement can be extended to include cases where the characteristic equation possesses multiple zero roots.

III. $e(t) = \epsilon^{\gamma t}$.

$$(D^2 + b_1 D + b_0)w_p = \epsilon^{\gamma t}. \tag{2-93}$$

Since the derivatives of an exponential function $\epsilon^{\gamma t}$ will all contain the same function $\epsilon^{\gamma t}$, it is certainly reasonable to choose

$$w_p = A\epsilon^{\gamma t}. \tag{2-94}*$$

Substituting w_p in Eq. (2–93) and cancelling the factor $\epsilon^{\gamma t}$ on both sides, we have

$$A(\gamma^2 + b_1\gamma + b_0) = 1.$$

Hence

$$A = \frac{1}{\gamma^2 + b_1\gamma + b_0} \tag{2-95}$$

and

$$w_p = \frac{\epsilon^{\gamma t}}{\gamma^2 + b_1\gamma + b_0}. \tag{2-96}$$

This method can be applied to find the particular integral for Eq. (2–43) or Eq. (2–44) in Example 2–3, where the given equation was

$$(D^2 + \tfrac{3}{2}D + \tfrac{1}{2})w = 5\epsilon^{-3t}.$$

Here, in view of Eq. (2–96), we obtain directly, without integration,

$$w_p = \frac{5\epsilon^{-3t}}{(-3)^2 + \tfrac{3}{2}(-3) + \tfrac{1}{2}} = \epsilon^{-3t},$$

as in Eq. (2–47).

---

* We note here that case I is also a special case of III with $\gamma = 0$.

IV. $e(t) = \cos \gamma t$ or $e(t) = \sin \gamma t$.

It is convenient to examine the cosine and sine functions in terms of their exponential components, since we now know how to handle an excitation function of the exponential type:

$$\cos \gamma t = \frac{1}{2} (\epsilon^{j\gamma t} + \epsilon^{-j\gamma t}), \tag{2-97}$$

$$\sin \gamma t = \frac{1}{2j} (\epsilon^{j\gamma t} - \epsilon^{-j\gamma t}). \tag{2-98}$$

For every type of exponential excitation function we must assume that an exponential function of the same type exists in the particular integral. Thus, for either $\cos \gamma t$ or $\sin \gamma t$, we choose

$$w_p = B_1 \epsilon^{j\gamma t} + B_2 \epsilon^{-j\gamma t}. \tag{2-99}$$

Equation (2–99) can be put in another form which avoids the combination and manipulation of exponential functions and complex coefficients:

$$\begin{aligned} w_p &= B_1 (\cos \gamma t + j \sin \gamma t) + B_2 (\cos \gamma t - j \sin \gamma t) \\ &= (B_1 + B_2) \cos \gamma t + j (B_1 - B_2) \sin \gamma t \\ &= A_1 \cos \gamma t + A_2 \sin \gamma t, \end{aligned} \tag{2-100}$$

where A_1 has been written for $(B_1 + B_2)$ and A_2 for $j(B_1 - B_2)$. Let us illustrate this with an example.

EXAMPLE 2–7. Find the general solution of the equation

$$\frac{d^2 w}{dt^2} - w = 3 \sin \omega t. \tag{2-101}$$

Solution. We note that the left side of Eq. (2–101) is the same as that of Eq. (2–79) in Example 2–5. Therefore these two equations have the same characteristic equation and hence also the same complementary function.

$$\begin{aligned} w_c &= c_1 \epsilon^{-t} + c_2 \epsilon^t \\ &= c_1 (\cosh t - \sinh t) + c_2 (\cosh t + \sinh t) \\ &= C_1 \cosh t + C_2 \sinh t, \end{aligned} \tag{2-102}$$

where C_1 has been written for $(c_1 + c_2)$ and C_2 for $(c_2 - c_1)$. To find the particular integral, we first choose, as suggested in Eq. (2–99),

$$w_p = B_1 \epsilon^{j\omega t} + B_2 \epsilon^{-j\omega t}. \tag{2-103}$$

Substituting w_p for w in Eq. (2–101), we have

$$-B_1(\omega^2 + 1)\epsilon^{j\omega t} - B_2(\omega^2 + 1)\epsilon^{-j\omega t} = \frac{3}{2j}(\epsilon^{j\omega t} - \epsilon^{-j\omega t}).$$

Equating the coefficients on both sides of the equation for like exponential terms, we determine the two constants B_1 and B_2 directly:

$$B_1 = -\frac{3}{2j(\omega^2 + 1)}, \tag{2–104}$$

$$B_2 = +\frac{3}{2j(\omega^2 + 1)}. \tag{2–105}$$

Hence, from Eqs. (2–103), (2–104), and (2–105),

$$w_p = -\frac{3}{2j(\omega^2 + 1)}(\epsilon^{j\omega t} - \epsilon^{-j\omega t})$$

$$= -\frac{3}{\omega^2 + 1}\sin \omega t, \tag{2–106}$$

and the general solution of Eq. (2–101) is

$$w = w_c + w_p = C_1 \cosh t + C_2 \sinh t - \frac{3}{\omega^2 + 1}\sin \omega t. \tag{2–107}$$

If we had chosen to use a trigonometric form instead of Eq. (2–103), we would write, according to Eq. (2–10C),*

$$w_p = A_1 \cos \omega t + A_2 \sin \omega t. \tag{2–108}$$

Substituting back into Eq. (2–101), we now have

$$-A_1(\omega^2 + 1)\cos \omega t - A_2(\omega^2 + 1)\sin \omega t = 3 \sin \omega t.$$

Equating the coefficients of cosine and sine terms on both sides, we find

$$A_1 = 0, \tag{2–109}$$

$$A_2 = -\frac{3}{\omega^2 + 1}, \tag{2–110}$$

and

$$w_p = -\frac{3}{\omega^2 + 1}\sin \omega t;$$

as before.

---

* Note that even though the excitation function is a sine function we must include *both* the cosine *and* the sine terms in the trial solution. Sometimes the cosine term will finally drop out because $A_1 = 0$, as happens to be true in this example; but this is a special case. In general, this will not be true, and there is no easy way to tell from the given equation whether A_1 will be zero.

To recapitulate, the method of undetermined coefficients affords us a simple way of finding the particular integrals of linear differential equations with constant coefficients when the excitation function is composed of the sum or the product of functions of certain special types. These special types of functions and their corresponding forms of particular integrals are listed in Table 2–1.

TABLE 2–1

METHOD OF UNDETERMINED COEFFICIENTS*

Excitation function	Form of particular integral
K (constant)	A (constant)
Kt^k (k a positive integer)	$A_0 + A_1t + A_2t^2 + \cdots + A_kt^k$
$K\epsilon^{\gamma t}$	$A\epsilon^{\gamma t}$
$K \cos \gamma t$	$A_1 \cos \gamma t + A_2 \sin \gamma t$
$K \sin \gamma t$	$A_1 \cos \gamma t + A_2 \sin \gamma t$

* Modifications are necessary (1) when the characteristic equation has a zero root (see Example 2–6), and (2) when the excitation function contains a term which appears in the complementary function (see Example 2–9).

If the excitation function is the sum or the product of two or more terms of the types listed in Table 2–1, then the particular integral will also be the sum or the product of two or more forms corresponding to the given excitation functions. Thus for

$$e(t) = Kt^3\epsilon^{\gamma t} \sin kt \qquad (2\text{–}111)$$

the correct form for the particular integral would be

$$w_p = (A_0 + A_1t + A_2t^2 + A_3t^3)\epsilon^{\gamma t} \cos kt$$
$$+ (B_0 + B_1t + B_2t^2 + B_3t^3)\epsilon^{\gamma t} \sin kt. \qquad (2\text{–}112)$$

In general, when the excitation function is of such a type that the method of undetermined coefficients is applicable, then this method is the simplest way of finding the particular-integral part of the solution.

2–5 Equations with multiple roots. Suppose now that we wish to solve the following homogeneous differential equation of the second order:

$$\frac{d^2w}{dt^2} + b_1 \frac{dw}{dt} + b_0w = 0, \qquad (2\text{–}113)$$

where the coefficients b_1 and b_0 are related in the following way:

$$b_1^2 = 4b_0. \qquad (2\text{–}114)$$

Let us first write Eq. (2–113) in operator notation:

$$(D^2 + b_1 D + b_0)w = 0. \tag{2–115}$$

Equation (2–115) can be factored to give

$$(D - s_1)(D - s_1)w = 0 \tag{2–116}$$

because, by virtue of Eq. (2–114), the characteristic equation has a double root, that is, the two roots are not distinct:

$$s_1 = s_2 = -\frac{b_1}{2}. \tag{2–117}$$

If we tried to follow the method outlined in Section 2–3 without modification, we would obtain the following solution:

$$w = c_1 \epsilon^{s_1 t} + c_2 \epsilon^{s_1 t}. \tag{2–118}$$

But the two terms in Eq. (2–118) can be combined:

$$w = (c_1 + c_2)\epsilon^{s_1 t} = c\epsilon^{s_1 t}. \tag{2–119}$$

The solution then actually contains only one arbitrary constant, and hence cannot be the general solution of the second-order equation (2–113).
  Going back to Eq. (2–116), let us first set

$$(D - s_1)w = r \tag{2–120}$$

and write Eq. (2–116) as

$$(D - s_1)r = 0, \tag{2–121}$$

which has as its solution

$$r = c_2 \epsilon^{s_1 t}. \tag{2–122}$$

Substituting this value of r in Eq. (2–120), we have

$$(D - s_1)w = c_2 \epsilon^{s_1 t}. \tag{2–123}$$

Equation (2–123) is a nonhomogeneous first-order linear differential equation, and its solution is composed of two parts, namely,

  a complementary function:

$$w_c = c_1 \epsilon^{s_1 t}$$

and a particular integral:

$$w_p = \epsilon^{s_1 t} \int \epsilon^{-s_1 t}(c_2 \epsilon^{s_1 t})\, dt = c_2 t \epsilon^{s_1 t}.$$

Hence the general solution of Eq. (2–123) or Eq. (2–113) is

$$w = w_c + w_p = (c_1 + c_2 t)\epsilon^{s_1 t}, \tag{2–124}$$

which now correctly contains two arbitrary constants.

By extending the above procedure, we can easily establish that if the characteristic equation possesses a root s of multiplicity k, then the solution corresponding to that root is

$$w_s = \epsilon^{st}(c_1 + c_2 t + c_3 t^2 + \cdots + c_k t^{k-1}). \tag{2–125}$$

Of course, if the characteristic equation has other roots, they should be added to w_s to form the complementary function.

EXAMPLE 2–8. Find the general solution of the following differential equation:

$$\frac{d^3 w}{dt^3} - 3\frac{dw}{dt} - 2w = \epsilon^{j\omega t}. \tag{2–126}$$

Solution. We first write Eq. (2–126) in operator form:

$$(D^3 - 3D - 2)w = \epsilon^{j\omega t}. \tag{2–127}$$

The characteristic equation is $(s^3 - 3s - 2) = 0$, or

$$(s + 1)^2(s - 2) = 0. \tag{2–128}$$

Hence it has a double root -1 and a single root $+2$.

(1) *Complementary function.*

$$w_c = (c_1 + c_2 t)\epsilon^{-t} + c_3\epsilon^{2t}. \tag{2–129}$$

(2) *Particular integral.* Let us use the method of successive integration. Specializing Eq. (2–69) for

$$s_1 = s_2 = -1 \quad \text{and} \quad s_3 = 2,$$

we have

$$w_p = \epsilon^{s_1 t} \iint \epsilon^{(s_3 - s_1)t} \int \epsilon^{-s_3 t}(\epsilon^{j\omega t})\,(dt)^3$$

$$= \epsilon^{-t} \iint \epsilon^{3t} \int \epsilon^{-2t}(\epsilon^{j\omega t})\,(dt)^3$$

$$= -\frac{1}{2 + j\omega(3 + \omega^2)}\,\epsilon^{j\omega t}. \tag{2–130}$$

We could have determined the particular integral by the method of undetermined coefficients as described in the preceding section without integrations. We start by assuming the form of the particular integral (see Table 2–1) to be

$$w_p = A \epsilon^{j\omega t}$$

and substituting it directly in Eq. (2–126) for w:

$$A[(j\omega)^3 - 3(j\omega) - 2]\epsilon^{j\omega t} = \epsilon^{j\omega t}.$$

Hence

$$A = -\frac{1}{2 + j\omega(3 + \omega^2)},$$

which, after being combined with the $\epsilon^{j\omega t}$ factor, checks with the answer found in Eq. (2–130).

The general solution of Eq. (2–126) is then

$$w = w_c + w_p = (c_1 + c_2 t)\epsilon^{-t} + c_3 \epsilon^{2t} - \frac{\epsilon^{j\omega t}}{2 + j\omega(3 + \omega^2)}. \qquad (2\text{–}131)$$

Having solved Eq. (2–126), we have in fact also solved the following two equations:

$$\frac{d^3w}{dt^3} - 3\frac{dw}{dt} - 2w = \cos \omega t \qquad (2\text{–}132)$$

and

$$\frac{d^3w}{dt^3} - 3\frac{dw}{dt} - 2w = \sin \omega t. \qquad (2\text{–}133)$$

The reason is twofold: (1) Since the characteristic equations for Eqs. (2–132) and (2–133) are the same as that for Eq. (2–126), the complementary function in both cases would remain the same as that given in Eq. (2–129). (2) Since $\cos \omega t = \mathrm{Re}\ (\epsilon^{j\omega t})$ and $\sin \omega t = \mathrm{Im}\ (\epsilon^{j\omega t}),$* then, by virtue of the linear property of the equations,

Particular integral for Eq. (2–132) is

$$\mathrm{Re}\left[-\frac{\epsilon^{j\omega t}}{2 + j\omega(3 + \omega^2)}\right] = -\frac{2 \cos \omega t + \omega(3 + \omega^2) \sin \omega t}{4 + \omega^2(3 + \omega^2)^2},$$
$$(2\text{–}134)$$

Particular integral for Eq. (2–133) is

$$\mathrm{Im}\left[-\frac{\epsilon^{j\omega t}}{2 + j\omega(3 + \omega^2)}\right] = -\frac{2 \sin \omega t - \omega(3 + \omega^2) \cos \omega t}{4 + \omega^2(3 + \omega^2)^2}.$$
$$(2\text{–}135)$$

---

* $\mathrm{Re}\ (\epsilon^{j\omega t})$ stands for "the real part of" $\epsilon^{j\omega t}$. $\mathrm{Im}\ (\epsilon^{j\omega t})$ stands for "the imaginary part of" $\epsilon^{j\omega t}$.

EXAMPLE 2–9. Find the general solution of the following differential equation:

$$\frac{d^2w}{dt^2} - w = \epsilon^{-t}. \tag{2–136}$$

Solution. The given differential equation in operator form is

$$(D^2 - 1)w = \epsilon^{-t}. \tag{2–137}$$

The roots of the characteristic equation $s^2 - 1 = 0$ are

$$s_1 = +1 \qquad \text{and} \qquad s_2 = -1.$$

The complementary function is then

$$w_c = c_1\epsilon^t + c_2\epsilon^{-t}. \tag{2–138}$$

If, for the particular integral, we followed Table 2–1 without modification and assumed

$$w_p = A\epsilon^{-t}, \tag{2–139}$$

then we would find

$$(D^2 - 1)w_p = (D^2 - 1)A\epsilon^{-t} = 0,$$

because w_p in Eq. (2–139) is of the same form as the $c_2\epsilon^{-t}$ term in w_c. The expression in Eq. (2–139) then does not satisfy the given equation and cannot be a particular integral.

We should treat this case, where the excitation function contains a term which appears in the complementary function, in the same way as if a double root for the characteristic equation existed. Thus, assume

$$w_p = At\epsilon^{-t}. \tag{2–140}$$

Substituting this back into the given equation, we find that it is satisfied when

$$A = -\tfrac{1}{2}. \tag{2–141}$$

Hence

$$w_p = -\tfrac{1}{2}t\epsilon^{-t}, \tag{2–142}$$

and the general solution of Eq. (2–136) or (2–137) is

$$w = w_c + w_p = c_1\epsilon^t + (c_2 - \tfrac{1}{2}t)\epsilon^{-t}. \tag{2–143}$$

2–6 Simultaneous differential equations. In engineering we sometimes have to solve simultaneous linear differential equations with constant coefficients that contain one independent variable but several dependent variables. A typical problem of this type in electrical engineering would be the determination of loop currents in an electrical network of several interconnected loops in response to some given source of excitation. The most straightforward way of solving simultaneous equations is to reduce the given system of equations to one with a single dependent variable, which can then be solved with methods we already know. We shall illustrate the procedure with an example.

EXAMPLE 2–10. Find the general solution of the following pair of simultaneous equations:

$$\frac{dx}{dt} + x - 2y = \epsilon^{-t}, \tag{2-144a}$$

$$6x - 3\frac{dy}{dt} + 3y = -t. \tag{2-144b}$$

Solution. We shall try to solve this problem in systematic steps.

Step 1. Obtain an equation with one dependent variable only, say x. It is convenient to use operator notation and write D for d/dt.

$$(D + 1)x - 2y = \epsilon^{-t} \tag{2-145a}$$

$$6x - 3(D - 1)y = -t. \tag{2-145b}$$

To eliminate y and its derivatives from Eqs. (2–145), we operate on Eq. (2–145a) with the operator $3(D - 1)$ and multiply Eq. (2–145b) by -2:

$$3(D^2 - 1)x - 6(D - 1)y = 3(D - 1)\epsilon^{-t} = -6\epsilon^{-t}, \tag{2-146a}$$

$$-12x + 6(D - 1)y = 2t. \tag{2-146b}$$

Adding Eqs. (2–146a) and (2–146b), we obtain a second-order equation in x:

$$3(D^2 - 5)x = 2t - 6\epsilon^{-t},$$

or

$$(D^2 - 5)x = \tfrac{2}{3}t - 2\epsilon^{-t}. \tag{2-147}$$

Step 2. We next solve this equation for x. The characteristic equation for Eq. (2–147) is

$$s^2 - 5 = 0,$$

which has two distinct roots: $s_1 = +\sqrt{5}$ and $s_2 = -\sqrt{5}$. Hence the complementary function is

$$x_c = c_1\epsilon^{\sqrt{5}t} + c_2\epsilon^{-\sqrt{5}t}. \tag{2-148}$$

Next we assume, according to Table 2–1, that the particular integral of Eq. (2–147) is of the form

$$x_p = A_0 + A_1t + A_2\epsilon^{-t}. \tag{2-149}$$

Substituting x_p in Eq. (2–147) for x and carrying out the differentiations, we have

$$-5A_0 - 5A_1 t - 4A_2 \epsilon^{-t} = \tfrac{2}{3}t - 2\epsilon^{-t}.$$

This will hold if

$$-5A_0 = 0, \qquad -5A_1 = \tfrac{2}{3}, \qquad -4A_2 = -2,$$

or

$$A_0 = 0, \qquad A_1 = -\tfrac{2}{15}, \qquad A_2 = \tfrac{1}{2}.$$

Hence

$$x_p = -\tfrac{2}{15}t + \tfrac{1}{2}\epsilon^{-t}$$

and

$$x = x_c + x_p = c_1 \epsilon^{\sqrt{5}t} + c_2 \epsilon^{-\sqrt{5}t} - \tfrac{2}{15}t + \tfrac{1}{2}\epsilon^{-t}. \tag{2–150}$$

Step 3. We now solve for y. Of course, we could solve for y from the given Eqs. (2–144) in the same way as x has been solved. If we did this, we would obtain the following equation for y:

$$(D^2 - 5)y = \tfrac{1}{3}(1 + t) + 2\epsilon^{-t}. \tag{2–151}$$

But it is unnecessary to solve this second-order differential equation. Moreover, if we proceeded to solve Eq. (2–151) for y, two additional arbitrary constants would result which would have to be expressed in terms of the c_1 and c_2 in Eq. (2–150) through the original Eqs. (2–144). We can solve for y much more simply by noting from Eq. (2–144a) that

$$y = \frac{1}{2}\left(\frac{dx}{dt} + x - \epsilon^{-t}\right)$$
$$= \frac{c_1}{2}(\sqrt{5} + 1)\epsilon^{\sqrt{5}t} - \frac{c_2}{2}(\sqrt{5} - 1)\epsilon^{-\sqrt{5}t} - \frac{1}{15}(1 + t) - \frac{1}{2}\epsilon^{-t}. \tag{2–152}$$

Equations (2–150) and (2–152) constitute the general solution for the pair of simultaneous differential equations (2–144).

We are now ready to make several important observations from a pair of general simultaneous linear differential equations with constant co-efficients. We consider two equations as follows:

$$f_1(D)x + g_1(D)y = e_1(t), \tag{2–153a}$$

$$f_2(D)x + g_2(D)y = e_2(t), \tag{2–153b}$$

where $f_1(D)$, $f_2(D)$, $g_1(D)$, and $g_2(D)$ are operators with constant coefficients. To obtain an equation for the dependent variable x, we operate on Eq. (2–153a) with $g_2(D)$ and on Eq. (2–153b) with $g_1(D)$, and then subtract:

$$[g_2(D)f_1(D) - g_1(D)f_2(D)]x = g_2(D)e_1(t) - g_1(D)e_2(t). \tag{2–154}$$

Similarly, to obtain an equation for the dependent variable y, we operate on Eq. (2–153a) with $f_2(D)$ and on Eq. (2–153b) with $f_1(D)$, and then subtract:

$$[f_1(D)g_2(D) - f_2(D)g_1(D)]y = f_1(D)e_2(t) - f_2(D)e_1(t). \quad (2\text{–}155)$$

We note that the operator in brackets operating on x in Eq. (2–154) is exactly the same as that operating on y in Eq. (2–155), both being equal to the expansion of the determinant formed by the operator coefficients in the original Eqs. (2–153a) and (2–153b):

$$\Delta(D) = \begin{vmatrix} f_1(D) & g_1(D) \\ f_2(D) & g_2(D) \end{vmatrix} \quad (2\text{–}156)$$

By extending this development to simultaneous equations with more than two dependent variables, we are able to draw the following general conclusions about the solutions of simultaneous linear differential equations with constant coefficients:

1. The characteristic equations for all dependent variables are the same: $\Delta(s) = 0$. Consequently, the complementary functions of the solutions for all dependent variables have the same number and types of terms.

2. The *total* number of independent arbitrary constants in the general solution is the same as the order of the resulting differential equation in one dependent variable only, after all the other dependent variables and their derivatives have been eliminated. This number is equal to the highest degree of D in the expansion of the determinant $\Delta(D)$.

3. The coefficients for the terms in the complementary functions for different dependent variables are definitely related through the original equations. Hence, when the coefficients of the terms in the complementary function of any one dependent variable are chosen arbitrarily, the coefficients for the terms in the complementary functions of all other dependent variables can no longer be chosen at will.

PROBLEMS

Find the general solution of each of the following equations:

2-1. $L\dfrac{di}{dt} + Ri = E \sin \omega t$ $(L, R, E,$ and ω are constants$)$

2-2. $x^2\dfrac{dy}{dx} = \epsilon^x - 2xy$

2-3. $2t^2\dfrac{dw}{dt} = \sin \beta t - 4tw$

2-4. $\sec \theta \dfrac{d\alpha}{d\theta} + \alpha = 1$

2-5. $(x + 1)\, dy = [y + (x - 1)]\, dx$

2-6. $\dfrac{dy}{dx} = \dfrac{1}{x + 3\epsilon^{2y}}$

2-7. Show that the nonlinear *Bernoulli's equation*

$$\frac{dy}{dx} + a(x)y = b(x)y^n$$

can be reduced to a linear first-order equation by the substitution $z = y^{1-n}$.

2-8. Solve the equation: $\dfrac{dy}{dx} + \dfrac{y}{x} + 3x^2y^3 = 0$

2-9. Solve the equation: $\dfrac{dw}{dt} + 3\dfrac{w}{t} = \sqrt{3w}$

2-10. Prove that $D^n(\epsilon^{mt}w) = \epsilon^{mt}(D + m)^n w$

Find the general solution of each of the following equations:

2-11. $2\dfrac{d^2y}{dx^2} - \dfrac{dy}{dx} - 10y = x^2\epsilon^{-x}$ 2-12. $\dfrac{d^2y}{dx^2} + y = \sin x$

2-13. $\dfrac{d^3y}{dx^3} - 4\dfrac{dy}{dx} = 8$ 2-14. $\dfrac{d^3y}{dx^3} - 8y = 3x + \epsilon^x$

2-15. $4\dfrac{d^2y}{dx^2} + 4\dfrac{dy}{dx} + y = \epsilon^{-x/2} + x$

2-16. $3\dfrac{d^2i}{dt^2} + 4\dfrac{di}{dt} + 2i = 20\epsilon^{-t/5}\sin (t/5)$

2-17. $\dfrac{d^3w}{dt^3} + 6\dfrac{d^2w}{dt^2} + 11\dfrac{dw}{dt} + 6w = t^2$

2-18. $\dfrac{d^3w}{dx^3} + 2\dfrac{d^2w}{dx^2} + \dfrac{dw}{dx} = 3 + \sin x$

2-19. (a) Show that a linear differential equation with variable coefficients of the following type (known as *Euler's* or *Cauchy's differential equation*):

$$a_nx^n\frac{d^ny}{dx^n} + a_{n-1}x^{n-1}\frac{d^{n-1}y}{dx^{n-1}} + \cdots + a_1x\frac{dy}{dx} + a_0y = f(x)$$

can be reduced to an equation with constant coefficients by means of the substitution $x = \epsilon^z$.

(b) Solve the equation

$$x^2 \frac{d^2 y}{dx^2} + 3x \frac{dy}{dx} + 2y = \log x$$

2–20. Find the general solution of

$$\frac{d^2 y}{dx^2} + 4 \frac{dy}{dx} + 4y = \frac{\epsilon^{-2x}}{x^2}$$

2–21. Find the general solution of

$$\frac{d^3 w}{dt^3} + 3 \frac{d^2 w}{dt^2} + 3 \frac{dw}{dt} + w = 3t\epsilon^{-t/2}$$

Find the general solutions of the following simultaneous differential equations:

2–22. $\dfrac{d^2 y}{dx^2} = y - z$ $\qquad\qquad$ $\dfrac{d^2 z}{dx^2} = z - y$

2–23. $\dfrac{d^2 u}{dx^2} = 2v$ $\qquad\qquad$ $\dfrac{d^2 v}{dx^2} = 2u$

2–24. $\dfrac{dy}{dx} + w = -1$ $\qquad$ $\dfrac{dz}{dx} - 3w = 5$ $\qquad$ $\dfrac{dw}{dx} + y - z = 2x - 3$

2–25. $L \dfrac{d^2 i_1}{dt^2} + R \dfrac{di_1}{dt} - R \dfrac{di_2}{dt} = E\omega \cos \omega t,$

$\qquad - R \dfrac{di_1}{dt} + R \dfrac{di_2}{dt} + \dfrac{1}{C} i_2 = 0,$

where L, R, C, E, and ω are given constants, $R = \frac{1}{2}\sqrt{L/C}$, and $\omega^2 = 1/LC$.

CHAPTER 3

LUMPED-ELEMENT ELECTRICAL SYSTEMS

3–1 Introduction. The analytical determination of the response of a physical system to a given excitation involves two essential steps, namely, the setting up of the mathematical equations that describe the system and the solution of these equations subject to appropriate initial or boundary conditions. In the preceding chapters we have discussed the general properties of linear ordinary differential equations and their solution by classical procedures. We shall now consider methods for setting up the equations that describe a physical system in accordance with physical laws.

In the present chapter we shall deal with lumped-element electrical systems. A lumped-element electrical system or network is one in which the individual components can be identified by *lumped* parameters at their terminals. From a mathematical viewpoint we can say that the voltage and current observables in a linear lumped-element electrical system satisfy linear *ordinary* differential equations of finite orders. In contrast to this, in systems with distributed parameters the component elements are distributed in character. Typical examples of distributed elements are transmission lines, wave guides, and the like. Partial differential equations are needed to describe systems with distributed parameters, the discussion of which is reserved for Chapter 11.

For systems other than electrical, the next chapter on analogous systems discusses in detail methods of drawing electrical circuits analogous to linear mechanical and electromechanical systems. Once the circuit diagram of the analogous electrical system is determined, it is possible to visualize, and sometimes even to predict, system behaviors by inspection. In the following sections we shall start with the fundamental voltage-current relationships of basic electrical network elements and show the techniques of setting up network equations for lumped-element systems, using Kirchhoff's laws. Typical circuit problems will be solved and the concepts of impedance, resonance, and time constants introduced. We shall also discuss the principle of duality and develop the important Thévenin's and Norton's theorems.

3–2 Network elements. Electrical network elements can be classified into two kinds: *active elements* and *passive elements*. Active elements are sources of electric energy. For purposes of analysis, it is convenient to consider two types of ideal sources: *voltage sources* and *current sources*. A

43

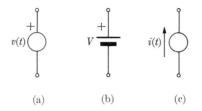

(a) (b) (c)

FIG. 3–1. Symbolic representation of ideal voltage and current sources. (a) A general voltage source. (b) A d-c voltage source. (c) A general current source.

voltage source is one which maintains a potential difference of a given magnitude and a prescribed waveform across its terminals independently of the current in it. Symbolically we represent a general voltage source as a two-terminal device in Fig. 3–1(a). Note that *it is essential to always indicate the reference polarity of a voltage source with a plus sign on one of its terminals.* A plus sign alongside a given terminal does not mean that the particular terminal has a positive potential with respect to the other terminal at all instants of time, since $v(t)$ may be an arbitrary function of t; it serves as a *reference polarity,* without which unambiguous network equations cannot be written. When a voltage source maintains a constant potential difference across its terminals, it is customary to represent it by a battery, as shown in Fig. 3–1(b). It is obvious that a voltage source such as we have defined is an ideal device which cannot be realized (although it can be approximated closely) physically because all physical sources have some internal resistance; the potential difference appearing across the terminals of a physical source will depend upon the current through it, since the internal voltage drop will vary with the current. Nevertheless, the ideal voltage source is an extremely useful device in analysis. All physical voltage sources may be represented as a combination of an ideal voltage source and some passive elements, the latter to describe the internal circuit properly.

A current source is one which maintains a current of a given magnitude and a prescribed waveform at its terminals independently of the potential difference appearing across its terminals. A general current source is represented symbolically in Fig. 3–1(c). No separate symbol is commonly used for an ideal source of direct current mainly because there exists no physical device which approximates a d-c constant current generator. *It is essential to always indicate the reference current direction of a current source by an arrow.* A reference current direction is necessary in order that unambiguous equations for a network containing the current source can be written. All physical current sources may be represented as a combination of an ideal current source and some passive elements, the latter to describe the internal circuit properly.

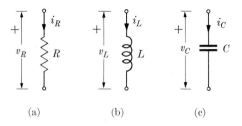

Fig. 3–2. Symbolic representation of passive electrical network elements. (a) A resistance. (b) An inductance. (c) A capacitance.

The ideal voltage and current sources described above are *independent* sources because the terminal voltage and current of these sources are, respectively, independent of any other voltage or current. In the analysis of circuits containing vacuum tubes or transistors, we find that they can be considered as equivalent sources whose terminal voltage or current is a function of other voltages or currents. Voltage and current sources of this type can be classified as *dependent sources*.

There are three kinds of linear, bilateral, passive electrical network elements: *resistance, inductance,* and *capacitance.* They are symbolically represented in Fig. 3–2. We shall confine our attention to those elements whose values do not vary with time. An electrical network is linear when its element values are dependent upon neither the voltage nor the current. The relations between the current in and the voltage drop across (in the direction of the current) the three kinds of passive elements are as follows:

(a) *Resistance, R.* (Unit: ohm, Ω)

$$v_R(t) = Ri_R(t), \tag{3-1}$$

$$i_R(t) = \frac{1}{R}v_R(t). \tag{3-2}$$

(b) *Inductance, L.* (Unit: henry, h)

$$v_L(t) = L\frac{di_L(t)}{dt}. \tag{3-3}*$$

---

* The self-inductance L is defined as the flux linkage per unit current. If ϕ_L represents the flux linkage which is the product of the number of turns and the flux linking with each turn of the coil, we have for a linear element

$$L = \frac{d\phi_L}{di_L} = \frac{\phi_L}{i_L}.$$

Faraday's law of electromagnetic induction tells us that the voltage drop across an inductive element is equal to the time rate of change of flux linkage. Thus

$$v_L(t) = \frac{d\phi_L}{dt} = \frac{d\phi_L}{di_L}\frac{di_L}{dt} = L\frac{di_L}{dt}.$$

The inverse relationship of Eq. (3–3) is usually written as an indefinite integral, as follows:

$$i_L(t) = \frac{1}{L} \int v_L(t) \, dt. \tag{3–4}$$

However, an *indefinite integral*, or *antiderivative*, of a function, say $v_L(t)$, is defined in calculus as *any* function whose derivative with respect to the independent variable, t, yields the given function, $v_L(t)$. Since two functions differing by a constant have the same derivative, the indefinite integral of a function is not unique. In fact, the function

$$\phi_L(t) = \int_{t_0}^{t} v_L(t) \, dt + K, \tag{3–5}^*$$

where both t_0 and K are arbitrary constants, is an indefinite integral of $v_L(t)$, since the derivative of $\phi_L(t)$ given in Eq. (3–5) is equal to $v_L(t)$. That is, if we write

$$i_L(t) = \frac{1}{L} \phi_L(t) = \frac{1}{L} \left[\int_{t_0}^{t} v_L(t) \, dt + K \right], \tag{3–6}$$

the voltage-current relationship in Eq. (3–3) is still satisfied. In a physical problem, where all physical conditions are prescribed, leaving no freedom in the choice of constants, the indefiniteness implied in the arbitrary constants t_0 and K in Eq. (3–6) will lead to confusion.

In the analysis of linear systems, the response of a system *after* the application of an excitation or the occurrence of a disturbance in the system is usually desired. The behavior of the system at the instant when the excitation or disturbance is applied is then of particular importance. For convenience, we use this instant as the reference time and call it the instant of $t = 0$. Hence, by setting $t_0 = 0$ and $K = \phi_L(0)$ in Eq. (3–6) we write, instead of Eq. (3–4), the relation between the current in and the voltage across an inductance as a combination of a definite integral (from 0 to t) and an initial condition:

$$i_L(t) = \frac{1}{L} \left[\int_{0}^{t} v_L(t) \, dt + \phi_L(0) \right] \tag{3–7}$$

or

$$i_L(t) = \frac{1}{L} \int_{0}^{t} v_L(t) \, dt + i_L(0). \tag{3–8}$$

---

* The t in $v_L(t)$ under the integral sign on the right side of Eq. (3–5) is a *dummy variable;* any other symbol could have been used for it without affecting the value of the integral. Thus, instead of Eq. (3–5), we can write

$$\phi_L(t) = \int_{t_0}^{t} v_L(x) \, dx + K.$$

In Eq. (3–8), $i_L(0)$ is the initial current in the inductance, which is equal to the initial flux linkage $\phi_L(0)$ divided by L. By convention, $i_L(0)$ is considered positive when it flows in the direction of the reference voltage drop v_L.

(c) *Capacitance, C.* (Unit: farad, f)

$$i_C(t) = \frac{dq_C(t)}{dt} = C\,\frac{dv_C(t)}{dt}. \tag{3–9}*$$

Following the preceding arguments, we write the inverse relationship, $v_C(t)$ in terms of $i_C(t)$, as a definite integral plus an initial voltage:

or

$$v_C(t) = \frac{1}{C}\left[\int_0^t i_C(t)\,dt + q_C(0)\right] \tag{3–10}$$

$$v_C(t) = \frac{1}{C}\int_0^t i_C(t)\,dt + v_C(0), \tag{3–11}$$

where the reference polarity of $v_C(0)$ is the same as that of $v_C(t)$.

Resistances dissipate power; that is, they irreversibly transform electric energy into heat. The rate of energy dissipation, or power dissipated, in a resistance R is given by

$$P_R = i_R^2 R = v_R i_R = v_R^2/R. \tag{3–12}$$

Inductances store energy in a magnetic field. Energy is not dissipated in inductances, which serve only as energy reservoirs. The energy stored in an inductance L is given by

$$W_L = \tfrac{1}{2}Li_L^2 = \tfrac{1}{2}\phi_L i_L = \phi_L^2/2L. \tag{3–13}$$

Capacitances store energy in an electric field; they are reservoirs of electric energy. The energy stored in a capacitance C is given by

$$W_C = \tfrac{1}{2}Cv_C^2 = \tfrac{1}{2}q_C v_C = q_C^2/2C. \tag{3–14}$$

Since neither magnetic nor electric energy in a physical system can change instantaneously even for sudden but finite changes in power, *the*

---

* The capacitance C is defined as the ratio of the charge on and the voltage across a capacitor:

$$C = \frac{q_C(t)}{v_C(t)}$$

or

$$q_C(t) = Cv_C(t),$$

from which Eq. (3–9) follows immediately.

TABLE 3-1

VOLTAGE-CURRENT RELATIONSHIPS OF ELECTRICAL NETWORK ELEMENTS

Element	Symbol	Voltage	Current
Voltage source	$v(t)$	$v(t)$ Independent of current	Depending on network connection
Current source	$i(t)$	Depending on network connection	$i(t)$ Independent of voltage
Resistance	R, i_R, v_R	$v_R = i_R R$	$i_R = v_R/R$
Inductance	L, i_L, v_L	$v_L = L \dfrac{di_L}{dt}$	$i_L = \dfrac{1}{L}\displaystyle\int_0^t v_L\,dt + i_L(0)$
Capacitance	C, i_C, v_C	$v_C = \dfrac{1}{C}\displaystyle\int_0^t i_C\,dt + v_C(0)$	$i_C = C \dfrac{dv_C}{dt}$

flux linkage of and current in inductances and the charge on and voltage across capacitances cannot change instantaneously; they take time to change (no sudden jumps) and will be continuous functions of time. In particular, at and near $t = 0$,

$$\text{for inductances:} \quad \phi_L(0+) = \phi_L(0-), \tag{3–15}$$

$$i_L(0+) = i_L(0-); \tag{3–16}$$

$$\text{for capacitances:} \quad q_C(0+) = q_C(0-), \tag{3–17}$$

$$v_C(0+) = v_C(0-). \tag{3–18}$$

The symbols $0-$ and $0+$ are used to denote instants immediately before and after $t = 0$, respectively. Equations (3–15) through (3–18) are of extreme importance in network analysis involving initial conditions.*

Table 3–1 is a tabulation of electrical network elements and their voltage-current relationships.

It is to be noted that, just as voltage and current sources are ideal active elements, pure resistance, inductance, and capacitance are ideal passive elements. All physical passive elements possess to some extent a combination of the properties ascribed to the three pure elements. For instance, all inductors have some resistance, although in some cases the resistance may be so small that its effect is negligible in comparison with the inductive effect. In other cases an inductor must be considered as a combination of a resistance and an inductance. At high frequencies the distributed capacitance between turns may become important, and an inductor should then be considered as a complex structure consisting of resistance, inductance, and capacitance. Idealized pure elements afford great convenience in analysis.

3–3 Magnetic coupling. Energy can be transferred from one circuit to another, not connected to the first, through magnetic flux linking with both circuits. The two circuits are then said to be magnetically coupled. The basis of this magnetic action is *Faraday's law of electromagnetic induction*, which states that an electromotive force (a voltage) is induced in a closed circuit when the magnetic flux linking with the circuit is changed; the induced electromotive force (emf) is always in such a direction as to

---

* An electrical network is initially *at rest*, or *relaxed*, if it contains no stored energy at the time an external excitation is applied. To put it in another way: there is no current in the inductances and no voltage across the capacitances in the network at $t = 0$.

Equations (3–15) through (3–18) would no longer be true under certain idealized conditions. These conditions will be discussed in Section 8–2.

oppose the flux change and its magnitude is numerically equal to the time rate of change of flux linkage. Two things are worthy of note here. First, there must be flux linkage; that is, some of the magnetic flux produced by the current in one circuit must *link* with the other circuit. Second, the magnetic flux must be *changing* with time; otherwise there will be no induced emf.

A typical situation of two magnetically coupled circuits is shown in Fig. 3–3(a), where two coils are wound on a common iron core that provides a path of low reluctance for the magnetic flux. It is a picture of a *transformer.* A source connected to terminals 1–1′ of the first coil (the primary winding) produces a current i_1 which varies with time. The current i_1 gives rise to a varying magnetic flux ϕ linking with the second coil (the secondary winding) to which a load is connected. By virtue of Faraday's law, the varying ϕ induces an emf in both windings, which tends to oppose the flux change. The induced emf in the primary winding must therefore be a voltage rise from terminal 1′ to terminal 1, seeking to "buck" the source voltage and oppose the change in i_1 and ϕ. Similarly, the induced emf in the secondary winding makes terminal 2 positive relative to terminal 2′, producing a current i_2 in the secondary circuit which opposes the change in ϕ according to the right-hand rule.* The relative polarities of the terminal pairs are indicated in Fig. 3–3(a).

Had the sense of the secondary winding been reversed, the direction of the induced emf in the secondary winding would also reverse, making terminal 2 negative relative to terminal 2′. Since it is rather inconvenient to draw the core and the associated windings in electrical networks, and since the conventional symbol for coils does not indicate the sense of winding, *it is customary to put dots on those terminals of the windings whose*

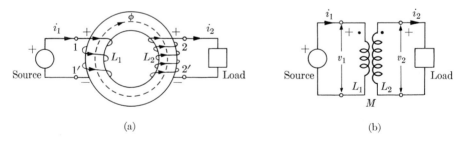

(a)   (b)

FIG. 3–3. Magnetically coupled circuits. (a) Arrangement showing relative senses of windings. (b) Schematic representation of the arrangement in (a).

* The right-hand rule relating the direction of magnetic flux and the current in a coil tells us that if the fingers of the right hand follow the direction of the current encircling the coil, then the thumb points in the direction of the flux *inside* the coil.

potentials due to electromagnetic induction rise and fall together. The arrangement in Fig. 3–3(a) can therefore be represented by the simple schematic diagram of Fig. 3–3(b) without ambiguity as to the relative senses of the windings.

The tightness of the magnetic coupling between two coils can be measured by the *mutual inductance,* which is defined as the flux linkage with coil 1 per unit current in coil 2, or vice versa. Quantitatively the mutual inductance M between two coils of self-inductances L_1 and L_2 respectively can be expressed as

$$M = k\sqrt{L_1 L_2}, \tag{3–19}$$

where k is called the *coefficient of coupling.* M, like L_1 and L_2, is in henrys, and k is dimensionless. Because of the existence of *leakage flux,* which links with one coil only, the coefficient of coupling of practical transformers is always less than unity, although in cases where an iron core is used it can be made very close to unity.

The voltages v_1 and v_2 at the terminal pairs 1–1' and 2–2' consist not only of the voltage drops across the self-inductances due to current changes in the coils themselves, but also of the induced emf's due to current changes in the other coil by mutual coupling. With the reference polarities in Fig. 3–3(b), we have

$$v_1 = L_1 \frac{di_1}{dt} - M \frac{di_2}{dt}, \tag{3–20}$$

$$v_2 = M \frac{di_1}{dt} - L_2 \frac{di_2}{dt}, \tag{3–21}$$

$$\underset{\substack{\text{due to a} \\ \text{changing} \\ i_1}}{\phantom{M\frac{di_1}{dt}}} \quad \underset{\substack{\text{due to a} \\ \text{changing} \\ i_2}}{\phantom{L_2\frac{di_2}{dt}}}$$

where *the mutual inductance M is itself always a positive quantity.* Resistances of the coils have been neglected in Eqs. (3–20) and (3–21).

An *ideal transformer* is one which has zero winding resistance, unity coefficient of coupling, and an infinite number of turns in each winding while maintaining finite turns ratios. It is obviously a hypothetical device. However, the consideration of ideal transformers is important because it leads to very simple voltage and current relationships among the windings. Good practical transformers with iron cores can approximate the conditions of an ideal transformer rather closely. With the assumption of an infinite number of turns in the windings, equations based upon self- and mutual inductances, such as (3–20) and (3–21), are no longer useful because the inductances become infinite.

Let us consider a two-winding ideal transformer. First of all, since unity coefficient of coupling is assumed, there is no leakage flux; all magnetic fluxes are mutual, that is, they link with both windings. The voltages appearing across the primary and secondary windings must therefore be proportional to the numbers of turns in the windings:

$$\frac{v_1}{v_2} = \frac{N_1}{N_2} = a. \tag{3-22}$$

We note that although both N_1 and N_2 are infinitely large, their ratio, a, is assumed finite for an ideal transformer. Secondly, with infinite N_1 and N_2, the induced voltages v_1 and v_2 can be finite only if the mutual magnetic flux approaches zero. In other words, the net total flux-producing ampere-turns around the transformer core must approach zero. For the schematic representation in Fig. 3–3(b) (disregarding the symbols L_1, L_2, and M), we have

$$N_1 i_1 - N_2 i_2 = 0. \tag{3-23}$$

The negative sign appears in Eq. (3–23) because i_1 is *entering* a dotted terminal and i_2 is *leaving* a dotted terminal; these currents have opposing effects. Rewriting Eq. (3–23), we have

$$\frac{i_1}{i_2} = \frac{N_2}{N_1} = \frac{1}{a}. \tag{3-24}$$

We see that the voltage and current relationships for an ideal transformer as given in Eqs. (3–22) and (3–24) are indeed very simple. This is why we often use them for rough calculations relative to nonideal transformers with closely coupled windings. We note further that *Eqs. (3–22) and (3–24) hold for an ideal transformer regardless of what is connected to its terminals.*

3–4 Kirchhoff's voltage law and loop analysis. Two fundamental laws form the basis of network analysis and enable us to write the equations describing electrical systems: Kirchhoff's voltage law and Kirchhoff's current law. These two laws were originally based upon experimental observations, but we can consider them as postulates, just like the laws of conservation of energy and conservation of charge. As a matter of fact, the two laws of Kirchhoff are restatements of the latter two laws in the terminology of electrical circuits.

In applying Kirchhoff's laws, reference polarities must be assigned to voltages across network elements and reference current directions must be assigned to currents in the elements. These are *algebraic* quantities and do not necessarily represent the actual voltage polarities and current directions. The actual voltages and currents may be functions of time, either positive or negative, based upon the set of chosen reference polarities and directions.

Kirchhoff's voltage law may be stated as follows:

Around any closed path (loop, mesh) in an electrical network the alge-braic sum of voltage drops is zero. Or, around a loop,

$$\sum_k v_k(t) = 0, \tag{3–25}$$

which is true at all instants of time. This law can be stated in other ways. For example, we may say that around any closed loop in an electrical net-work the sum of voltage rises is equal to the sum of voltage drops; this will lead to the same equations.

In taking the algebraic sum of voltage drops around a loop, we can traverse the loop in either a clockwise or a counterclockwise direction. Hence we must assign an orientation to the loop. Kirchhoff's voltage law is the basis of *loop analysis* of electrical networks. Let us apply it to the network in Fig. 3–4. Three closed loops are in evidence, for which we choose clockwise loop orientations, as indicated by the arrows. Applying Kirchhoff's voltage law to each of the three loops, we get

$$\text{Loop } a: \quad v_1 + v_2 - v_3 = 0, \tag{3–26a}$$

$$\text{Loop } b: \quad v_3 + v_4 - v_5 = 0, \tag{3–26b}$$

$$\text{Loop } c: \quad v_1 + v_2 + v_4 - v_5 = 0. \tag{3–26c}$$

Examination reveals that Eq. (3–26c) can be obtained by adding Eqs. (3–26a) and (3–26b); hence these three equations are not all inde-pendent. The question then arises: How do we determine the total number

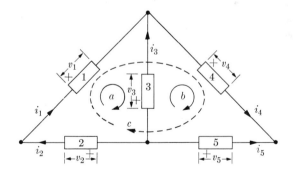

Fig. 3–4. Application of Kirchhoff's voltage law.

of independent loop equations that completely describe a given network? The answer can be obtained from the theory of linear graphs and will be given below without proof. *Let N_b, N_j, and N_s be the number of branches,*[*]

[*] A branch is any combination of elements grouped between two terminals (or junctions). In applying Eq. (3–27), each network element (active or passive) can be counted as a branch and each point connecting two or more elements as a junction.

number of junctions, and number of separate parts, respectively, of an elec-
trical network; then the number of independent loop equations, N_l, obtained
by the application of Kirchhoff's voltage law is given by

$$N_l = N_b - N_j + N_s. \tag{3-27}$$

For the network in Fig. 3–4, we have $N_b = 5$, $N_j = 4$, and $N_s = 1$;
hence $N_l = 5 - 4 + 1 = 2$, as we already know. Which two of the
three loops we choose is immaterial, so long as *every* element in the given
network is traced by *at least one* of the chosen loops.

Let us now go back to Eqs. (3–26a) and (3–26b), which are independent
equations describing the network in Fig. 3–4. We have here five voltage
terms, v_1, v_2, v_3, v_4, and v_5, related by only two equations. However, we
do not have the dilemma of having to solve for more unknowns than the
number of available equations. As we have discussed in Section 3–2, all
network elements have definite voltage-current relationships; hence if the
current in an element is known, the voltage across it is also determined.
In loop analysis, we assign a fictitious loop current circulating in each in-
dependent loop in the reference direction. The current in each element can
then be obtained by a proper combination of the loop currents. We have
already emphasized that the loops should be chosen in such a way that at
least one loop current flows through each element in the network.

For the network in Fig. 3–4, we have two independent loops *a* and *b*
for which we can assign circulating loop currents i_a and i_b. All element
currents can be expressed in terms of i_a and i_b:

$$i_1 = i_a, \quad i_2 = i_a, \quad i_3 = -i_a + i_b, \quad i_4 = i_b, \quad i_5 = -i_b. \tag{3-28}$$

Equations (3–26a) and (3–26b) can then be written in terms of the *two*
unknown loop currents i_a and i_b.

Loop analysis based on Kirchhoff's voltage law yields voltage equations
in which loop currents are unknown quantities. The following is a formalized procedure for loop analysis.

1. Determine the number of independent loops of the given network:
$N_l = N_b - N_j + N_s$.

2. Choose a total of N_l loops and assign clockwise directions to the N_l
loop currents. Make sure that at least one loop current flows through every
branch of the network.* (The clockwise assignment of loop current reference directions is not essential in this procedure.)

3. Write Kirchhoff's voltage law for each loop, using voltage-current
relationships of the elements in terms of loop currents.

---

* A sufficient (although not necessary) condition for the N_l loops to be all
independent is that, in assigning loops to the network, each additional loop
covers at least one branch which has not been covered by other loops.

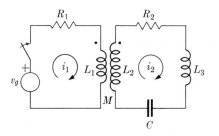

FIG. 3–5. Electrical network for Example 3–1.

4. Solve the system of N_l simultaneous equations for the loop currents desired.

5. Find the required branch currents or branch voltages.

EXAMPLE 3–1.   The network shown in Fig. 3–5 is initially relaxed. Write the loop equations that completely describe its electrical behavior after the switch is closed at $t = 0$.

Solution. First we determine the number of independent loops for the given network:
$$N_b = 7, \qquad N_j = 7, \qquad N_s = 2.$$
Hence
$$N_l = 7 - 7 + 2 = 2.$$

Two loops are chosen and reference directions for the two loop currents i_1 and i_2 are assigned as shown. We are now ready to write Kirchhoff's voltage law for the two loops:

Loop 1: $$R_1 i_1 + L_1 \frac{di_1}{dt} - M \frac{di_2}{dt} = v_g. \qquad (3\text{–}29)$$

Loop 2: $$-M \frac{di_1}{dt} + (L_2 + L_3) \frac{di_2}{dt} + R_2 i_2 + \frac{1}{C} \int_0^t i_2 \, dt = 0. \quad (3\text{–}30)$$

Equations (3–29) and (3–30) are two simultaneous equations from which the loop currents i_1 and i_2 can be solved, the exact solution being dependent upon the given v_g function and the network parameters. Equation (3–30) is seen to be an integro-differential equation, the solution of which was not discussed in Chapter 2. This equation could be tackled in either of two ways. We could differentiate the whole equation with respect to t once, which will yield

$$-M \frac{d^2 i_1}{dt^2} + (L_2 + L_3) \frac{d^2 i_2}{dt^2} + R_2 \frac{di_2}{dt} + \frac{1}{C} i_2 = 0. \qquad (3\text{–}31)$$

Equation (3–31) is now a second-order differential equation and can be solved simultaneously with Eq. (3–29) by the techniques of Section 2–6. Alternatively, we could substitute charge q for $\int_0^t i_2\,dt$, dq/dt for i_2, and d^2q/dt^2 for di_2/dt. Equations (3–29) and (3–30) then become

$$R_1 i_1 + L_1 \frac{di_1}{dt} - M \frac{d^2q}{dt^2} = v_g \qquad (3\text{–}32)$$

and

$$-M \frac{di_1}{dt} + (L_2 + L_3) \frac{d^2q}{dt^2} + R_2 \frac{dq}{dt} + \frac{1}{C} q = 0, \qquad (3\text{–}33)$$

which can be solved for i_1 and q. Once the charge q is found, i_2 can be obtained by a simple differentiation.

EXAMPLE 3–2. Write the loop equations for the network in Fig. 3–5, assuming that the transformer is ideal and has a primary-to-secondary turns ratio $N_1/N_2 = a$.

Solution. For an ideal transformer, winding inductances approach infinity, and the symbols L_1, L_2, and M are no longer useful. We redraw the network in Fig. 3–6. Now the voltages appearing across the primary and secondary windings and the currents in the windings of the ideal transformer are definitely related by the turns ratio according to Eqs. (3–22) and (3–24):

$$\frac{v_1}{v_2} = \frac{i_2}{i_1} = a. \qquad (3\text{–}34)$$

The two loop equations are:

Loop 1: $$R_1 i_1 + v_1 = v_g. \qquad (3\text{–}35)$$

Loop 2: $$R_2 i_2 + L_3 \frac{di_2}{dt} + \frac{1}{C} \int_0^t i_2\,dt = v_2. \qquad (3\text{–}36)$$

If we write v_1/a for v_2 and $a i_1$ for i_2 in Eq. (3–36) and substitute it

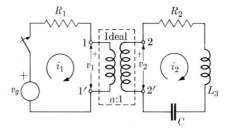

FIG. 3–6. Electrical network with ideal transformer.

into Eq. (3–35), we obtain

$$R_1 i_1 + a \left[a R_2 i_1 + a L_3 \frac{di_1}{dt} + \frac{a}{C} \int_0^t i_1 \, dt \right] = v_g$$

or

$$(R_1 + a^2 R_2) i_1 + (a^2 L_3) \frac{di_1}{dt} + \frac{1}{C/a^2} \int_0^t i_1 \, dt = v_g. \qquad (3\text{–}37)$$

Equation (3–35) contains only one dependent variable, i_1, which can be solved without difficulty. Examination of Eq. (3–37) reveals that it describes the single-loop network in Fig. 3–7 exactly. Hence, so far as the primary circuit of the network in Fig. 3–6 is concerned, Fig. 3–7 is an equivalent circuit which gives the correct i_1 and a correct v_1 between terminals 1–1′. To obtain the equivalent circuit, we transferred all the elements in the secondary circuit of the ideal transformer to the primary circuit such that resistances and inductances were multiplied by a^2 and capacitances were divided by a^2, where a is the turns ratio. Of course, when i_1 is found, i_2 is simply $a i_1$, and the original network in Fig. 3–6 is completely solved.

Instead of substituting Eq. (3–36) into Eq. (3–35), we could write $a v_2$ for v_1 and i_2/a for i_1 in Eq. (3–35) and substitute it into Eq. (3–36). If we did this we would obtain an equation in i_2:

$$R_2 i_2 + L_3 \frac{di_2}{dt} + \frac{1}{C} \int_0^t i_2 \, dt = \frac{1}{a} \left(v_g - R_1 \frac{i_2}{a} \right)$$

or

$$\left(R_2 + \frac{R_1}{a^2} \right) i_2 + L_3 \frac{di_2}{dt} + \frac{1}{C} \int_0^t i_2 \, dt = \frac{v_g}{a}. \qquad (3\text{–}38)$$

Equation (3–38) describes the single-loop network in Fig. 3–8 exactly. Hence, so far as the secondary circuit of the network in Fig. 3–6 is concerned, Fig. 3–8 is an equivalent circuit which gives the correct i_2 and a correct v_2 between terminals 2–2′. In obtaining the equivalent circuit, we

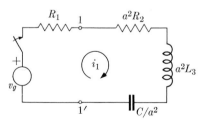

Fig. 3–7. An equivalent circuit for network in Fig. 3–6 referred to the primary.

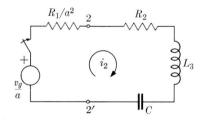

Fig. 3–8. An equivalent circuit for network in Fig. 3–6 referred to the secondary.

transferred all the elements in the primary circuit of the ideal transformer to the secondary circuit such that resistances were divided by a^2 and source voltages were divided by a, where a is the turns ratio. The ideal transformer is absent in the equivalent circuit, its effect having been taken care of properly through the transference of network elements.

The conclusion obtained in Example 3–2 in connection with the transfer of elements from the secondary circuit of an ideal transformer to the primary, and vice versa, are important techniques in the analysis of electrical networks containing ideal transformers. The same technique is useful in analyzing mechanical systems containing coupling devices, as we shall see in Chapter 4.

We can make the following general statements: The primary-circuit responses of an electrical network containing an *ideal* transformer of turns ratio $N_1:N_2 = a$ can be found by transferring all the elements (active and passive) in the secondary circuit to the primary circuit such that source voltages are multiplied by a, source currents by $1/a$, resistances and inductances by a^2, and capacitances by $1/a^2$. Similarly, the secondary-circuit responses of an electrical network containing an *ideal* transformer of turns ratio $N_1:N_2 = a$ can be found by transferring all the elements (active and passive) in the primary circuit to the secondary circuit such that source voltages are multiplied by $1/a$, source currents by a, resistances and inductances by $1/a^2$, and capacitances by a^2.

3–5 Kirchhoff's current law and node analysis. Kirchhoff's current law is a restatement of the law of conservation of charge:

At any junction (node) in an electrical network the algebraic sum of currents in a direction away from the junction is zero. Or, at a node,

$$\sum_k i_k(t) = 0, \tag{3–39}$$

which is true at all instants of time. This law can also be stated in other ways. For example, we may say that at any node in an electrical network the sum of the currents leaving is equal to the sum of the currents entering. The same equations will result when we apply either form of this law to a node.

Kirchhoff's current law is the basis of *node analysis* of electrical networks. Let us apply it to the network in Fig. 3–9. Four junctions or nodes are in evidence: a, b, c, and d. *Reference directions* have been assigned to the branch currents. We emphasize again that the assignment of reference direction is arbitrary. Reference directions do not necessarily correspond

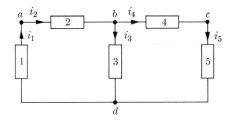

FIG. 3–9. Application of Kirchhoff's current law.

to actual current directions; they are algebraic quantities. Reference directions are assigned to currents so that network equations can be written with no ambiguities in sign. The application of Kirchhoff's current law to each of the four nodes yields the following equations:

$$\text{Node } a: \quad -i_1 + i_2 = 0, \tag{3–40a}$$

$$\text{Node } b: \quad -i_2 + i_3 + i_4 = 0, \tag{3–40b}$$

$$\text{Node } c: \quad -i_4 + i_5 = 0, \tag{3–40c}$$

$$\text{Node } d: \quad i_1 - i_3 - i_5 = 0. \tag{3–40d}$$

We note that Eq. (3–40d) is the negative of the sum of the first three equations; hence the four equations above are not all independent. Here again we need to determine the total number of independent node equations that completely describe a given network. Looking at Fig. 3–9, we could have predicted that the equation we wrote for node d would not be independent of the equations for nodes a, b, and c, since the other ends of branches 1, 3, and 5 that come together at node d also go respectively to nodes a, b, and c. Restrictions have been imposed on the currents i_1, i_3, and i_5 in Eqs. (3–40a), (3–40b), and (3–40c) so that the law of conservation of charge is satisfied; an equation connecting i_1, i_3, and i_5 at node d then would not give anything new.

In general, if N_j is the total number of junctions and N_s is the number of separate parts in an electrical network, then the number of independent node equations, N_n, obtained by the application of Kirchhoff's current law is given by

$$N_n = N_j - N_s. \tag{3–41}$$

For the network in Fig. 3–9, we have $N_j = 4$ and $N_s = 1$; hence $N_n = 4 - 1 = 3$. Which three of the four nodes we choose is immaterial; the remaining node is called the *datum node* or the *reference node*.

Obviously, we cannot solve for the five branch currents i_1, i_2, i_3, i_4, and i_5 directly from three independent node equations. In node analysis, we write current equations in which N_n node-pair voltages between N_n

independent nodes and the datum node are unknown quantities. Since the voltage across each element can be expressed as a combination of the node-pair voltages referred to the datum node, and since each element has a definite voltage-current relationship, the analysis of the network is essentially complete when the N_n simultaneous node equations are solved.

A complete description of an electrical network with N_b branches requires the knowledge of all the N_b branch currents or, alternatively, of all the N_b branch voltages which are definitely related to the branch currents. In using loop analysis, we solve N_l independent loop equations and express all branch currents in terms of N_l loop currents. When node analysis is used, we solve N_n independent node equations and express all branch voltages in terms of N_n node-pair voltages. N_l and N_n for a given network are not necessarily equal. (The network in Fig. 3–9, for example, has $N_l = 2$ and $N_n = 3$.) Hence the decision of whether to use loop analysis or node analysis depends upon whether N_l or N_n is the smaller. When $N_l = N_n$, the choice rests upon which method is more convenient for the particular problem at hand; factors such as the type of sources and the required response may deserve careful consideration.

The following is a formalized procedure for node analysis.

1. Determine the number of independent nodes of the given network: $N_n = N_j - N_s$.

2. Choose any one node as the datum or reference node. (The criterion for this choice is that the simplest set of equations will result. For most networks, the choice is obvious.) The reference polarities for all independent nodes are taken as positive with respect to the datum node.

3. Write Kirchhoff's current law for each independent node, using voltage-current relationships of the elements in terms of node-pair voltages.

4. Solve the system of N_n simultaneous equations for the node-pair voltages desired.

5. Find the required branch voltages or branch currents.

EXAMPLE 3–3. The network shown in Fig. 3–10 is initially at rest. Write the node equations that completely describe its electrical behavior after the switch is opened at $t = 0$.

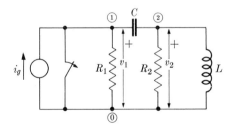

FIG. 3–10. Electrical network for Example 3–3.

Solution. For the given network,

$$N_b = 5, \qquad N_j = 3, \qquad N_s = 1.$$

Hence

$$N_n = 3 - 1 = 2 \qquad (N_l = 5 - 3 + 1 = 3 > N_n).$$

Nodes 1 and 2 have been chosen as the independent nodes, and reference polarities of node-pair voltages v_1 and v_2 with respect to the datum node 0 are assigned as shown. Applying Kirchhoff's current law to the two independent nodes, we obtain

Node 1: $$\frac{1}{R_1} v_1 + C \frac{d}{dt} (v_1 - v_2) = i_g. \tag{3-42}$$

Node 2: $$C \frac{d}{dt} (v_2 - v_1) + \frac{1}{R_2} v_2 + \frac{1}{L} \int_0^t v_2 \, dt = 0. \tag{3-43}$$

Equations (3–42) and (3–43) are two simultaneous equations from which the node-pair voltages v_1 and v_2 can be solved, the exact solution being dependent upon the given i_g function and the network parameters. The integro-differential equation (3–43) could be tackled in either of two ways. We could differentiate the whole equation with respect to t once, which will yield

$$C \frac{d^2}{dt^2} (v_2 - v_1) + \frac{1}{R_2} \frac{dv_2}{dt} + \frac{1}{L} v_2 = 0. \tag{3-44}$$

Equation (3–44) is now a second-order differential equation and can be solved simultaneously with Eq. (3–42). Alternatively, we could substitute flux linkage ϕ for $\int_0^t v_2 \, dt$, $d\phi/dt$ for v_2, and $d^2\phi/dt^2$ for dv_2/dt. Equations (3–42) and (3–43) then become

$$\frac{1}{R} v_1 + C \frac{dv_1}{dt} - C \frac{d^2\phi}{dt^2} = i_g \tag{3-45}$$

and

$$C \frac{d^2\phi}{dt^2} - C \frac{dv_1}{dt} + \frac{1}{R_2} \frac{d\phi}{dt} + \frac{1}{L} \phi = 0, \tag{3-46}$$

which can be solved for v_1 and ϕ. Once the flux linkage function ϕ is found, v_2 can be obtained by a simple differentiation.

EXAMPLE 3–4. The network in Fig. 3–11 is initially under steady-state conditions. At $t = 0$ the switch across the current source $i_0(t)$ is opened. Obtain the integro-differential equation for the voltage across the inductance, v_L.

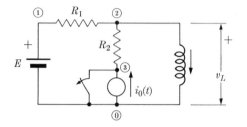

Fig. 3–11. Electrical network for Example 3–4.

Solution. This example is designed to illustrate two things: (1) the method of setting up network equations when the electrical network contains both voltage and current sources, and (2) the method of setting up network equations when the network is initially not at rest. Let us use node analysis. We always write the equations which hold *after* the network is disturbed, that is, for $t > 0$. With the switch open,

$$N_n = N_j - N_s = 4 - 1 = 3.$$

The given network then has three independent nodes, 1, 2, and 3 as indicated in Fig. 3–11, for which three independent node equations can be written. We recall that node equations are current equations in which the unknown quantities are node-pair voltages. Let us denote the node-pair voltages between the independent nodes and the datum node by v_1, v_2, and v_3 respectively. We find that we cannot conveniently write a current equation for node 1 because we have no way of expressing the current in the voltage source in terms of E and its zero internal resistance. However, this situation should be a cause for joy, not one for frustration! For node 1, we write simply

$$v_1 = E. \tag{3–47}$$

Equation (3–47) is true at all times; hence v_1 is not really an unknown quantity.

For node 2, the application of Kirchhoff's current law yields

$$\frac{1}{R_1}(v_2 - E) + \frac{1}{R_2}(v_2 - v_3) + \frac{1}{L}\int_0^t v_2 \, dt + i_L(0) = 0. \tag{3–48}$$

For node 3, we have

$$\frac{1}{R_2}(v_3 - v_2) - i_0(t) = 0. \tag{3–49}$$

Substituting Eq. (3–49) into Eq. (3–48) and writing v_L for v_2, we have

$$\frac{1}{R_1}(v_L - E) - i_0(t) + \frac{1}{L}\int_0^t v_L \, dt + i_L(0) = 0, \tag{3–50}$$

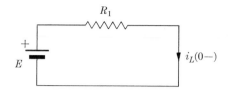

FIG. 3–12. Circuit for evaluating $i_L(0-)$ at $t = 0-$.

which contains only a single unknown, v_L. Equation (3–50) is actually the only equation needed in this problem; it could have been written directly for node 2. The current in R_2 (or any other passive element in its place) must be $i_0(t)$, flowing toward node 2, since it is connected in series with the current generator. We note that R_2 does not even appear in Eq. (3–50)!

We must now determine $i_L(0)$ from the conditions at $t = 0-$ when the current source is short-circuited by the switch. Since an inductance is effectively a short circuit to the d-c voltage source E under steady-state conditions, the circuit in Fig. 3–12 can be used to evaluate $i_L(0-)$. We have

$$i_L(0-) = \frac{E}{R_1} = i_L(0+) \qquad (3\text{--}51)$$

by virtue of Eq. (3–16). The $+$ and $-$ signs after the zero are then not needed. Substituting $i_L(0) = E/R_1$ in Eq. (3–50) and rearranging terms, we have the required equation for v_L:

$$\frac{1}{R_1} v_L + \frac{1}{L} \int_0^t v_L \, dt = i_0(t). \qquad (3\text{--}52)$$

3–6 Dual networks. In preceding sections we have discussed the application of Kirchhoff's voltage and current laws in setting up loop and node equations for an electrical network in which loop currents and node-pair voltages, respectively, are the unknown quantities. We have certainly sensed a degree of parallelism in these two methods of analysis. This feeling is enhanced if we also recall the similarity between the voltage-current relationship of an inductance and the current-voltage relationship of a capacitance, and vice versa (see Table 3–1). It is therefore not unreasonable to expect that two analogous electrical networks exist which, when analyzed by the two methods, will yield the same types of equations. Two electrical networks that are governed by the same set of equations are called *dual networks*. Dual electrical networks are a special type of analogous systems which will be discussed in more general terms in the next chapter.

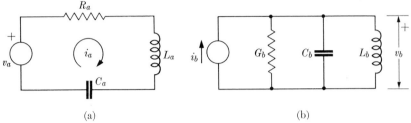

(a)   (b)

FIG. 3–13. Dual electrical circuits.

Let us consider the series R_a-L_a-C_a circuit fed by a voltage source v_a as shown in Fig. 3–13(a), and the parallel G_b-C_b-L_b circuit* fed by a current source i_b as shown in Fig. 3–13(b). The governing equations are:

For Fig. 3–13(a),

$$L_a \frac{di_a}{dt} + R_a i_a + \frac{1}{C_a} \left[\int_0^t i_a \, dt + q_a(0) \right] = v_a. \tag{3-53}$$

For Fig. 3–13(b),

$$C_b \frac{dv_b}{dt} + G_b v_b + \frac{1}{L_b} \left[\int_0^t v_b \, dt + \phi_b(0) \right] = i_b. \tag{3-54}$$

Comparing Eqs. (3–53) and (3–54), we note the similarity immediately. In fact, the solution of Eq. (3–53) will be identical to the solution of Eq. (3–54) when the following changes of symbols are made:

$$L_a \rightarrow C_b, \qquad R_a \rightarrow G_b, \qquad C_a \rightarrow L_b,$$

$$v_a \rightarrow i_b, \qquad i_a \rightarrow v_b, \qquad q_a \rightarrow \phi_b.$$

The series circuit in Fig. 3–13(a) and the parallel circuit in Fig. 3–13(b) are therefore *duals* of each other, and there is no need to analyze both types of circuits, inasmuch as the solution of one automatically gives the solution of the other with a suitable change of symbols for the physical quantities. Table 3–2 lists the corresponding quantities for dual electrical circuits.

It is very important to note that *dual circuits are not the same as equivalent circuits.†* The parallel circuit in Fig. 3–13(b) is not equivalent to the series circuit in Fig. 3–13(a) because not only the individual elements but also the entire configuration of the circuit has been altered. The voltage

---

* The symbol G is used to denote the *conductance*, or the reciprocal of the resistance, of a resistive element. Its unit is mho, the reciprocal of ohm.

† See Section 3–8 for a more detailed discussion of equivalent circuits.

TABLE 3–2

TABLE OF CONVERSION FOR DUAL ELECTRICAL CIRCUITS

Loop basis	Node basis
A loop comprising several branches	A node joining the same number of branches
Voltage sources	Current sources
Loop currents	Node-pair voltages (referring to datum node)
Inductances	Capacitances
Resistances	Conductances
Capacitances	Inductances

appearing across L_b, for example, is certainly not the same as that appearing across L_a; and there is no relation whatever between L_a and L_b in these two circuits.

Only planar networks (networks which can be laid flat on a plane without branches crossing one another) have duals. For these networks there is a convenient graphical method of obtaining the duals, based on the correspondence between the two columns of quantities for dual electrical circuits listed in Table 3–2. It is called the *dot method* and may be carried out according to the following procedure:

1. Put a dot in each independent loop of the given network. These internal dots correspond to the independent (nondatum) nodes in the dual network.

2. Put a dot outside the given network. This external dot corresponds to the datum node in the dual network.

3. Connect all internal dots in adjacent loops by dashed lines crossing the common branches. Elements which are the duals of the common branches will form the branches connecting the corresponding independent nodes in the dual network. For example, if a common branch in the original network has R-L-C in series, then a parallel combination of G-C-L should be put between the corresponding nodes in the dual network.

4. Connect all internal dots to the external dot by dashed lines crossing all external branches. Duals of these external branches will form the branches connecting independent nodes and the datum node.

A convention should be established between the reference voltage polarities in the given network and the reference current directions in its dual, and vice versa, in order to avoid ambiguity. We shall agree then that (1) a clockwise current in a loop corresponds to a positive polarity (with respect to the datum node) at the dual independent node, and

(2) a voltage rise in the direction of a clockwise loop current corresponds to a current flowing toward the dual independent node. These additional rules* should be carefully checked in constructing duals for networks that contain active elements.

The above procedure can be tested readily on the circuits in Fig. 3–13, by constructing the dual in (b) from the given circuit in (a), or by constructing the dual in (a) from the given circuit in (b).

EXAMPLE 3–5. Construct the dual electrical circuit for the network of Fig. 3–14.

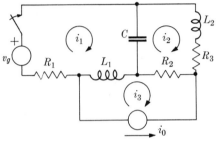

FIG. 3–14. An electrical network.

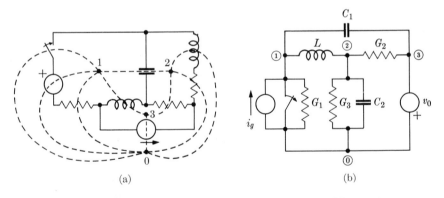

FIG. 3–15. Construction of the dual for network in Fig. 3–14.

Solution. Figure 3–14 is redrawn in Fig. 3–15(a) with three internal nodes 1, 2, and 3 (corresponding to the three independent loops) and an external node 0 in place. Dashed lines connecting the nodes and crossing *each element* in the common and external branches are also shown. The

---

* These rules conform with the convention that reference loop currents are chosen as clockwise in loop analysis and that node-to-datum voltages are chosen as positive at independent nodes in node analysis.

dual network obviously has three independent nodes and is drawn in Fig. 3–15(b); all dual elements carry the same subscripts as the original elements given in Fig. 3–14. Note that i_g flows *toward* node 1 because the original v_g is a voltage rise in loop 1 (in the direction of the clockwise i_1) and that the reference polarity for v_0 is *negative* on node 3 because the original i_0 is counterclockwise (counter to the reference loop current i_3). Note further that a closing switch in series with a voltage source corresponds to an opening switch in parallel with a current source in the dual network.

The reader can readily check the duality of the networks in Figs. 3–14 and 3–15(b) by writing the integro-differential equations for these two networks. Figure 3–15(b) can be similarly converted into its dual (the original network in Fig. 3–14) by applying the same dot method.

3–7 Solutions of some typical problems. So far in this chapter we have confined our discussion to the methods of setting up equations that describe the behavior of an electrical network. Of course the analysis is not complete until these equations have been solved for the desired response. To this end we have already been provided with the necessary techniques in Chapter 2, where classical solutions of linear differential equations were developed. In this section we shall analyze some typical electrical network problems completely and, in the process, also bring out such important concepts as time constant, resonance, impedance, etc.

Example 3–6. The switch in the series R-L circuit in Fig. 3–16 is closed at $t = 0$. Determine the current for (a) $v_g = V_0$ (a constant), and for (b) $v_g = V_g \sin \omega t$.

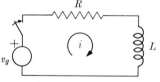

Fig. 3–16. A series R-L circuit.

Solution. Obviously, it is simplest to use loop analysis in this problem since it has only one loop and since the desired response is the loop current. Application of Kirchhoff's voltage law yields the following equation:

$$L \frac{di}{dt} + Ri = v_g. \tag{3–55}$$

The solution of Eq. (3–55) consists of two parts, namely, the complementary function and the particular integral. The *complementary function* is the general solution of the homogeneous equation with the right side of Eq. (3–55) set to zero; *it corresponds to the transient solution of the physical problem.* The characteristic equation is

$$Ls + R = 0, \tag{3-56}$$

which has a single root $s = -R/L$. Thus, writing i_{tr} as the transient current, we have

$$i_{tr} = K\epsilon^{-Rt/L}, \tag{3-57}$$

where K is an arbitrary constant to be determined from the initial condition *after the complete solution has been obtained.* We see that i_{tr} eventually dies out as t increases indefinitely.

The *particular integral* is the particular solution of the nonhomogeneous equation (3–55) containing no arbitrary constants; *it corresponds to the steady-state solution of the physical problem.* We shall use the symbol i_{ss} for this part of the solution.

(a) $v_g = V_0$ (*a constant*).

From Section 2–4, we know that the particular integral is a constant, and

$$i_{ss} = \frac{V_0}{R}. \tag{3-58}$$

Combining Eqs. (3–57) and (3–58), we have

$$i = K\epsilon^{-Rt/L} + \frac{V_0}{R}. \tag{3-59}$$

The constant K is determined from the condition that the circuit is initially at rest, that is,

$$i = 0 \quad \text{at} \quad t = 0.$$

Hence, from Eq. (3–59),

$$0 = K + \frac{V_0}{R}$$

or

$$K = -\frac{V_0}{R}, \tag{3-60}$$

and the complete solution is then

$$i = \frac{V_0}{R}(1 - \epsilon^{-Rt/L}). \tag{3-61}$$

A typical i versus t curve represented by Eq. (3–61) is plotted in Fig. 3–17. The current builds up from its zero initial value at $t = 0$ toward

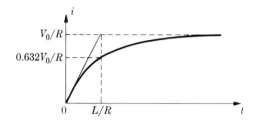

FIG. 3–17. Typical i versus t curve for a series $R\text{-}L$ circuit when $v_g = V_0$ (a constant).

the final value V_0/R. At $t = L/R$, i becomes

$$i = \frac{V_0}{R}(1 - \epsilon^{-1}) = 0.632\frac{V_0}{R}, \tag{3–62}$$

or 63.2 percent of its final value. The ratio L/R has the dimension of time and is called the *time constant* of the series $R\text{-}L$ circuit. It is a convenient measure of the rate at which the current builds up to its steady-state or final value, and is used frequently in the jargon of electrical engineers. We note that the time constant is a characteristic of the transient response of the circuit and is independent of the magnitude of the applied voltage V_0. We further note that there is no direct relationship between the time constant L/R and the general slope of the i versus t curve, the latter being a variable equal to

$$\frac{di}{dt} = \frac{V_0}{L}\epsilon^{-Rt/L}. \tag{3–63}$$

The initial slope is equal to V_0/L; hence the tangent line at $t = 0$ intersects the V_0/R line at $t = L/R$, as shown.

(b) $v_g = V_g \sin \omega t$.

The nonhomogeneous equation is now

$$L\frac{di}{dt} + Ri = V_g \sin \omega t. \tag{3–64}$$

To find the particular integral, let us use the method of undetermined coefficients and assume a solution of the following form in accordance with Eq. (2–99):

$$i_{ss} = B_1\epsilon^{j\omega t} + B_2\epsilon^{-j\omega t}. \tag{3–65}$$

Substituting Eq. (3–65) in Eq. (3–64) and equating the coefficients, we obtain

$$B_1 = \frac{V_g}{2j(R + j\omega L)} = \frac{V_g}{2jZ} \tag{3–66}$$

and

$$B_2 = -\frac{V_g}{2j(R - j\omega L)} = -\frac{V_g}{2jZ^*}, \tag{3-67}$$

where the complex quantity

$$Z = R + j\omega L = |Z|\epsilon^{j\theta_z}, \tag{3-68}$$

called the *impedance* of the series *R-L* circuit, has a magnitude $|Z|$ and a phase angle θ_z:

$$|Z| = \sqrt{R^2 + \omega^2 L^2}, \tag{3-69}$$

$$\theta_z = \tan^{-1}(\omega L/R). \tag{3-70}$$

The quantity $Z^* = |Z|\epsilon^{-j\theta_z}$ is the complex conjugate of Z. Putting B_1 and B_2 obtained in Eqs. (3-66) and (3-67) into Eq. (3-65), we get

$$\begin{aligned}
i_{ss} &= \frac{V_g}{2j}\left[\frac{\epsilon^{j\omega t}}{Z} - \frac{\epsilon^{-j\omega t}}{Z^*}\right] \\
&= \frac{V_g}{2j|Z|}[\epsilon^{j(\omega t - \theta_z)} - \epsilon^{-j(\omega t - \theta_z)}] \\
&= \frac{V_g}{|Z|}\sin(\omega t - \theta_z).
\end{aligned} \tag{3-71}$$

Combining the i_{tr} in Eq. (3-57) and the i_{ss} in Eq. (3-71), we have

$$i = i_{tr} + i_{ss} = K\epsilon^{-Rt/L} + \frac{V_g}{|Z|}\sin(\omega t - \theta_z). \tag{3-72}$$

To determine K, we set $i = 0$ at $t = 0$, and obtain

$$K = \frac{V_g}{|Z|}\sin\theta_z. \tag{3-73}$$

Finally, we have

$$i = \frac{V_g}{|Z|}[\epsilon^{-Rt/L}\sin\theta_z + \sin(\omega t - \theta_z)], \tag{3-74}$$

where $|Z|$ and θ_z are given in Eqs. (3-69) and (3-70) respectively.

The exact i versus t curve will depend upon V_g, ω, R, and L. Figure 3-18 represents a typical sketch for the current curves.

In electrical engineering a *sinor*† notation is employed to simplify manipulation of sinusoidally varying quantities. A sinor is a time-inde-

---

† Also called a *phasor*.

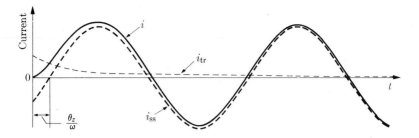

FIG. 3–18. Typical current curves for a series R-L circuit when $v_g = V_g \sin \omega t$.

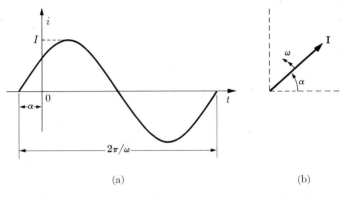

(a)   (b)

FIG. 3–19. The sinor notation. (a) $i = I \sin(\omega t + \alpha)$. (b) Sinor $\mathbf{I} = I\epsilon^{j\alpha}$ symbolizing the instantaneous variation of i in (a).

pendent complex coefficient which together with the factor $\epsilon^{j\omega t}$ yields a complex time function. A sinor such as $\mathbf{I} = I\epsilon^{j\alpha}$ is representable graphically by an arrow of magnitude I and initial phase angle α which, when rotating with an angular frequency ω, has a projection on the vertical axis whose instantaneous value is equal to $i = I \sin(\omega t + \alpha)$. This is depicted in Fig. 3–19. The advantage of employing such a notation is that sinusoidally varying quantities can be handled as complex quantities in determining the magnitude and phase angle of a sinusoidally varying response without tedious manipulations of trigonometric functions of time. Thus, in the present problem, we can use $\mathbf{V}_g = V_g\epsilon^{j0}$ to *symbolize* $v_g = V_g \sin(\omega t + 0)$. The complex impedance $Z = |Z|\epsilon^{j\theta_z}$ is the ratio of sinor voltage $\mathbf{V}_g$ and sinor steady-state current $\mathbf{I}_{ss}$. (Impedance Z, although a complex quantity, is not a sinor because it does not symbolize a quantity that varies sinusoidally with time.)

$$\mathbf{I}_{ss} = \frac{\mathbf{V}_g}{Z} = \frac{V_g}{|Z|\epsilon^{j\theta_z}} = \frac{V_g}{|Z|}\epsilon^{-j\theta_z}, \tag{3–75}$$

which, according to our definition, *symbolizes* the instantaneous current i_{ss} in Eq. (3–71). It is very important to remember that *sinor notations can be used to symbolize only sinusoidally varying quantities under steady-state conditions.* One *must not* mix sinors and instantaneous, time-varying quantities in the same equation.

It is to be noted that, by virtue of the duality principle discussed in the preceding section, in solving for the current in the series R-L circuit given in Fig. 3–16, we have essentially also determined the node-pair voltage v for the parallel G-C circuit in Fig. 3–20, where the switch across the current source i_g is opened at $t = 0$.

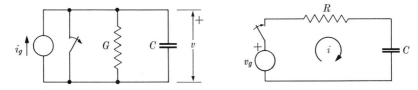

Fig. 3–20. Dual electrical circuit for   Fig. 3–21. A series R-C circuit.
Fig. 3–16.

EXAMPLE 3–7.   The switch in the series R-C circuit in Fig. 3–21 is closed at $t = 0$. Determine the current for (a) $v_g = V_0$ (a constant), and for (b) $v_g = V_g \sin \omega t$.

Solution. Application of Kirchhoff's voltage law to the given circuit yields the following equation:

$$Ri + \frac{1}{C} \int_0^t i\, dt = v_g. \tag{3–76}$$

In Eq. (3–76) we have assumed that there is no initial voltage across the capacitance. If we make use of the relation

$$i = \frac{dq}{dt}, \tag{3–77}$$

Eq. (3–76) can be converted to

$$R \frac{dq}{dt} + \frac{1}{C} q = v_g, \tag{3–78}$$

which is seen to be of exactly the same form as Eq. (3–55) in the previous example. Hence the solutions for i of Eq. (3–55) in Example 3–6 could be borrowed in entirety as those for q of Eq. (3–78) save for the following changes of symbols:

$$i \rightarrow q, \qquad L \rightarrow R, \qquad R \rightarrow 1/C.$$

(a) $v_g = V_0$ (*a constant*).

From Eq. (3–61) with the changes of symbols:

$$q = CV_0(1 - \epsilon^{-t/CR}), \qquad (3\text{–}79)$$

which, after differentiation with respect to t, yields the desired current:

$$i = \frac{V_0}{R} \epsilon^{-t/CR}. \qquad (3\text{–}80)$$

A typical i versus t curve represented by Eq. (3–80) is plotted in Fig. 3–22. The current decays from an initial value V_0/R at $t = 0$ toward zero. We see clearly that the current in an R-C circuit can have sudden jumps. (In this example, it jumps from 0 to V_0/R at $t = 0$.) At $t = CR$, i becomes

$$i = \frac{V_0}{R} \epsilon^{-1} = 0.368 \frac{V_0}{R}, \qquad (3\text{–}81)$$

or 36.8 percent of its initial value, and

$$q = CV_0(1 - \epsilon^{-1}) = 0.632\, CV_0, \qquad (3\text{–}82)$$

or 63.2 percent of its final value. The product RC has the dimension of time and is called the *time constant* of the series R-C circuit. It is a convenient measure of the rate at which the capacitor is being charged. Again we note that the time constant is a characteristic of the circuit and is independent of the magnitude of the applied voltage V_0. The slope of the current curve is

$$\frac{di}{dt} = -\frac{V_0}{CR^2} \epsilon^{-t/CR}, \qquad (3\text{–}83)$$

which has an initial value $-V_0/CR^2$. The tangent line at $t = 0$ intersects the time axis at $t = CR$, as shown.

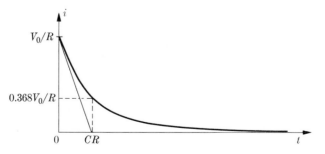

Fig. 3–22. Typical i versus t curve for a series R-C circuit when $v_g = V_0$ (a constant).

(b) $v_g = V_g \sin \omega t$.

From Eqs. (3–74), (3–69), and (3–70), with the proper changes of symbols, we have

$$q = \frac{V_g}{\sqrt{(1/C)^2 + \omega^2 R^2}} [\epsilon^{-t/CR} \sin (\tan^{-1} \omega C R) + \sin (\omega t - \tan^{-1} \omega C R)]$$

$$(3\text{–}84)$$

and

$$i = \frac{V_g}{\omega \sqrt{R^2 + (1/\omega C)^2}}$$

$$\times \left[-\frac{\epsilon^{-t/CR}}{CR} \sin (\tan^{-1} \omega C R) + \omega \cos (\omega t - \tan^{-1} \omega C R) \right] \cdot \quad (3\text{–}85)$$

But

$$\sin (\tan^{-1} \omega C R) = \frac{\omega C R}{\sqrt{1 + \omega^2 C^2 R^2}} = \frac{R}{\sqrt{R^2 + (1/\omega C)^2}} \quad (3\text{–}86)$$

and

$$\cos (\omega t - \tan^{-1} \omega C R) = \sin [\omega t + (90° - \tan^{-1} \omega C R)]$$

$$= \sin \left(\omega t + \tan^{-1} \frac{1}{\omega C R} \right) \cdot \quad (3\text{–}87)$$

Substituting Eqs. (3–86) and (3–87) in Eq. (3–85), we obtain

$$i = \frac{V_g}{\sqrt{R^2 + (1/\omega C)^2}}$$

$$\times \left[-\frac{\epsilon^{-t/CR}}{\omega C \sqrt{R^2 + (1/\omega C)^2}} + \sin \left(\omega t + \tan^{-1} \frac{1}{\omega C R} \right) \right]$$

$$= i_{\text{tr}} + i_{\text{ss}}, \quad (3\text{–}88)$$

where

$$i_{\text{tr}} = -\frac{V_g \epsilon^{-t/CR}}{\omega C [R^2 + (1/\omega C)^2]} = -\frac{\omega C V_g}{1 + (\omega C R)^2} \epsilon^{-t/CR} \quad (3\text{–}89)$$

and

$$i_{\text{ss}} = \frac{V_g}{\sqrt{R^2 + (1/\omega C)^2}} \sin \left(\omega t + \tan^{-1} \frac{1}{\omega C R} \right) \cdot \quad (3\text{–}90)$$

If we use sinor notation for the steady-state part of the solution, we have, as in Eq. (3–75),

$$\mathbf{I}_{\text{ss}} = \frac{\mathbf{V}_g}{Z} = \frac{V_g}{|Z|} \epsilon^{-j\theta_z}. \quad (3\text{–}75)$$

Comparing the symbolic notation in Eq. (3–75) with the expression for i_{ss} in Eq. (3–90), we see that the complex impedance at angular frequency ω for the series R-C circuit is

$$Z = R + \frac{1}{j\omega C} = |Z|\epsilon^{j\theta_z}, \qquad (3\text{–}91)$$

where

$$|Z| = \sqrt{R^2 + (1/\omega C)^2} \qquad (3\text{–}92)$$

and

$$\theta_z = -\tan^{-1}\left(\frac{1}{\omega C R}\right). \qquad (3\text{–}93)$$

It is to be noted that, as we commonly define it, the "impedance" of a circuit containing energy-storage elements (L or C) refers to a definite frequency and is meaningful only for sinusoidal excitation under steady-state conditions.* Under steady-state conditions the impedance of an inductance L to a sinusoidal current of angular frequency ω is $j\omega L$, and that of a capacitance C is $1/j\omega C$.† The total impedance of a series or parallel combination of R, $j\omega L$, and $1/j\omega C$ can be computed in the conventional manner.

Figure 3–23 represents a typical sketch for the current curves in an R-C circuit with a sinusoidal applied voltage.

By virtue of the principle of duality, solution for the current in the series R-C circuit shown in Fig. 3–21 gives at the same time the solution for the node-pair voltage v for the parallel G-L circuit in Fig. 3–24, where the switch across the current source i_g is opened at $t = 0$.

---

* It is possible to extend the impedance concept to exponential (instead of sinusoidal) excitations. Thus, to an excitation of the ϵ^{st} type, the impedance of a series R-L circuit would be $Z_{R\text{-}L}(s) = R + sL$, and that of a series R-C circuit would be $Z_{R\text{-}C}(s) = R + (1/sC)$. However, the impedance concept is not meaningful in the discussion of transient responses; nor is it correct to speak of the impedance of a circuit to, say, a square wave. (A periodic square wave can be resolved into sinusoidal components, and we can talk about the impedance of a circuit to *each* of these components. See Chapter 5.)

† Since pure inductances and capacitances do not dissipate power, their impedances are purely imaginary and are usually written as j times a *reactance*. Thus,

$$Z_L = j\omega L = jX_L \qquad \text{where} \qquad X_L = \omega L,$$

and

$$Z_C = \frac{1}{j\omega C} = jX_C \qquad \text{where} \qquad X_C = -\frac{1}{\omega C}.$$

The reactance of an inductance is positive (ωL) and the reactance of a capacitance is negative ($-1/\omega C$).

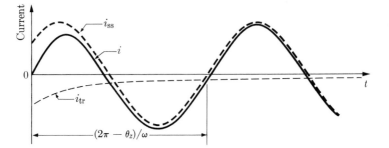

FIG. 3-23. Typical current curves for a series R-C circuit when $v_g = V_g \sin \omega t$.

EXAMPLE 3-8. The switch in the series R-L-C circuit in Fig. 3-25 is closed at $t = 0$. Determine the current for (a) $v_g = V_0$ (a constant), and for (b) $v_g = V_g \sin \omega t$.

Solution. The loop equation for the given circuit, assuming zero initial conditions, is

$$L \frac{di}{dt} + Ri + \frac{1}{C} \int_0^t i \, dt = v_g. \tag{3-94}$$

The complementary function, or the transient part of the solution, for Eq. (3-94) is independent of the excitation and can be obtained by solving the following homogeneous equation in operator notation:*

$$\left(LD^2 + RD + \frac{1}{C}\right) i_{tr} = 0. \tag{3-95}$$

---

* The integro-differential equation (3-94) can be considered as a second-order differential equation in q:

$$L \frac{d^2q}{dt^2} + R \frac{dq}{dt} + \frac{1}{C} q = v_g,$$

which has a characteristic equation

$$Ls^2 + Rs + \frac{1}{C} = L(s - s_1)(s - s_2) = 0.$$

The transient solution for q is then

$$q_{tr} = c_1 \epsilon^{s_1 t} + c_2 \epsilon^{s_2 t}.$$

The corresponding current,

$$i_{tr} = \frac{dq_{tr}}{dt} = c_1 s_1 \epsilon^{s_1 t} + c_2 s_2 \epsilon^{s_2 t} = K_1 \epsilon^{s_1 t} + K_2 \epsilon^{s_2 t},$$

is the same as Eq. (3-99).

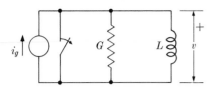

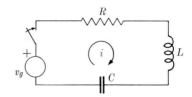

FIG. 3–24. Dual electrical circuit for Fig. 3–21.   FIG. 3–25. A series R-L-C circuit.

The characteristic equation corresponding to Eq. (3–95) is

$$Ls^2 + Rs + \frac{1}{C} = 0, \tag{3-96}$$

which has two roots:

$$s_1 = -\frac{R}{2L} + \sqrt{\left(\frac{R}{2L}\right)^2 - \frac{1}{LC}} = -\alpha + b, \tag{3-97}$$

$$s_2 = -\frac{R}{2L} - \sqrt{\left(\frac{R}{2L}\right)^2 - \frac{1}{LC}} = -\alpha - b. \tag{3-98}$$

Hence

$$i_{tr} = K_1 \epsilon^{s_1 t} + K_2 \epsilon^{s_2 t}, \tag{3-99}$$

where K_1 and K_2 are two arbitrary constants to be determined from two initial conditions after the complete solution for i has been obtained.

Before we take up the steady-state solution, let us pause and examine the behavior of the transient solution in Eq. (3–99). Depending upon the relative values of R, L, and C, three different situations are possible.

(i) $(R/2L)^2 > (1/LC)$, or $R > 2\sqrt{L/C}$; b is real.

$$i_{tr} = \epsilon^{-\alpha t}(K_1 \epsilon^{bt} + K_2 \epsilon^{-bt}). \tag{3-100}$$

Since $\alpha > b$, the transient current consists of two exponentially decreasing components with different rates of decay. This is called the *overdamped* case and occurs when the resistance in the circuit is large. The exponential functions with real exponents can be expressed in terms of hyperbolic sines and cosines:

$$\epsilon^{bt} = \cosh bt + \sinh bt, \tag{3-101}$$

$$\epsilon^{-bt} = \cosh bt - \sinh bt. \tag{3-102}$$

An alternative form for an overdamped i_{tr} is then

$$i_{tr} = \epsilon^{-\alpha t}(A_1 \cosh bt + A_2 \sinh bt), \tag{3-103}$$

where the constants A_1 and A_2 stand for $(K_1 + K_2)$ and $(K_1 - K_2)$, respectively.

(ii) $(R/2L)^2 = (1/LC)$, or $R = 2\sqrt{L/C}$; $b = 0$.

Under this condition the characteristic equation (3–96) has a double root, and the transient solution becomes

$$i_{\mathrm{tr}} = (K_1 + K_2 t)\epsilon^{-\alpha t}. \qquad (3\text{–}104)$$

The two terms in Eq. (3–104) are both nonoscillatory. The first term is a simple exponential function starting from a value K_1 at $t = 0$ and decaying at a rate equal to the *damping coefficient* α; the second term starts from zero at $t = 0$, increases toward a maximum, and then decreases toward zero when t is very large. This is called the *critically damped* case.

(iii) $(R/2L)^2 < (1/LC)$, or $R < 2\sqrt{L/C}$; b *is imaginary.*

$$b = j\sqrt{\frac{1}{LC} - \left(\frac{R}{2L}\right)^2} = j\beta, \qquad (3\text{–}105)$$

Now

$$i_{\mathrm{tr}} = \epsilon^{-\alpha t}(K_1 \epsilon^{j\beta t} + K_2 \epsilon^{-j\beta t}). \qquad (3\text{–}106)$$

$$\epsilon^{j\beta t} = \cos \beta t + j \sin \beta t, \qquad (3\text{–}107)$$

and

$$\epsilon^{-j\beta t} = \cos \beta t - j \sin \beta t. \qquad (3\text{–}108)$$

Substituting Eqs. (3–107) and (3–108) in Eq. (3–106), we have

$$i_{\mathrm{tr}} = \epsilon^{-\alpha t}(B_1 \cos \beta t + B_2 \sin \beta t) \qquad (3\text{–}109)$$

$$= \sqrt{B_1^2 + B_2^2}\, \epsilon^{-\alpha t} \sin\left(\beta t + \tan^{-1}\frac{B_1}{B_2}\right), \qquad (3\text{–}110)$$

where B_1 has been written for $(K_1 + K_2)$ and B_2 for $j(K_1 - K_2)$. In Eq. (3–110) i_{tr} is a damped sine wave oscillating with an angular frequency β. This is called the *underdamped* case and occurs when the resistance in the circuit is small.

Now let us consider the steady-state solution for Eq. (3–94).

(a) $v_g = V_0$ (*a constant*).

The steady-state current in this case is apparently zero, since a series circuit containing a capacitance cannot sustain a direct current:

$$i_{\mathrm{ss}} = 0. \qquad (3\text{–}111)$$

Therefore the transient current is the total current, and it assumes one of the expressions derived above, depending upon whether the circuit is overdamped, critically damped, or underdamped. Typical i versus t

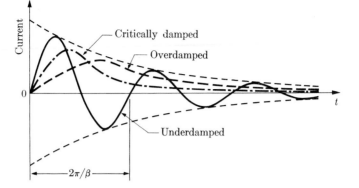

FIG. 3–26. Typical current curves for a series R-L-C circuit when $v_g = V_0$ (a constant).

curves are sketched in Fig. 3–26. The curves for overdamped and critically damped cases are similar in form.

(b) $v_g = V_g \sin \omega t$.

It is simplest to use the impedance concept to obtain the steady-state current for an applied sinusoidal voltage. The impedance of a series R-L-C circuit to a current of angular frequency ω is

$$Z = R + j\omega L + \frac{1}{j\omega C} = |Z|\epsilon^{j\theta_z}, \qquad (3\text{–}112)$$

where

$$|Z| = \sqrt{R^2 + \left(\omega L - \frac{1}{\omega C}\right)^2} \qquad (3\text{–}113)$$

and

$$\theta_z = \tan^{-1}\frac{\omega L - (1/\omega C)}{R}. \qquad (3\text{–}114)$$

In view of the general relation given in Eq. (3–75), we have for the expression of the steady-state current

$$\begin{aligned}
i_{\text{ss}} &= \frac{V_g}{|Z|}\sin(\omega t - \theta_z) \\
&= \frac{V_g}{\sqrt{R^2 + [\omega L - (1/\omega C)]^2}}\sin\left[\omega t - \tan^{-1}\frac{1}{R}\left(\omega L - \frac{1}{\omega C}\right)\right].
\end{aligned}$$
$$(3\text{–}115)$$

The total current is then the sum of the i_{ss} in Eq. (3–115) and one of the expressions for i_{tr} discussed above. The determination of the arbitrary constants from given initial conditions is routine but tedious; we shall not carry it out here. For an initially relaxed circuit, the initial conditions to be used are $i(0) = 0$ and $q(0) = 0$.

The variation of the magnitude of the steady-state current in a series R-L-C circuit with respect to the frequency of the applied sinusoidal voltage is of particular interest. From Eq. (3–115), we have

$$|i_{ss}| = \frac{V_g}{\sqrt{R^2 + [\omega L - (1/\omega C)]^2}}. \tag{3–116}$$

It is clear that $|i_{ss}|$ is maximum at an angular frequency ω_r such that

$$\omega_r L - \frac{1}{\omega_r C} = 0$$

or

$$\omega_r = \frac{1}{\sqrt{LC}}. \tag{3–117}$$

In terms of frequency, we have

$$f_r = \frac{\omega_r}{2\pi} = \frac{1}{2\pi\sqrt{LC}}. \tag{3–118}$$

When Eq. (3–117) or (3–118) is satisfied, we say that the series R-L-C circuit is in *resonance; f_r* in Eq. (3–118) is the *resonant frequency*. Let us denote the maximum value of $|i_{ss}|$ by I_m. Then, at resonance,

$$|Z| = R, \qquad \theta_z = 0, \qquad I_m = \frac{V_g}{R}. \tag{3–119}$$

In Fig. 3–27 are plotted typical curves for normalized current magnitudes, $|i_{ss}|/I_m$, for two values of R. The magnitude of the steady-state current decreases as the frequency of the applied voltage deviates from the resonant frequency, but the rate of decrease is slower for a circuit with a higher resistance. In circuit applications the range of frequency

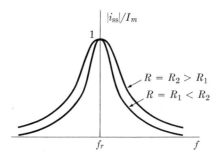

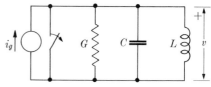

FIG. 3–27. Resonance current curves for a series R-L-C circuit.

FIG. 3–28. Dual electrical circuit for Fig. 3–25.

within which $|i_{ss}|/I_m \geq 1/\sqrt{2}$ or 0.707 is called the *bandwidth* of the circuit. Thus, a circuit with a larger resistance has a wider bandwidth but a smaller maximum current. Besides the change in current magnitude, the change in the phase angle θ_z is also important, but we shall not delve into that in this book.

We note again that by virtue of the principle of duality, solution for the current in the series R-L-C circuit shown in Fig. 3–25 gives at the same time the solution for the node-pair voltage v for the parallel G-C-L circuit in Fig. 3–28, where the switch across the current source i_g is opened at $t = 0$.

3–8 Thévenin's and Norton's theorems; equivalent circuits. There are a number of network theorems that are useful in the analysis of electrical networks. These theorems are usually statements which can be proved in a general way; once proved, they can be applied to simplify the analysis of actual systems. Among the most important are Thévenin's theorem and Norton's theorem, which lead us to the idea of *equivalent circuits*. The equivalence of two electrical circuits must be discussed in terms of their external behavior, which in turn is described in terms of the voltages and currents appearing at the terminals of these circuits. Thus if two circuits are to be equivalent, we must require that the voltages and currents at the stated terminals of one circuit be the same as those at the corresponding terminals of the other when the same arbitrary external circuit is connected to the terminals.

Very often in the analysis of electrical networks, the voltage or current in a particular part of the network is required. Because straightforward application of loop or node analysis yields simultaneous equations which involve other voltages and currents in addition to the one required, it would be highly desirable if we could find a way to get at the required quantity without having to obtain essentially a complete solution for the whole network. Thévenin's and Norton's theorems are very useful in this respect.

Thévenin's theorem. Consider the situation in Fig. 3–29(a), where an electrical network is drawn in two parts, N_A and N_B, connected at terminals 1–1'. Let us assume that N_B is a passive network containing no sources, initial voltages, or initial currents. In an actual problem, it may be the voltage or the current at terminals 1–1' of N_B that we wish to determine; N_A may in general be a complex network containing voltage and current sources. Suppose an external voltage source $v(t)$ is introduced in the circuit as shown in Fig. 3–29(b), such that the current between the two networks N_A and N_B is made zero. (The $v(t)$ required to do this will be determined presently.) Since there is now no current between N_A and N_B, the voltage across terminals 1–1' will be zero. Consequently we

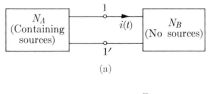

(a)

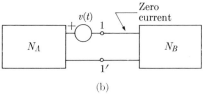

(b)

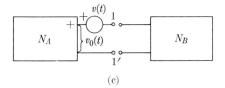

(c)

FIG. 3–29. Steps used for establishing Thévenin's theorem.

can break the circuit off at 1–1′ as shown in Fig. 3–29(c) without disturbing any voltages or currents.

For the network on the left in Fig. 3–29(c), the voltage across terminals 1–1′ (which is now zero) is the superposition of the externally applied voltage $v(t)$ and the *open-circuit voltage* $v_0(t)$ at 1–1′ due to the sources contained within N_A; that is,

$$v_0(t) - v(t) = 0$$

or

$$v(t) = v_0(t). \qquad (3\text{--}120)$$

Thus the voltage $v(t)$ of the external source which, when connected in the manner shown in Fig. 3–29(b), will produce a current just the negative of that due to network N_A is equal to the voltage at the terminals of N_A when N_B is disconnected. Consequently, if we reverse the reference polarities of $v(t)$ and remove all the sources in N_A, the original current $i(t)$ will be set up between N_A and N_B; there will also be no change in the voltage appearing across terminals 1–1′, since N_B is a passive network. In other words, the circuit shown in Fig. 3–30 to the left of terminals 1–1′ is an equivalent circuit of N_A with its sources. This is the essence of *Thévenin's theorem*, which may be stated as follows:

   Insofar as the external characteristics are concerned, a two-terminal electrical network containing sources and passive elements is equivalent to

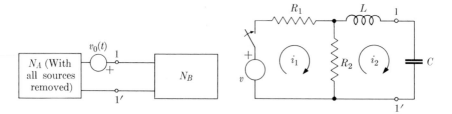

FIG. 3–30.   Thévenin's equivalent   FIG. 3–31.   Network for Example 3–9.
for Fig. 3–29(a).

a voltage source in series with the network with all sources removed; the
voltage of the equivalent voltage source has the same magnitude and reference polarity as those of the voltage appearing at the open-circuit terminals
of the original network.*

Removal of a voltage source should result in zero voltage across its
terminals; hence a voltage source is removed by replacing it with a short
circuit. Similarly, removal of a current source should result in zero current; a current source is removed by replacing it with an open circuit. In the
above it has been tacitly implied that N_A is not magnetically coupled to
external circuits. This should be regarded as a restriction on the type of
two-terminal networks to which Thévenin's theorem can be applied.

EXAMPLE 3–9. The switch in the network shown in Fig. 3–31 is closed
at $t = 0$. Determine the current in the capacitance C.

Solution. If we solve this problem in the conventional way by loop
analysis, we would obtain two equations as follows:

$$(R_1 + R_2)i_1 - R_2 i_2 = v, \tag{3–121}$$

$$-R_2 i_1 + R_2 i_2 + L \frac{di_2}{dt} + \frac{1}{C} \int_0^t i_2 \, dt = 0. \tag{3–122}$$

Equations (3–121) and (3–122) can then be solved simultaneously for i_2,
which is the desired current in C.

Now let us replace the network to the left of terminals 1–1' by its
Thévenin's equivalent. The part of the given network shown in Fig.
3–32(a) is network N_A in Fig. 3–29. The open-circuit voltage v_0 at 1–1'

* If N_A contains dependent sources as well as independent sources, the determination of the equivalent passive network to be connected in series with the
equivalent voltage source is a more involved matter. Briefly, it should be found
as the effective network appearing at the terminals of N_A with all its independent
sources removed.

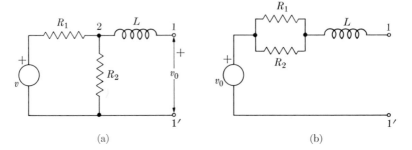

(a)   (b)

FIG. 3–32. Development of Thévenin's equivalent. (a) Original network to the left of terminals 1–1'. (b) Thévenin's equivalent for the network in (a).

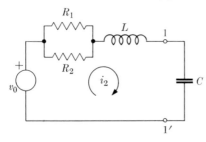

FIG. 3–33.   Equivalent circuit to find current in C.

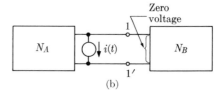

(a)

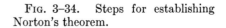

(b)

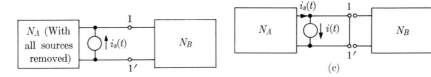

FIG. 3–35.   Norton's equivalent for Fig. 3–34(a).

(c)

FIG. 3–34.   Steps for establishing Norton's theorem.

is the same as that appearing across the resistance R_2, which can be found most easily by writing the node equation for node 2:

$$\left(\frac{1}{R_1} + \frac{1}{R_2}\right) v_0 - \frac{1}{R_1} v = 0$$

or

$$v_0 = \frac{R_2}{R_1 + R_2} v, \tag{3–123}$$

which could have been written from Fig. 3–32(a) by inspection.   By

Thévenin's theorem, we can then replace the network in Fig. 3–32(a) by its equivalent circuit in Fig. 3–32(b). The equation for the current i_2 in C is obtained from the single-loop series circuit shown in Fig. 3–33:

$$L\frac{di_2}{dt} + \Big(\frac{R_1 R_2}{R_1 + R_2}\Big) i_2 + \frac{1}{C}\int_0^t i_2\, dt = v_0 = \frac{R_2 v}{R_1 + R_2}. \quad (3\text{–}124)$$

The solution of Eq. (3–124) depends upon the given v.

———————

Norton's theorem.   Consider the situation in Fig. 3–34(a), where an electrical network is drawn in two parts, N_A and N_B, connected at terminals 1–1′. This is the same situation that we considered in Fig. 3–29(a). A voltage $v(t)$ appears across the N_B network. Suppose an external current source $i(t)$ is introduced in the circuit as shown in Fig. 3–34(b), such that the voltage across N_B is made zero. Since the voltage is now zero, terminals 1–1′ can be short-circuited as in Fig. 3–34(c). There will be no current in the short circuit, and the circuit is not disturbed even when N_B is disconnected.

For the network on the left in Fig. 3–34(c), the current through the short circuit at terminals 1–1′ (which is now zero) is the superposition of the externally applied current $i(t)$ and the *short-circuit current* $i_s(t)$ due to the sources contained within N_A; that is,

$$i_s(t) - i(t) = 0$$

or

$$i(t) = i_s(t). \quad (3\text{–}125)$$

If we reverse the reference direction of $i(t)$ and remove all the sources in N_A, the original voltage $v(t)$ will appear across N_B; there will also be no change in the current between N_A and N_B, since N_B is a passive network. In other words, the circuit shown in Fig. 3–35 to the left of terminals 1–1′ is an equivalent circuit of N_A with its sources. This is the essence of *Norton's theorem*, which may be stated as follows:

> *Insofar as the external characteristics are concerned, a two-terminal electrical network containing sources and passive elements is equivalent to a current source in parallel with the network with all sources removed; the current of the current source has the same magnitude and reference direction as those of the current which would exist at the terminals in the original network if the terminals were short-circuited.**

———————

* The footnote on p. 83 applies when N_A contains dependent as well as independent sources.

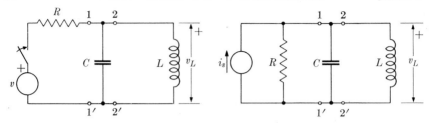

FIG. 3–36. Network for Example 3–10.   FIG. 3–37. Norton's equivalent for
network in Fig. 3–36.

Again we note that a voltage source is removed by replacing it with a short circuit and that a current source is removed by replacing it with an open circuit. Network N_A is assumed to have no magnetic coupling with external circuits.

EXAMPLE 3–10. The switch in the network shown in Fig. 3–36 is closed at $t = 0$. Determine the voltage appearing across the inductance L by applying Norton's theorem.

Solution. We shall solve this problem in two ways. (a) Replace the network to the left of terminals 1–1′ by its Norton's equivalent. In this case, N_A is simply the given voltage source v in series with the resistance R. If we short-circuit terminals 1–1′, the short-circuit current will be

$$i_s = \frac{v}{R}. \tag{3–126}$$

From Eq. (3–126) we draw Norton's equivalent for the network to the left of terminals 1–1′ as a parallel combination of a current source i_s and the resistance R, as shown in Fig. 3–37.

(b) Replace the network to the left of terminals 2–2′ by its Norton's equivalent. Now N_A is composed of three elements, v, R, and C, in a series-parallel combination. If we short-circuit terminals 2–2′, we get the same short-circuit current as given in Eq. (3–126), since the capacitance C in no way affects i_s. Hence the Norton's equivalent consists of the current source i_s in parallel with the parallel combination of R and C, resulting in a situation exactly the same as in the previous case, shown in Fig. 3–37.

The rest of this problem is the solution of v_L from the following node equation for a given v:

$$C \frac{dv_L}{dt} + \frac{1}{R} v_L + \frac{1}{L} \int_0^t v_L \, dt = \frac{v}{R}. \tag{3–127}$$

We note that in Example 3–10 we essentially changed a voltage source in series with a passive element to a current source in parallel with the same passive element. This *exchange of sources* can be discussed in more general terms if we refer to Figs. 3–30 and 3–35. By comparing the networks to the left of terminals 1–1' in these two figures, we see that *a voltage source in series with a passive network can be replaced by a current source in parallel with the same passive network, the current of the current source being equal to the short-circuit current of the original combination.* Conversely, *a current source in parallel with a passive network can be replaced by a voltage source in series with the same network, the voltage of the voltage source being equal to the open-circuit voltage appearing across the original combination.*

It is important to remember that *two circuits are equivalent only with respect to their external behaviors.* We *cannot* use equivalent circuits to determine the internal characteristics (voltages or currents associated with elements behind the terminals to which equivalence applies). For example, it is quite obvious that the voltage across the resistance R in Fig. 3–36 would in general not be equal to that across the capacitance C and inductance L, as the "equivalent circuit" in Fig. 3–37 seems to indicate!

Thévenin's and Norton's theorems can also be used to establish the equivalence between a charged capacitance and the suitable combination of a source and the uncharged capacitance, and the equivalence between an inductance with an initial current and the suitable combination of a source and the inductance without an initial current:

(1) A capacitance C with an initial voltage $v_c(0) = V_0$. From Fig. 3–38(a) it is obvious that the open-circuit voltage of the charged capacitance at terminals 1–1' is V_0; the Thévenin's equivalent circuit in Fig. 3–38(b) therefore follows directly. Thus, *a charged capacitance with an initial voltage V_0 is equivalent to the series combination of a d-c voltage source with voltage V_0 and the uncharged capacitance.*

Norton's theorem can also be applied to find another equivalent circuit for Fig. 3–38(a). However, the short-circuit current in this case is $C[dv_c(0)/dt]$, which would lead to an impulsive current source. Since we shall not take up the discussion of impulse functions until Chapter 8, and since there is no particular advantage in introducing the Norton's equiv-

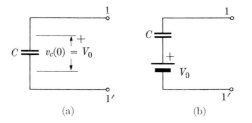

(a)   (b)

Fig. 3–38. Equivalence of initial voltage and series voltage source. (a) A capacitance with an initial voltage. (b) Thévenin's equivalent for the charged capacitance in (a).

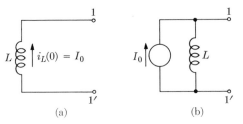

FIG. 3–39. Equivalence of initial current and parallel current source. (a) An inductance with an initial current. (b) Norton's equivalent for the inductance with initial current in (a).

alent of a charged capacitance, we shall not pursue the above remark further.

(2) An inductance L with an initial current $i(0) = I_0$. The equivalence of the circuits in parts (a) and (b) of Fig. 3–39 can be established immediately by an application of Norton's theorem. Hence an inductance with an initial current I_0 is equivalent to the parallel combination of a current source with direct current I_0 and the inductance with no initial current. A correct Thévenin's equivalent for the circuit in Fig. 3–39(a) would involve an impulsive voltage source, and is not of enough importance to be discussed here.

In introducing Thévenin's and Norton's theorems, we merely said that network N_A in Figs. 3–29 and 3–34 contained sources; no mention was made of initial conditions. In view of what we have just discussed, it should now be clear that initial voltages on capacitances and initial currents in inductances can be converted into d-c voltage and current sources; there should be no trouble in applying Thévenin's and Norton's theorems to an N_A with sources as well as nonzero initial conditions. The passive network to be connected in series with the voltage source in the Thévenin's equivalent or in parallel with the current source in the Norton's equivalent should then be initially relaxed, inasmuch as the effects of initial voltages on capacitances and initial currents in inductances have been accounted for in the equivalent sources.

EXAMPLE 3–11. The network shown in Fig. 3–40 is initially under steady-state conditions. The switch is opened at $t = 0$. Determine the voltage across the resistance R_2, assuming $R_2 = 2R_1$, (a) by applying Thévenin's theorem and (b) by applying Norton's theorem to the left of points 2–2'.

Solution. First of all, we note from Fig. 3–41 that the initial voltage across the capacitance C is

$$v_c(0+) = v_c(0-) = (\tfrac{2}{3} - \tfrac{1}{2})V = \tfrac{1}{6}V, \qquad (3\text{–}128)$$

with reference polarity as indicated.

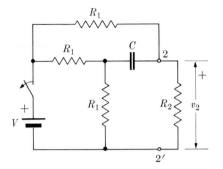
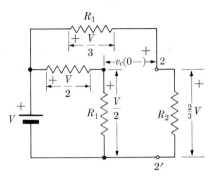

FIG. 3–40. Network for Example 3–11.   FIG. 3–41. Condition of network in Fig. 3–40 at $t = 0-$.

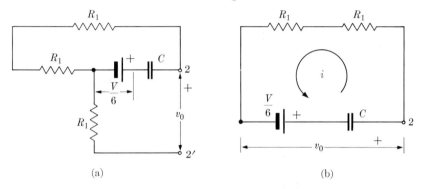

FIG. 3–42. For the determination of v_0.

(a) *Thévenin's equivalent, $t > 0$*. The circuit to the left of points 2–2' is as shown in Fig. 3–42, where the d-c voltage source is to account for the initial voltage across C. The open-circuit voltage v_0 at terminals 2–2' determines the Thévenin's equivalent voltage source. The equivalence of the circuits in parts (a) and (b) of Fig. 3–42 for the determination of v_0 should be obvious. From Fig. 3–42(b), we have

$$2R_1 i + \frac{1}{C} \int_0^t i \, dt = \frac{V}{6}. \tag{3–129}$$

Note that C is now being treated as an uncharged capacitance. Since

$$2R_1 i = v_0, \tag{3–130}$$

Eq. (3–129) can be written as

$$v_0 + \frac{1}{2CR_1} \int_0^t v_0 \, dt = \frac{V}{6}, \tag{3–131}$$

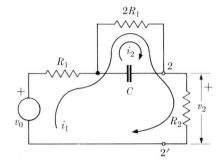

FIG. 3–43. Thévenin's equivalent of network in Fig. 3–40.

from which we find

$$v_0 = \frac{V}{6} \epsilon^{-t/2CR_1}. \tag{3-132}$$

The Thévenin's equivalent of the original network for $t > 0$ can now be drawn as shown in Fig. 3–43. Let us apply loop analysis to this equivalent network:

$$(3R_1 + R_2)i_1 + 2R_1 i_2 = v_0 = \frac{V}{6} \epsilon^{-t/2CR_1}, \tag{3-133}$$

$$2R_1 i_1 + 2R_1 i_2 + \frac{1}{C}\int_0^t i_2 \, dt = 0. \tag{3-134}$$

Differentiating Eq. (3–134) with respect to t once and combining it with Eq. (3–133) to eliminate i_2, we obtain (remembering that $R_2 = 2R_1$),

$$\frac{di_1}{dt} + \frac{5}{6CR_1} i_1 = 2R_1 \frac{dv_0}{dt} + \frac{v_0}{C} = 0. \tag{3-135}$$

The solution for Eq. (3–135) is

$$i_1 = K\epsilon^{-5t/6CR_1}. \tag{3-136}$$

At $t = 0+$, the entire v_0 is applied across R_1 and R_2 in series, since the voltage across C cannot change instantaneously:

$$i_1(0+) = \frac{v_0(0+)}{R_1 + R_2} = \frac{1}{3R_1}\left(\frac{V}{6}\right) = K. \tag{3-137}$$

Hence

$$i_1 = \frac{V}{18R_1} \epsilon^{-5t/6CR_1}, \tag{3-138}$$

and finally,

$$v_2 = R_2 i_1 = 2R_1 i_1 = \frac{V}{9} \epsilon^{-5t/6CR_1}. \tag{3-139}$$

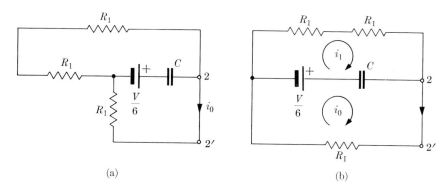

FIG. 3–44. For the determination of i_0.

(b) *Norton's equivalent, $t > 0$.* To find the Norton's equivalent current source, we place a short circuit between terminals 2–2' in Fig. 3–42(a) and get Fig. 3–44(a), which has been rearranged in Fig. 3–44(b). Using loop analysis, we obtain

$$2R_1 i_1 + \frac{1}{C} \int_0^t i_1 \, dt - \frac{1}{C} \int_0^t i_0 \, dt = -\frac{V}{6}, \qquad (3\text{–}140)$$

$$-\frac{1}{C} \int_0^t i_1 \, dt + \frac{1}{C} \int_0^t i_0 \, dt + R_1 i_0 = \frac{V}{6}. \qquad (3\text{–}141)$$

The sum of Eqs. (3–140) and (3–141) yields the obviously correct result

$$i_1 = -\frac{i_0}{2}. \qquad (3\text{–}142)$$

Substitution of Eq. (3–142) in the derivative of either Eq. (3–140) or Eq. (3–141) results in a first-order homogeneous differential equation in i_0:

$$\frac{di_0}{dt} + \frac{3}{2CR_1} i_0 = 0. \qquad (3\text{–}143)$$

The solution of Eq. (3–143) is

$$i_0 = \frac{V}{6R_1} \epsilon^{-3t/2CR_1}. \qquad (3\text{–}144)$$

The Norton's equivalent of the original network for $t > 0$ is therefore as shown in Fig. 3–45, where C is uncharged. v_2 can be solved by writing the node equations for nodes 1 and 2, using 2' as the reference node:

For node 1: $\left(\dfrac{1}{R_1} + \dfrac{1}{2R_1} \right) v_1 + C \dfrac{dv_1}{dt} - \dfrac{1}{2R_1} v_2 - C \dfrac{dv_2}{dt} = 0.$ (3–145)

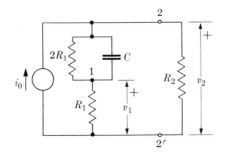

FIG. 3–45. Norton's equivalent of network in Fig. 3–40.

For node 2: $\quad -\dfrac{1}{2R_1} v_1 - C \dfrac{dv_1}{dt} + \left(\dfrac{1}{2R_1} + \dfrac{1}{R_2}\right) v_2 + C \dfrac{dv_2}{dt} = i_0.$ (3–146)

Eliminating v_1 from Eqs. (3–145) and (3–146), we get

$$\frac{dv_2}{dt} + \frac{5}{6CR_1} v_2 = 0. \tag{3–147}$$

The solution for Eq. (3–147) is

$$v_2 = A\epsilon^{-5t/6CR_1}. \tag{3–148}$$

At $t = 0+$, the capacitance can be viewed as a short circuit, since its voltage remains at zero and $i_0(0+)$ effectively flows through R_1 and R_2 in parallel, yielding

$$v_2(0+) = i_0(0+) \frac{R_1 R_2}{R_1 + R_2} = \frac{V}{6R_1} \left(\frac{2}{3} R_1\right) = \frac{V}{9} = A. \tag{3–149}$$

Hence

$$v_2 = \frac{V}{9} \epsilon^{-5t/6CR_1}, \tag{3–150}$$

which should, of course, be the same as the answer in Eq. (3–139) if both are correct.

The above example illustrates the applications of Thévenin's and Norton's theorems to a network with nonquiescent initial conditions. They are not necessarily simpler than the straightforward application of loop or node analysis to the original network.

PROBLEMS

3-1. Write the equations that completely describe the bridge network in Fig. 3-46, (a) on a loop basis, and (b) on a node basis.

3-2. Write the equations that completely describe the twin T network in Fig. 3-47, (a) on a loop basis, and (b) on a node basis.

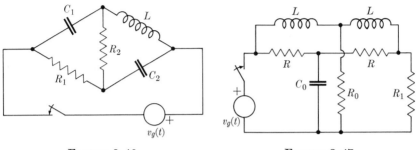

FIGURE 3-46   FIGURE 3-47

3-3. The network in Fig. 3-48 is under steady-state conditions. At $t = 0$, the switch shunting the current source is opened. Write the equations that completely describe the network for $t \geq 0$, (a) on a loop basis, and (b) on a node basis.

3-4. Write the loop equations for the network in Fig. 3-49.

3-5. Write the loop equations for the network in Fig. 3-50 assuming (a) finite L_1 and L_2, and $k = 0.9$, and (b) infinite L_1 and L_2 but $L_1/L_2 = n$ (finite), and $k = 1.0$.

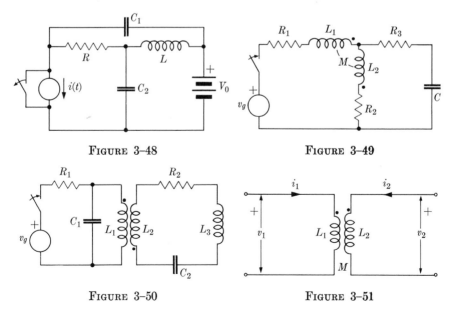

FIGURE 3-48   FIGURE 3-49

FIGURE 3-50   FIGURE 3-51

3–6. Draw the dual for the network in Fig. 3–46.

3–7. Draw the dual for the network in Fig. 3–47.

3–8. Draw the dual for the network in Fig. 3–48.

3–9. Draw the dual for the network in Fig. 3–50 for finite L_1, L_2, and M.

3–10. In order to write the node equations of a network containing magnetically coupled elements, it is necessary to express the currents in the coupled coils in terms of the voltages appearing across them. For the two-winding transformer shown in Fig. 3–51, express i_1 and i_2 in terms of v_1 and v_2, assuming that the initial currents in the primary and secondary windings are $i_1(0)$ and $i_2(0)$ respectively.

3–11. A d-c voltage V is applied to an initially relaxed series R-L-C circuit at $t = 0$. Determine the maximum instantaneous current and the time at which this maximum current occurs for (a) the overdamped case, (b) the critically damped case, and (c) the underdamped case.

3–12. Assuming that $v_g = V_g \cos(\omega t + \alpha)$ in Fig. 3–49, determine the instantaneous expression for the steady-state current in the capacitance C.

3–13. The switch in the circuit of Fig. 3–52 is closed at $t = 0$. Determine the voltage across the inductance L, (a) for $v_g = 10$ volts, and (b) for $v_g = 10 \sin(100t + \alpha)$ volts. (c) Find the value of the angle α in part (b) for which there will be no transient.

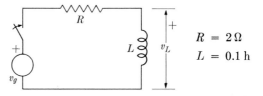

$$R = 2\,\Omega$$
$$L = 0.1\,\text{h}$$

FIGURE 3–52

3–14. When the switch in the circuit of Fig. 3–53 is closed at $t = 0$, there is an initial voltage of 5 volts across the capacitance C such that the lower plate is positive relative to the upper plate. Determine the voltage across C as a function of time, (a) for $v_g = 10$ volts, and (b) for $v_g = 10 \sin(10t + \alpha)$ volts. (c) Find the value of the angle α in part (b) for which there will be no transient.

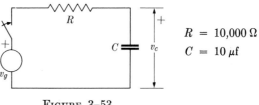

$$R = 10{,}000\,\Omega$$
$$C = 10\,\mu\text{f}$$

FIGURE 3–53

3–15. The circuit in Fig. 3–54 is initially under steady-state conditions. The switch is opened at time $t = 0$. Find the voltage across the capacitance C as a function of time.

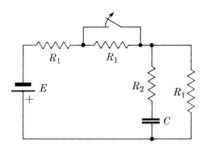

FIGURE 3–54

3–16. The circuit in Fig. 3–55 is a "relaxation oscillator," in which the capacitor charges and discharges alternately, depending upon whether the gas diode is nonconducting or conducting. Assume that the gas diode conducts (breaks down) when the voltage across it reaches 100 volts and extincts when the voltage decreases to 50 volts, and that when it conducts the gas diode can be represented by a 10-ohm resistance. The switch is closed at $t = 0$. (a) Find the expression for $v(t)$ for t from 0 to the time when the gas diode first breaks down. (b) Find a general expression for $v(t)$ which applies when the gas diode is conducting. (c) Sketch $v(t)$ as a function of t. (d) What are the time constants of the circuit for charging and for discharging?

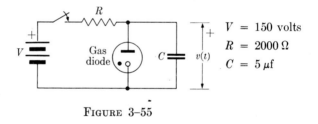

FIGURE 3–55

3–17. The capacitance C in Fig. 3–56 has an initial voltage $v_c(0) = 10$ volts when the switch is closed. Find $v_c(t)$ (a) for $v_g = 100$ volts, and (b) for $v_g = 100 \sin 7000t$ volts. (c) What is the steady-state expression for $v_c(t)$ in response to a sinusoidal v_g with amplitude V_g and angular frequency ω? What is the resonant frequency of the given circuit? What happens to $v_c(t)$ at resonance?

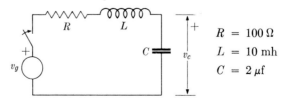

FIGURE 3–56

3–18. The circuit in Fig. 3–57 is initially under steady-state conditions. The switch is opened at $t = 0$. (a) Use node analysis to find $v(t)$ for $I = 10$ ma. (b) Replace the network to the left of terminals 1–1′ by its Thévenin's equivalent.

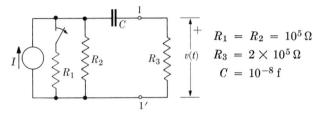

$$R_1 = R_2 = 10^5 \, \Omega$$
$$R_3 = 2 \times 10^5 \, \Omega$$
$$C = 10^{-8} \, \text{f}$$

FIGURE 3–57

3–19. The circuit in Fig. 3–58 is initially at rest. The switch is closed at $t = 0$. (a) Find the voltage across the capacitance. (b) Replace the network to the left of terminals 1–1′ by its Norton's equivalent.

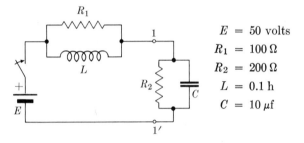

$$E = 50 \text{ volts}$$
$$R_1 = 100 \, \Omega$$
$$R_2 = 200 \, \Omega$$
$$L = 0.1 \text{ h}$$
$$C = 10 \, \mu\text{f}$$

FIGURE 3–58

3–20. The switch in the circuit shown in Fig. 3–59 is initially at position A. After the steady state has been reached, the switch is quickly thrown to position B. Determine the current in R_2 by first finding (a) the Thévenin's equivalent and (b) the Norton's equivalent of the network to the left of points 1–1′.

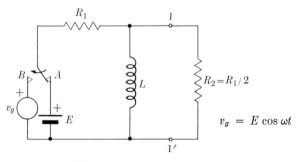

$$R_2 = R_1/2$$
$$v_g = E \cos \omega t$$

FIGURE 3–59

CHAPTER 4

ANALOGOUS SYSTEMS

4–1 Introduction. In the analysis of linear systems the mathematical procedure for obtaining the solutions to a given set of equations does not depend upon what physical system the equations represent. Hence if the response of one physical system to a given excitation is determined, the responses of all other systems which can be described by the same set of equations are known for the same excitation function. Systems which are governed by the same types of equations are called *analogous systems*.

Analogous systems may have entirely different physical appearances. For example, a given electrical circuit consisting of resistances, inductances, and capacitances may be analogous to a mechanical system consisting of a suitable combination of dashpots, weights, and springs; or it may be analogous to an acoustical device consisting of an appropriate arrangement of fine-mesh screens, tubes, and cavities. Dual electrical circuits that are governed by the same differential equations are a special type of analogous system. As we discussed in Section 3–6, the series R_a-L_a-C_a circuit fed by a voltage source v_a, as shown in Fig. 4–1(a), is a dual of the parallel G_b-C_b-L_b circuit fed by a current source i_b as shown in Fig. 4–1(b). The equations describing these two circuits are:

For Fig. 4–1(a),

$$L_a \frac{di_a}{dt} + R_a i_a + \frac{1}{C_a} \left[\int_0^t i_a \, dt + q_a(0) \right] = v_a. \qquad (4\text{–}1)$$

For Fig. 4–1(b),

$$C_b \frac{dv_b}{dt} + G_b v_b + \frac{1}{L_b} \left[\int_0^t v_b \, dt + \phi_b(0) \right] = i_b. \qquad (4\text{–}2)$$

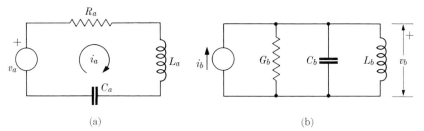

FIG. 4–1. Dual electrical circuits.

97

Equations (4–1) and (4–2) are identical if the following changes of symbols are made:

$$L_a \leftrightarrow C_b, \qquad R_a \leftrightarrow G_b, \qquad C_a \leftrightarrow L_b,$$

$$v_a \leftrightarrow i_b, \qquad i_a \leftrightarrow v_b, \qquad q_a \leftrightarrow \phi_b.$$

Mathematically, dual circuits are governed by the same types of equations. Physically, an electrical circuit can be changed to its dual by making appropriate conversions in accordance with Table 3–2 or by using the dot method described in Section 3–6.

When we deal with systems other than electrical, there are distinct advantages if we can reduce the systems under consideration to their analogous electrical circuits. First of all, electrical engineers have developed a set of convenient symbols for circuit elements that permits a complex system to be set down in the form of a circuit diagram from which the behavior of the system can be readily analyzed. Once the circuit diagram of the analogous electrical system is determined, it is possible to visualize and even predict system behaviors (resonances, passbands, damping coefficients, time constants, etc.) by inspection. Second, electrical circuit-theory techniques, such as the use of the impedance concept and the various network theorems, can be applied in the actual analysis of the system. Third, the ease of changing the values of electrical components, of connecting and disconnecting them in a circuit, and of measuring the voltages and currents all prove invaluable in model construction and testing.

In this chapter, we shall establish the analogy between linear mechanical and electrical systems, the ultimate objective being the ability to draw analogous electrical circuits (and solve as such) for given mechanical systems by inspection. This technique becomes even more valuable when we deal with electromechanical systems, in which electrical and mechanical phenomena are interrelated. Analogies can be extended beyond electrical and mechanical systems to acoustical, thermal, and even economic systems, but these require thorough knowledge of the parameters and system relationships in the respective fields and will not be treated here.

4–2 Linear mechanical elements. Three fundamental passive elements, or system parameters, will come into play in linear mechanical systems; they correspond to the coefficients in the expressions of three types of mechanical forces that resist motion. We must define the parameters separately for translational and for rotational systems.

A. *Translational systems.* There are three types of forces that resist motion:

1. Inertia force: Newton's second law of motion states that the inertia force is equal to mass times acceleration:

$$f_M = Ma = M\frac{du}{dt} = M\frac{d^2x}{dt^2},\tag{4-3}$$

where a denotes the acceleration, u the velocity, and x the displacement. The *mass* M of a body is thus the coefficient in the force equation and is the inertia force per unit acceleration. Symbolically, we represent it by a block, as shown in Fig. 4–2(a).

2. Damping force: In linear systems we assume the damping force to be proportional to the velocity. However, this is true only in the case of viscous friction; it is, in general, not a good approximation for dry friction. Coils moving in a uniform magnetic field experience a damping force which is proportional to the velocity. We have, for linear damping,

$$f_D = Du = D\frac{dx}{dt},\tag{4-4}$$

where D, the *damping coefficient*, is the damping force per unit velocity. Symbolically, we represent D by a dashpot, as shown in Fig. 4–2(b), signifying viscous damping between the piston and the cylinder.

3. Spring force: The restoring force of a spring is proportional to the displacement (amount of stretch or compression):

$$f_K = \frac{1}{K}x = \frac{1}{K}\int u\,dt = \frac{1}{K}\left[\int_0^t u\,dt + x(0)\right],\tag{4-5}$$

where K, the *compliance* of the spring, is the reciprocal of its *stiffness*. The stiffness of a spring is the restoring force per unit displacement. We represent the element by a coil spring, as shown in Fig. 4–2(c).

Note that Eqs. (4–3), (4–4), and (4–5) will hold in any self-consistent system of units.

In mechanical systems, it is convenient to define two kinds of ideal sources (active elements), corresponding to the ideal voltage and current

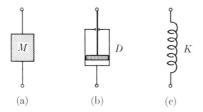

(a)   (b)   (c)

Fig. 4–2. Passive mechanical elements for translational motion.

generators in electrical systems. When the driving force $f(t)$ acting in a system is known, we can think of the system as being under the influence of a *force source*. The force developed by a force source is assumed to be independent of what is connected to it. When the driving velocity $u(t)$ at some point in a mechanical system is known and is controlled by external means, we can think of the system as being under the influence of a *velocity source*. The velocity developed by a velocity source is likewise assumed to be independent of what is connected to it.

B. *Rotational systems.* Corresponding to the three types of forces resisting translational motion, there are three types of torques resisting rotational motion. They are:

1. Inertia torque: The inertia torque, τ_I, is equal to the *moment of inertia* I_θ times the angular acceleration α:*

$$\tau_I = I_\theta \alpha = I_\theta \frac{d\Omega}{dt} = I_\theta \frac{d^2\theta}{dt^2}, \qquad (4\text{-}6)$$

where Ω denotes the angular velocity and θ the angular displacement.

2. Damping torque: The damping torque, τ_D, is equal to the *rotational damping coefficient* D_θ times the angular velocity Ω in a linear system:

$$\tau_D = D_\theta \Omega = D_\theta \frac{d\theta}{dt}. \qquad (4\text{-}7)$$

3. Spring torque: The restoring torque, τ_K, of a spring is equal to the angular displacement θ divided by the *torsional compliance* K_θ. The reciprocal of K_θ is the *torsional stiffness* of the spring.

$$\tau_K = \frac{1}{K_\theta} \theta = \frac{1}{K_\theta} \int \Omega \, dt = \frac{1}{K_\theta} \left[\int_0^t \Omega \, dt + \theta(0) \right]. \qquad (4\text{-}8)$$

Comparing Eqs. (4-6), (4-7), and (4-8) for rotational motion to Eqs. (4-3), (4-4), and (4-5), respectively, for translational motion, we see that these two systems are entirely similar mathematically. Here then is another example of analogous systems.

Table 4-1 tabulates the analogous quantities in translational and rotational mechanical systems.

Inasmuch as all the characteristics of a rotational system can be discussed in terms of the analogous translational system, no attempt will be made to assign special symbols to rotational mechanical elements as we have done for the translational system in Fig. 4-2.

---

* In rotational motion it is necessary to refer all quantities (torque, angular displacement, angular velocity, angular acceleration, moment of inertia, etc.) to an axis of rotation.

TABLE 4–1

ANALOGOUS QUANTITIES IN TRANSLATIONAL AND ROTATIONAL
MECHANICAL SYSTEMS

Translational	Rotational
Force, f	Torque, τ
Acceleration, a	Angular acceleration, α
Velocity, u	Angular velocity, Ω
Displacement, x	Angular displacement, θ
Mass, M	Moment of inertia, I_θ
Damping coefficient, D	Rotational damping coefficient, D_θ
Compliance, K	Torsional compliance, K_θ

It should be realized that expressions such as "force on" and "velocity of" a mechanical element, though conventionally used, are loose expressions. This becomes clear when we recall the element symbols in Fig. 4–2, where all three elements are represented as two-terminal devices. Which terminal do we refer to when we talk about the "force on" and the "velocity of" an element? In order to be more specific, let us consider the dashpot in Fig. 4–2(b). Obviously, the two terminals (the piston and the cylinder) of the dashpot can move with different velocities. In fact, the cylinder may remain stationary while the piston is moving. Furthermore, there would be no damping force if the piston and the cylinder moved with the same velocity. Hence the significant quantity is the *relative* velocity (velocity difference) of the two terminals. In other words, element relationships hold for velocity *across* a mechanical element. Similarly, relative to mechanical elements, we must imply displacement *across* and acceleration *across* the terminals. Motion of a rigid body (mass) refers to the stationary ground. The force, on the other hand, acts *through* mechanical elements. The same comments apply to torque *through* and angular displacement, velocity, or acceleration *across* mechanical elements in rotational motion.

4–3 D'Alembert's principle. D'Alembert's principle applies the conditions of static equilibrium to problems in dynamics by considering both the externally applied driving forces and the reaction forces of mechanical elements which resist motion. It is actually a slightly modified form of Newton's second law of motion, and can be stated as follows:

For any body, the algebraic sum of externally applied forces and the forces resisting motion in any given direction is zero.

D'Alembert's principle applies for all instants of time. A positive reference direction must first be chosen. Forces acting in the reference direction are then considered as positive and those against the reference direction as negative. D'Alembert's principle is as useful in writing the equations of motion for a mechanical system as Kirchhoff's laws are in writing the circuit equations for an electrical network. Let us apply it to the system in Fig. 4–3.

In Fig. 4–3, the mass M is attached to a fixed wall through a spring with compliance K. It is assumed that the contact between the mass and the floor offers viscous damping with damping coefficient D. (The problem would be exactly the same if we assumed that the floor was frictionless but that there existed a dashpot with damping coefficient D, in addition to the spring, between the mass and the fixed wall.) In the direction toward the right:

External force: f

Resisting forces: (1) inertia force, $f_M = -M \dfrac{du}{dt}$,

  (2) damping force, $f_D = -Du$,

  (3) spring force, $f_K = -\dfrac{1}{K}\left[\displaystyle\int_0^t u\, dt + x(0)\right].$

By d'Alembert's principle, we then have

$$f + f_M + f_D + f_K = 0, \qquad (4\text{–}9)$$

or

$$M \frac{du}{dt} + Du + \frac{1}{K}\left[\int_0^t u\, dt + x(0)\right] = f, \qquad (4\text{–}10)$$

which is the equation of motion for the system in Fig. 4–3.

A rotational system analogous to the translational system of Fig. 4–3 is shown in Fig. 4–4. Here a flywheel with moment of inertia I_θ is supported

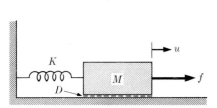

FIG. 4–3. A translational mechanical system.

FIG. 4–4. A rotational mechanical system which is analogous to the translational system in Fig. 4–3.

on a shaft with torsional compliance K_θ (each half having a torsional compliance $2K_\theta$). The shaft is securely clamped on both ends. The brake below the flywheel is to symbolize a linear (viscous) damping with rotational damping coefficient D_θ. D'Alembert's principle modified for a rotational system can be stated as follows:

For any body, the algebraic sum of externally applied torques and the torques resisting rotation about any axis is zero.

For Fig. 4–4:

    External torque: τ

    Resisting torques: (1) inertia torque, $\tau_I = -I_\theta \dfrac{d\Omega}{dt}$,

                             (2) damping torque, $\tau_D = -D_\theta\Omega$,

                             (3) spring torque, $\tau_K = -\dfrac{1}{K_\theta}\left[\displaystyle\int_0^t \Omega\,dt + \theta(0)\right]$.

Hence

$$\tau + \tau_I + \tau_D + \tau_K = 0, \tag{4–11}$$

or

$$I_\theta \frac{d\Omega}{dt} + D_\theta\Omega + \frac{1}{K_\theta}\left[\int_0^t \Omega\,dt + \theta(0)\right] = \tau, \tag{4–12}$$

which is, of course, entirely analogous to Eq. (4–10) for the translational case.

4–4 Force-voltage analogy. Comparison of Eq. (4–10) with Eq. (4–1) immediately reveals their close similarity. They therefore represent analogous systems. In other words, the behavior of the mechanical system of Fig. 4–3 can be completely predicted by what we know about the simple series R-L-C electrical circuit in Fig. 4–1(a) by making appropriate conversions of physical quantities, as listed in Table 4–2. Since force, f, in the mechanical system is set to be analogous to voltage, v, in the electrical system, we designate this type of analogy as the *force-voltage* (f-v) *analogy.*[*]

The usefulness of the electromechanical analogy lies in our ability to draw the analogous electrical circuit directly from a given mechanical system (without first writing the mechanical equations of motion) and to solve the problem entirely as an electrical one, with known techniques. For the simple system in Fig. 4–3, it is not too difficult to see that the circuit in Fig. 4–1(a) is its force-voltage electrical analog. However, with

---

[*] It is also referred to in the literature as the *direct analogy*, or the *impedance-type analogy*.

TABLE 4-2

TABLE OF CONVERSION FOR FORCE-VOLTAGE ANALOGY

Mechanical system	Electrical system (f-v analogy)
Force, f	Voltage, v
Velocity, u	Current, i
Displacement, x	Charge, q
Mass, M	Inductance, L
Damping coefficient, D	Resistance, R
Compliance, K	Capacitance, C

a complex system, it is not so easy to visualize. The following rule for drawing f-v analogous electrical circuits from mechanical systems will prove useful:

Each junction in the mechanical system corresponds to a closed loop which consists of electrical excitation sources and passive elements analogous to the mechanical driving sources and passive elements connected to the junction. All points on a rigid mass are considered as the same junction.

For the system in Fig. 4–3, since all points on the mass are considered as the same junction, there is only one junction, to which a driving force f and three passive mechanical elements, D, M, and K, are connected. This converts into one closed loop consisting of a voltage excitation source, v, and three passive electrical circuit elements, R, L, and C, as shown in Fig. 4–1(a). Let us now consider a more complex system.

EXAMPLE 4–1. Find the equations that describe the motion of the mechanical system of Fig. 4–5.

Solution. We shall solve this problem in two different ways. In part (a) the normal approach based upon d'Alembert's principle is used; in part (b) we make use of analogy.

(a) Using d'Alembert's principle.

It is clear that the system in Fig. 4–5 is a two-coordinate system, i.e., two variables, x_1 and x_2 as shown (measured from the equilibrium position before the external force is applied), are needed to describe the system completely. First consider the forces on mass M_1:

External force: f

Resisting forces:

(1) inertia force, $f_{M_1} = -M_1 \dfrac{d^2x_1}{dt^2} = -M_1 \dfrac{du_1}{dt},$

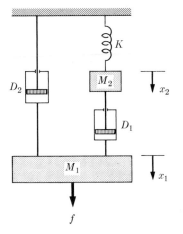

FIG. 4–5. A two-coordinate mechanical system.

(2) damping forces, $f_{D_1} = -D_1 \dfrac{d}{dt} (x_1 - x_2) = -D_1(u_1 - u_2),$

$$f_{D_2} = -D_2 \frac{dx_1}{dt} = -D_2 u_1.$$

Application of d'Alembert's principle yields

$$M_1 \frac{du_1}{dt} + (D_1 + D_2)u_1 - D_1 u_2 = f. \tag{4–13}$$

Next consider the forces on mass M_2:

External force: 0

Resisting forces:

(1) inertia force, $f_{M_2} = -M_2 \dfrac{d^2 x_2}{dt^2} = -M_2 \dfrac{du_2}{dt},$

(2) damping force, $f_{D_1} = -D_1 \dfrac{d}{dt} (x_2 - x_1) = -D_1(u_2 - u_1),$

(3) spring force, $f_K = -\dfrac{1}{K} x_2 = -\dfrac{1}{K}\left[\displaystyle\int_0^t u_2\, dt + x_2(0)\right].$

Hence

$$-D_1 u_1 + M_2 \frac{du_2}{dt} + D_1 u_2 + \frac{1}{K}\left[\int_0^t u_2\, dt + x_2(0)\right] = 0. \tag{4–14}$$

Equations (4–13) and (4–14) completely describe the motion of the system.

(b) Using f-v analogy.

Corresponding to the two coordinates x_1 and x_2, the mechanical system has two junctions. Hence, we will have two loops in the f-v analogous electrical circuit. The first loop consists of a voltage source v [f], an inductance L_1 [M_1], and two resistances R_1 [D_1] and R_2 [D_2], and the second loop consists of an inductance L_2 [M_2], a capacitance C [K], and a resistance R_1 [D_1], the last element being common to both loops. The analogous electrical circuit is shown in Fig. 4–6.

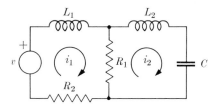

FIG. 4–6.   An electrical circuit analogous to that in Fig. 4–5, based upon f-v analogy.

It is now a simple matter to write the equations for the two-loop circuit of Fig. 4–6:

$$L_1 \frac{di_1}{dt} + (R_1 + R_2)i_1 - R_1 i_2 = v, \qquad (4\text{--}15)$$

$$-R_1 i_1 + L_2 \frac{di_2}{dt} + R_1 i_2 + \frac{1}{C}\left[\int_0^t i_2\, dt + q_2(0)\right] = 0. \quad (4\text{--}16)$$

Equations (4–15) and (4–16) are identical with Eqs. (4–13) and (4–14) except for the conversion of the parameters in accordance with Table 4–2, and the response of the given mechanical system can be determined by examining the behavior of the analogous electrical circuit. Note that Fig. 4–6 can be drawn directly from Fig. 4–5 by inspection; there is no need to write Eqs. (4–13) and (4–14) at all.

4–5 Force-current analogy. The force-voltage analogy described in the preceding section is based on the mathematical similarity between Eqs. (4–10) and (4–1). From the point of view of physical interpretation, it is not a natural analogy, because *forces* acting *through* mechanical elements are made to be analogous to *voltages across* the corresponding electrical elements, and *velocities across* (velocity differences between the terminals of) mechanical elements are made to be analogous to *currents through* the corresponding electrical elements. A direct consequence is that a *junction* in the mechanical system goes over to the analogous electrical circuit as a *loop.* Realizing these inherent imperfections, F. A. Firestone* advocated an analogy of the mobility type, which we shall call the force-current (f-i) analogy,† and which will be the subject matter of this section.

We arrived at the force-voltage analogy by noticing the similarity between Eq. (4–10) for the mechanical system in Fig. 4–3 and Eq. (4–1) for the electrical circuit in Fig. 4–1(a). Equation (4–10), of course, is also similar to Eq. (4–2) describing the circuit in Fig. 4–1(b), which is therefore also analogous to the mechanical system in Fig. 4–3. The corresponding electrical and mechanical quantities for this new analogy are listed in Table 4–3.

TABLE 4–3

TABLE OF CONVERSION FOR FORCE-CURRENT ANALOGY

Mechanical system	Electrical system (f-i analogy)
Force, f	Current, i
Velocity, u	Voltage, v
Displacement, x	Flux linkage, ϕ
Mass, M	Capacitance, C
Damping coefficient, D	Conductance, G
Compliance, K	Inductance, L

Since force, f, in the mechanical system is now set to be analogous to current, i, in the electrical system, we designate this type of analogy as the *force-current* (f-i) *analogy.* From the physical point of view, this analogy is more satisfactory than the force-voltage analogy, since force *through* is now analogous to current *through* and velocity *across* is now analogous to voltage *across.* Consequently, a *junction* in the mechanical system goes over to the electrical system as a *node* (junction), instead of as a loop. Furthermore, in mechanical devices we can measure velocity (or displacement) with a vibration pickup without disturbing the machine, just as we can measure voltage in electrical circuits with a voltmeter without disturbing the circuit; force and current, however, cannot be measured unless we break into the system.

The rule for drawing f-i analogous electrical circuits from mechanical systems can be stated as follows:

---

* F. A. Firestone, "The Mobility Method of Computing the Vibration of Linear Mechanical and Acoustical Systems: Mechanical-Electrical Analogies," *Journal of Applied Physics,* **9,** pp. 373–387; June 1938.

† It is also referred to as the *inverse analogy,* although it is in reality more direct than the force-voltage analogy.

Each junction in the mechanical system corresponds to a node (junction) which joins electrical excitation sources and passive elements analogous to the mechanical driving sources and passive elements connected to the junction. All points on a rigid mass are considered as the same junction and one terminal of the capacitance analogous to a mass is always connected to the ground.

The reason that one terminal of the capacitance analogous to a mass is always connected to the ground is that the velocity (or displacement, or acceleration) of a mass is always referred to the earth. Recognition of this fact greatly simplifies the drawing of analogous electrical circuits. Two or more masses rigidly connected go over to the analogous electrical circuit as two or more capacitances connected between the same node and the ground.

The mechanical system in Fig. 4–3 has only one junction, to which a driving force f and three passive mechanical elements, D, M, and K, are connected. This converts by f-i analogy into a single node consisting of a current source i and three passive electrical elements G, C, and L all connected to the ground (reference or datum node), as shown in Fig. 4–1(b).* It is quite clear that *the electrical circuits drawn from the f-v and f-i analogies are duals of each other.*

EXAMPLE 4–2. Draw the electrical circuit analogous to the mechanical system of Fig. 4–5, using force-current analogy.

Solution. Corresponding to the two coordinates x_1 and x_2 for the mechanical system, we will have two independent nodes in the f-i analogous electrical circuit. The first node joins a current source i [f], a capacitance C_1 [M_1] and two conductances G_1 [D_1] and G_2 [D_2]; the second node joins a capacitance C_2 [M_2], an inductance L [K] and a conductance G_1 [D_1], the last element being common to both nodes. This is shown in Fig. 4–7, which is clearly the dual of the circuit in Fig. 4–6 and can be derived from the latter by the dot method described in Section 3–6.

By applying Kirchhoff's current law, we can readily write the node equations from Fig. 4–7:

$$C_1 \frac{dv_1}{dt} + (G_1 + G_2)v_1 - G_1 v_2 = i, \qquad (4\text{–}17)$$

$$-G_1 v_1 + C_2 \frac{dv_2}{dt} + G_1 v_2 + \frac{1}{L}\left[\int_0^t v_2\, dt + \phi_2(0)\right] = 0. \qquad (4\text{–}18)$$

---

* Of course, in the actual electrical circuit, there is no difference between an element with a conductance G and an element with a resistance $R = 1/G$.

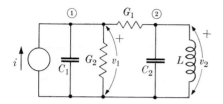

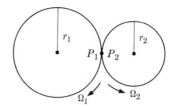

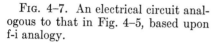

FIG. 4–7. An electrical circuit analogous to that in Fig. 4–5, based upon
f-i analogy.

FIG. 4–8. A pair of nonslipping friction wheels.

Equations (4–17) and (4–18) can be converted to the mechanical equations of motion (4–13) and (4–14) by using the conversions in Table 4–3. However, this conversion of parameters does not have to be done until after the desired response has been determined. As a matter of fact, the circuit elements in Fig. 4–7 could have been labeled with their analogous mechanical quantities; then no conversion of parameters in the solution would be necessary.

What we have said about translational mechanical systems would obviously apply equally well to analogous rotational mechanical systems except for changes in terminology, in accord with Table 4–1.

4–6 Mechanical coupling devices.　Common mechanical coupling devices, such as gears, friction wheels, and levers, also have electrical analogs.　Let us first consider the pair of nonslipping friction wheels shown in Fig. 4–8. (In function and in mathematical description, there is no difference between a pair of nonslipping friction wheels and a pair of positively engaged gears.)

At the point of contact, P_1 on wheel 1 and P_2 on wheel 2 must have the same linear velocity because they move together, and experience equal and opposite forces (action and reaction).　Since this is a rotational system, it is convenient to use angular velocities and torques.　The following relations between magnitudes hold:

$$\frac{\tau_1}{\tau_2} = \frac{r_1}{r_2}, \tag{4–19}$$

$$\frac{\Omega_1}{\Omega_2} = \frac{r_2}{r_1}. \tag{4–20}$$

Equations (4–19) and (4–20) remind us immediately of the relations that exist between the voltages and the currents in the primary and secondary windings of an ideal transformer. If $r_1:r_2$ is considered as the turns ratio $N_1:N_2$ of an ideal transformer, torque will then be analogous to voltage and angular velocity to current. This forms the basis of the f-v analogy in Fig. 4–9(a). If $r_2:r_1$ is considered as the turns ratio $N_1:N_2$, we have

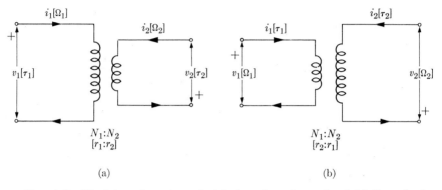

FIG. 4–9. Ideal transformer as electrical analog of a pair of friction wheels or meshed gears. (a) f-v analogy. (b) f-i analogy.

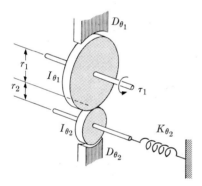

FIG. 4–10. A mechanical system with friction-wheel coupling.

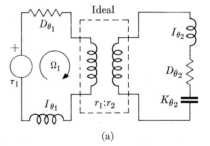

(a)

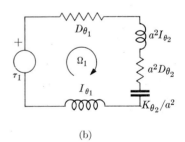

(b)

FIG. 4–11. Analogous electrical circuits for the system in Fig. 4–10 (on τ-v basis). (a) Circuit containing ideal transformer. (b) Circuit with ideal transformer removed, $a = r_1/r_2$.

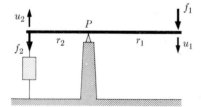

FIG. 4–12. A simple lever supported at rigid fulcrum P.

the analogous ideal transformer based on the f-i analogy in Fig. 4–9(b). The corresponding mechanical quantities are shown in brackets. The reversal of current directions and voltage polarities in the secondaries of these two figures is to show the reversal of the directions of both torque and angular velocity due to coupling; it is equivalent to putting dots on opposite ends of the primary and secondary windings of the transformer. In general, it is easy to determine the relative directions of motion of the coupled mechanical elements by inspection of the mechanical system, without elaborate notations in the electrical circuit.

In applying the electrical analogy in either Fig. 4–9(a) or Fig. 4–9(b) to mechanical systems containing friction wheels or gears, we must remember that the transformer involved is an ideal one. The voltage and current relationships in an ideal transformer are fixed by the turns ratio. Primary, secondary, and mutual inductances do not enter into the equations; in fact, in an ideal transformer they should all be infinitely large. The transformer in the analogous electrical circuit merely serves to properly transform the quantities in the secondary circuit to the primary circuit, and vice versa.

EXAMPLE 4–3. In the system shown in Fig. 4–10, a sinusoidally varying torque $\tau_1 = T_0 \sin \omega t$ is applied to wheel 1, which is engaged with wheel 2. Assume the shafts of both wheels to be inertialess and supported on frictionless bearings. Determine the steady-state angular velocity of wheel 1.

Solution. Let us use τ-v analogy in this example. The analogous electrical circuit is given in Fig. 4–11(a), where all electrical quantities have been written in terms of their analogous mechanical quantities. The ideal transformer can be removed when all the elements in the secondary circuit have been transformed to the primary circuit as shown in Fig. 4–11(b). Since the steady-state angular velocity Ω_1 is wanted, a simple application of the impedance concept yields the desired result:

$$\Omega_1 = \frac{T_0}{\sqrt{(D_{\theta_1} + a^2 D_{\theta_2})^2 + [\omega(I_{\theta_1} + a^2 I_{\theta_2}) - (a^2/\omega K_{\theta_2})]^2}} \sin(\omega t - \Psi),$$

where

$$\Psi = \tan^{-1}\left[\frac{\omega(I_{\theta_1} + a^2 I_{\theta_2}) - (a^2/\omega K_{\theta_2})}{D_{\theta_1} + a^2 D_{\theta_2}} \right].$$

The simple lever is another type of mechanical coupling device that is analogous to a transformer.* Consider the lever in Fig. 4–12, which rests

---

* L. L. Beranek, *Acoustics*, McGraw-Hill Book Company, New York, Chapter 3; 1954.

on a rigid fulcrum P. The lever is assumed to be massless but rigid, and its left end is connected to the ground through some mechanical element which resists motion. If a force f_1 applied to the right end makes it move with a velocity u_1, the following relations hold:

$$\frac{u_1}{u_2} = \frac{r_1}{r_2}, \tag{4–21}$$

$$\frac{f_1}{f_2} = \frac{r_2}{r_1}. \tag{4–22}$$

The similarity between these two equations and Eqs. (4–19) and (4–20) is obvious. We find that the velocities of the ends of the simple lever correspond to the torques on the gears, and the forces on the lever correspond to the angular velocities of the gears. Therefore, although the electrical analog of a simple lever is also an ideal transformer, the f-v analog for a lever will correspond to the f-i analog of a pair of meshed gears, and vice versa. This is shown in Figs. 4–13(a) and 4–13(b). The relative directions of motion of the two ends can be more readily determined from Fig. 4–12.

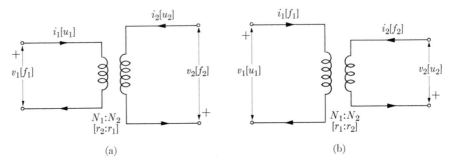

(a)   (b)

FIG. 4–13. Ideal transformer as electrical analog of a simple lever. (a) f-v analogy. (b) f-i analogy.

EXAMPLE 4–4. Find the f-i analogous electrical circuit of the mechanical system shown in Fig. 4–14. Assume that the bar is rigid but massless, and that the junctions are restricted to have vertical motion only.

Solution. Since this system has a lever-type coupling, the existence of an ideal transformer in the electrical analog is apparent. However, we do not have a simple lever here because the bar does not rest or pivot on a fixed fulcrum. The primary circuit in the electrical analog can be drawn without difficulty. By the rule of f-i analogy we know that the primary circuit has one independent node with a current source (f) and three elements (a capacitance M_1, a resistance $1/D_1$, and an inductance K_1) connected to it.

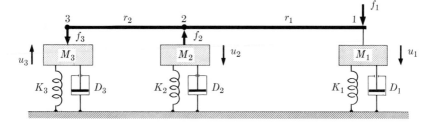

Fig. 4–14. A mechanical system with lever-type coupling.

To draw the secondary circuit, we apply the principle of superposition. First, consider junction 3 as fixed. We then have

$$\frac{u_1}{u_2} = \frac{r_1 + r_2}{r_2}, \tag{4–23}$$

$$\frac{f_1}{f_2} = \frac{r_2}{r_1 + r_2}. \tag{4–24}$$

Equations (4–23) and (4–24) indicate a primary to secondary turns ratio of $N_1 : N_2 = (r_1 + r_2):r_2$ in the f-i analogy. Next, considering junction 2 as fixed, we have

$$\frac{u_1}{u_3} = \frac{r_1}{r_2}, \tag{4–25}$$

$$\frac{f_1}{f_3} = \frac{r_2}{r_1}. \tag{4–26}$$

Equations (4–25) and (4–26) require a primary to secondary turns ratio of $N_1:N_3 = r_1:r_2$. In each of the two secondaries, three passive elements

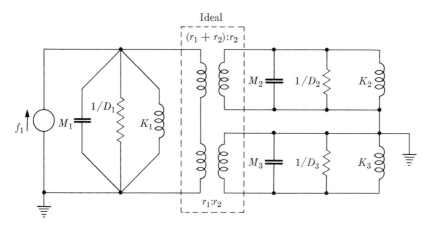

Fig. 4–15. Analogous electrical circuit for system in Fig. 4–14 (on f-i basis). Values beside resistive elements are resistances, not conductances.

(a mass, a spring, and a dashpot) meet at a common junction. This results in the f-i analogous electrical circuit of Fig. 4–15, in which all electrical quantities have been written in terms of their analogous mechanical quantities.

4–7 Electromechanical systems. Systems in which electrical and mechanical elements occur in combination and interact are called *electromechanical* systems. They are quite often referred to as *electromechanical transducers*, which convert electrical energy into mechanical energy or vice versa. Typical examples include microphones, loudspeakers, vibration pickups, and electrical machineries. We shall discuss in this section how an all-electrical analogous circuit can be drawn for an electromechanical transducer.

The basic relations can be derived from the two simple situations shown in Figs. 4–16(a) and 4–16(b), where a conductor of length l lies perpendicularly in a uniform magnetic field with flux density B. In Fig. 4–16(a), a current i through the conductor produces an upward force f on the conductor (left-hand rule):

$$f = Bli, \tag{4–27}$$

which can also be written as

$$i = \left(\frac{1}{Bl}\right)f. \tag{4–28}$$

In Fig. 4–16(b), the conductor moving with an upward velocity u will have induced in it an open-circuit voltage v with polarities as indicated (right-hand rule):

$$v = (Bl)u. \tag{4–29}$$

Equations (4–28) and (4–29) demonstrate the properties of an ideal transformer with turns ratio $Bl{:}1$; one side of the "transformer" carries electrical quantities v and i, while the other side carries mechanical quantities u and f. This is shown in Fig. 4–17. Since f and u on the mechanical side

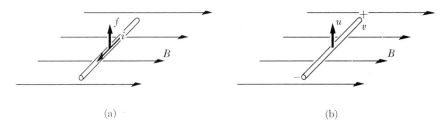

(a)   (b)

Fig. 4–16. A conductor situated in a uniform magnetic field. (a) Current i in conductor produces force f. (b) Motion of conductor with velocity u induces voltage v.

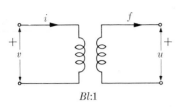

Fig. 4–17. Ideal transformer analogous to the simple electromechanical situation in Fig. 4–16.

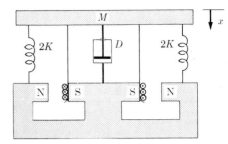

Fig. 4–18. An electromechanical system.

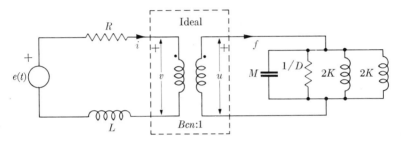

Fig. 4–19. Analogous circuit for the electromechanical system in Fig. 4–18 $(1/D$ is value of the analogous resistance).

are analogous to i and v respectively, this analogy is of the f-i type. We note that Eqs. (4–28) and (4–29) hold for all frequencies, including zero [d-c current in Fig. 4–16(a) and constant velocity in Fig. 4–16(b)]; hence we must also assume that the ideal "transformer" in Fig. 4–17 is operative at all frequencies, including zero.

EXAMPLE 4–5. The electromechanical system shown in Fig. 4–18 may represent a loudspeaker or an electromagnetic relay. The moving coil in the uniform magnetic field B has n turns of circumference c and its inductance and resistance are L and R respectively. A voltage $e(t)$ is applied to the coil. Determine the equation of motion of the mass M.

Solution. In this system we have electrical elements L and R of the coil, and mechanical elements M, D, $2K$, and $2K$. The coupling is provided by the moving coil in the magnetic field. Making use of the analogous ideal transformer in Fig. 4–17 and converting the mechanical elements into their electrical analogs on the f-i basis, we obtain the circuit of Fig. 4–19. The governing equations for the primary and the secondary are

$$L \frac{di}{dt} + Ri + v = e(t), \qquad (4\text{–}30)$$

$$M \frac{du}{dt} + Du + \frac{1}{K} \left[\int_0^t u \, dt + x(0) \right] = f. \tag{4-31}$$

Kirchhoff's voltage law has been used for the primary circuit, and current law for the secondary circuit. Now we have

$$v = (Bcn)u \tag{4-32}$$

and

$$f = (Bcn)i. \tag{4-33}$$

Hence Eqs. (4–30) and (4–31) become

$$L \frac{di}{dt} + Ri + (Bcn)u = e(t) \tag{4-34}$$

and

$$M \frac{du}{dt} + Du + \frac{1}{K} \left[\int_0^t u \, dt + x(0) \right] = (Bcn)i. \tag{4-35}$$

Equations (4–34) and (4–35) are simultaneous integro-differential equations in two unknowns, i and u. The correctness of the signs of the coupling terms $(Bcn)u$ and $(Bcn)i$ can be checked by noting that a positive u (downward motion) would induce an increased voltage drop in the electrical circuit and that current i in the indicated direction tends to pull M down (positive u).

If it is desired to find the current i in the coil, we can refer all elements to the primary circuit and remove the transformer, as shown in Fig. 4–20(a). On the other hand, if the motion (x, or u) of the mass M is to be determined, then it is more convenient to refer all quantities to the secondary, as in Fig. 4–20(b).

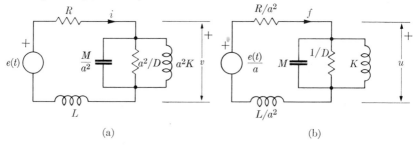

(a)   (b)

FIG. 4–20. Equivalent circuits for Fig. 4–19, $a = Bcn$. (a) Referred to the primary. (b) Referred to the secondary.

PROBLEMS

For each of the mechanical systems given in problems 4–1 through 4–10,
(a) draw the f-v (τ-v if system is rotational) analogous electrical circuit,
(b) draw the f-i (τ-i if system is rotational) analogous electrical circuit, and
(c) write the equations of motion in terms of the given mechanical quantities.

4–1.

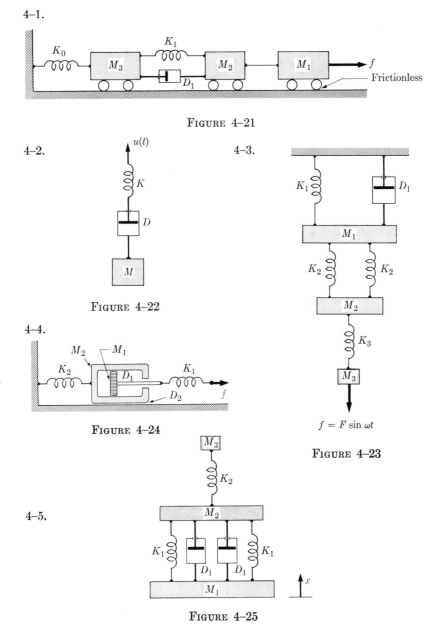

FIGURE 4–21

4–2.

4–3.

FIGURE 4–22

4–4.

FIGURE 4–24

$f = F \sin \omega t$

FIGURE 4–23

4–5.

FIGURE 4–25

4–6.

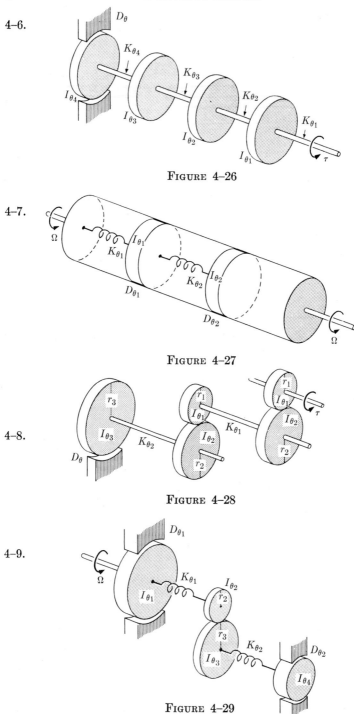

FIGURE 4–26

4–7.

FIGURE 4–27

4–8.

FIGURE 4–28

4–9.

FIGURE 4–29

4–10.

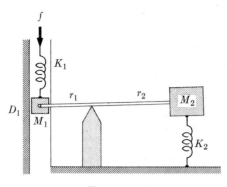

FIGURE 4–30

4–11. (a) Draw the f-v analogous mechanical system for the electrical circuit of Fig. 4–31. (b) Difficulty in drawing the f-v analogous mechanical system will arise if the circuit in Fig. 4–31 is rearranged as in Fig. 4–32. Why?

4–12. Draw the f-i analogous mechanical system for the electrical circuit of Fig. 4–33.

4–13. It has been suggested that it is possible to reduce the vibration of an electric shaver by attaching a mass and spring combination on the case, as illustrated schematically in Fig. 4–34. Comment on the merit of this suggestion and give your reasons analytically.

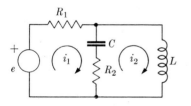

FIGURE 4–31

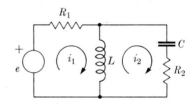

FIGURE 4–32

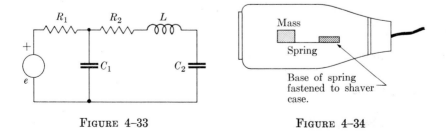

FIGURE 4–33

FIGURE 4–34

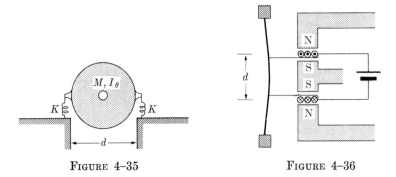

FIGURE 4–35   FIGURE 4–36

4–14. Figure 4–35 is a diagram of a rotary machine with mass M and moment of inertia I_θ about its shaft in the center. It is supported on two springs with identical compliance K, and translational motion is possible in the vertical direction only. Draw an analogous electrical circuit for this mechanical system with combined translational and rotational motion and express all elements in the circuit in terms of the given mechanical quantities.

4–15. Figure 4–36 is a schematic diagram of a moving-coil microphone. Sound waves impinging on the diaphragm make the coil move in the magnetic field, thereby producing a current change in the electrical circuit. Assuming that the diaphragm-coil combination has an effective mass M, a damping coefficient D, and a compliance K, and that the n-turn coil, which has an inductance L and a resistance R, moves in a uniform magnetic field of flux density B, draw an analogous circuit for this electromechanical device and write the equations that completely describe the system.

4–16. Figure 4–37 is a schematic diagram of a small d-c shunt motor. Assuming that the armature winding has an inductance L and a resistance R and that the entire rotor has a rotational damping coefficient D_θ and a moment of inertia I_θ about its axis, draw an analogous circuit for the motor after the switch is closed and write the equations that completely describe the system. The torque τ developed in a d-c motor is proportional to the armature current I and to the total flux per pole ϕ: $\tau = kI\phi$, where k is a constant for any one machine.

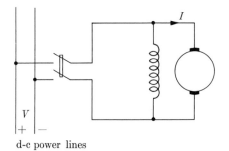

d-c power lines

FIGURE 4–37

CHAPTER 5

ANALYSIS BY FOURIER METHODS

5–1 Introduction. Sinusoidal functions occupy a unique position in engineering. They are easy to generate, and the steady-state response of a linear system to sinusoidal excitations can be found very easily by using the impedance concept. However, they are the favorite functions of engineers primarily because arbitrary periodic functions* with few restrictions can be expanded into Fourier series of harmonic sinusoidal components, and transient nonperiodic functions can be expressed as Fourier integrals. We can therefore discuss arbitrary waveforms in terms of their frequency spectra, and by superposition determine system response to an arbitrary excitation in terms of its responses to the various frequency components of the excitation.

We recall that a function $f(\theta)$ possessing derivatives of all orders at $\theta = 0$ may be expanded in a power series as follows:

$$f(\theta) = f(0) + f'(0)\theta + \frac{1}{2!} f''(0)\theta^2 + \frac{1}{3!} f'''(0)\theta^3 + \cdots$$

$$= \sum_{n=0}^{\infty} a_n \theta^n, \tag{5-1}$$

where the general coefficient a_n is

$$a_n = \frac{1}{n!} f^{(n)}(0). \tag{5-2}$$

Equation (5–1) is the famous *Maclaurin series expansion* of $f(\theta)$. The requirement that the given function $f(\theta)$ possess derivatives of all orders is, however, too restrictive to be of practical use; it excludes such important waveforms as rectangular, sawtooth, rectified sine, etc. Furthermore, there exist no useful common properties among the component terms of the Maclaurin series which would aid in the determination of the resultant response of a physical system. All terms except the first increase indefinitely with the independent variable θ, and none is periodic.

On the other hand, Fourier series expansion of a periodic function $f(\theta)$ is a trigonometric series consisting of sine and cosine terms. The restrictions that must be imposed on $f(\theta)$ are usually satisfied by the waveforms the engineer is likely to encounter. In this chapter we shall develop the

---

* A function $f(\theta)$ is said to be periodic in θ with period p ($\neq 0$) if $f(\theta + p) = f(\theta)$ for all θ.

Fourier series, discuss its important properties, transform it into the Fourier integral for nonperiodic functions, and apply Fourier methods to analysis of linear systems.

5–2 Fourier series expansion of periodic functions. Any periodic function $f(\theta)$ with period 2π that satisfies the *Dirichlet conditions*

(1) that it has at most a finite number of discontinuities in one period,

(2) that it has at most a finite number of maxima and minima in one period, and

(3) that the integral $\int_{-\pi}^{\pi} |f(\theta)|\, d\theta$ is finite

can be expanded into a Fourier series such that

$$f(\theta) = \frac{a_0}{2} + \sum_{n=1}^{\infty} (a_n \cos n\theta + b_n \sin n\theta), \qquad (5\text{–}3)$$

where the coefficients a_n and b_n are given by*

$$a_n = \frac{1}{\pi} \int_{-\pi}^{\pi} f(\theta) \cos n\theta\, d\theta, \qquad (5\text{–}4)$$

$$b_n = \frac{1}{\pi} \int_{-\pi}^{\pi} f(\theta) \sin n\theta\, d\theta. \qquad (5\text{–}5)$$

The proof that if the given function $f(\theta)$ satisfies the Dirichlet conditions as stated above then the right side of Eq. (5–3) converges to $f(\theta)$ is too involved to be given here.† However, the correctness of Eqs. (5–4) and (5–5) for the evaluation of the Fourier coefficients a_n and b_n can be readily verified by direct substitution. Thus, multiplying both sides of Eq. (5–3) by $\cos m\theta$ and integrating from $-\pi$ to π, we have

$$\int_{-\pi}^{\pi} f(\theta) \cos m\theta\, d\theta = \int_{-\pi}^{\pi} \left[\frac{a_0}{2} \cos m\theta + \sum_{n=1}^{\infty} a_n \cos n\theta \cos m\theta \right.$$

$$\left. + \sum_{n=1}^{\infty} b_n \sin n\theta \cos m\theta \right] d\theta. \qquad (5\text{–}6)$$

---

* The interval of integration in Eqs. (5–4) and (5–5) could just as well be from 0 to 2π or from θ_0 to $\theta_0 + 2\pi$ (for any θ_0), instead of from $-\pi$ to π, without affecting the values of a_n and b_n, since the function $f(\theta)$ repeats itself with a period 2π irrespective of the starting point of the interval. However, the choice of the fundamental interval $-\pi \leq \theta \leq \pi$, half in the negative region and half in the positive region about the vertical axis at the origin, has certain advantages when we discuss symmetry properties in Section 5–3.

† H. S. Carslaw, *Fourier's Series and Integrals*, 3rd edition, p. 207, Cambridge University Press; 1930. Reprinted by Dover Publications, New York, N. Y.; 1946.

The first integral on the right side of Eq. (5–6) obviously vanishes; the second and third integrals can be expanded as follows:

$$\int_{-\pi}^{\pi} a_n \cos n\theta \cos m\theta \, d\theta = \frac{a_n}{2} \int_{-\pi}^{\pi} \left[\cos (m+n)\theta + \cos (m-n)\theta \right] d\theta$$

$$= \begin{cases} 0, & m \neq n, \\ \pi a_n, & m = n, \end{cases} \tag{5–7}$$

$$\int_{-\pi}^{\pi} b_n \sin n\theta \cos m\theta \, d\theta = \frac{b_n}{2} \int_{-\pi}^{\pi} \left[\sin (m+n)\theta - \sin (m-n)\theta \right] d\theta = 0. \tag{5–8}$$

Hence Eq. (5–6) gives a nonzero value only when $m = n$ and

$$\int_{-\pi}^{\pi} f(\theta) \cos n\theta \, d\theta = \pi a_n, \tag{5–9}$$

from which the formula for a_n in Eq. (5–4) follows directly. An entirely similar procedure [multiplying both sides of Eq. (5–3) by $\sin m\theta$ and integrating] yields the formula for b_n in Eq. (5–5). Note that by putting $n = 0$ in Eq. (5–4) we obtain

$$a_0 = \frac{1}{\pi} \int_{-\pi}^{\pi} f(\theta) \, d\theta = 2 \left(\frac{1}{2\pi} \int_{-\pi}^{\pi} f(\theta) \, d\theta \right)$$

$$= 2 \times [\text{average value of } f(\theta) \text{ over a period}], \tag{5–10}$$

or

$$\frac{a_0}{2} = \text{average value of } f(\theta), \tag{5–11}$$

which of course is also correct. Note also that the simplicity of Eqs. (5–4) and (5–5) relies upon the so-called *orthogonality relations* between sine and cosine functions, as indicated in Eqs. (5–7) and (5–8).

If the given function is a function of time, as it often is in engineering problems, Eqs. (5–3), (5–4), and (5–5) can be modified accordingly. We should always remember that the arguments for sine and cosine functions must be angles; it would be an inexcusable mistake to write $\sin mt$ or $\cos nt$ when t represents time and m and n are dimensionless numbers. Let T be the period in time of the given function $f(t)$, then $\theta = (2\pi/T)t = \omega t$, where ω is the angular frequency and is always equal to $2\pi/T$. We obtain

$$f(t) = \frac{a_0}{2} + \sum_{n=1}^{\infty} (a_n \cos n\omega t + b_n \sin n\omega t), \tag{5–12}$$

with

$$a_n = \frac{2}{T} \int_{-T/2}^{T/2} f(t) \cos n\omega t \, dt, \qquad (n = 0, 1, 2, \ldots), \tag{5–13}$$

$$b_n = \frac{2}{T} \int_{-T/2}^{T/2} f(t) \sin n\omega t \, dt, \qquad (n = 1, 2, 3, \ldots). \tag{5–14}$$

(Note the necessity for changing the limits of the integrals properly when the variable of integration is changed from θ to t.) Equation (5–12) can also be expressed in an alternative form:

$$f(t) = \frac{a_0}{2} + \sum_{n=1}^{\infty} c_n \cos(n\omega t - \Psi_n),\qquad(5\text{–}15)$$

where

$$c_n = \sqrt{a_n^2 + b_n^2}\qquad(5\text{–}16)$$

and

$$\Psi_n = \tan^{-1}\left(\frac{b_n}{a_n}\right).\qquad(5\text{–}17)$$

Equation (5–15) shows clearly that an arbitrary periodic function satisfying the Dirichlet conditions can be decomposed into an average (d-c) value and harmonically related sinusoidal components whose frequencies are integral multiples of the fundamental frequency $(1/T)$. The amplitude and relative phase of the components are given by Eqs. (5–16) and (5–17).

If the magnitude of c_n, $|c_n|$, is plotted versus $n\omega$, we obtain a graph which may have the general appearance of Fig. 5–1. This is called the *frequency spectrum** of the given $f(t)$. The component at $\omega = 0$ is the average value of the given function; its magnitude does not bear a definite relationship to the magnitudes of the other components. In fact, this average value can be changed at will by moving the position of the horizontal axis (t-axis) up and down, without changing the shape of the given $f(t)$ or the rest of the frequency spectrum. The other frequency components are spaced ω apart, although some of the components may vanish (zero amplitude) in special cases. Since this spectrum consists of discrete lines, it is a *discrete frequency spectrum*. It gives no information about the

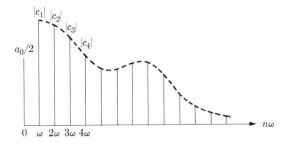

Fig. 5–1. A discrete frequency spectrum.

---

* Although it is really an *amplitude* spectrum, the term *frequency* spectrum is more commonly used.

relative phase Ψ_n of the components, which is usually of secondary importance unless one desires to synthesize $f(t)$. Of course, we can also plot a phase spectrum with Ψ_n versus $n\omega$ if we wish. It is clear that as T increases (ω decreases), the lines are closer to their neighbors. In the limit as T increases indefinitely, the discrete spectrum approaches a continuous one, and the Fourier series becomes a Fourier integral, as we shall discuss in more detail later.

In the above discussions we have considered the given function $f(\theta)$ as a periodic function of θ; but the formulas for the coefficients a_n and b_n in the Fourier series expansions use the values of $f(\theta)$ between $\theta = -\pi$ to π only. It follows that the *same* Fourier series of Eq. (5–3) can be used to represent an arbitrary function *within the interval from* $\theta = -\pi$ *to* π even though $f(\theta)$ does not continue or repeat outside this interval. Outside this interval the Fourier series expression will repeat faithfully from period to period irrespective of whether the given function continues. The same remarks apply to an arbitrary function $f(t)$ which is specified over any *finite* range, say from $t = 0$ to t_0. An infinite number of Fourier series expansions with fundamental periods $T \geq t_0$ can be found such that they all reproduce $f(t)$ *within the given range*. Outside this range, different expansions may have entirely different values, depending upon the choice of T as compared with t_0 and of the waveform in the interval from $t = t_0$ to T, which is entirely arbitrary except that the Dirichlet conditions must be satisfied.

EXAMPLE 5–1. Find the Fourier series expansion of the periodic rectangular wave shown in Fig. 5–2.

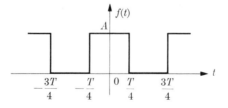

FIG. 5–2. A periodic rectangular wave.

Solution. First we write the expressions for the given waveform in one period $(-T/2 < t < T/2)$:

$$f(t) = \begin{cases} 0, & -T/2 < t < -T/4, \\ A, & -T/4 < t < T/4, \\ 0, & T/4 < t < T/2. \end{cases} \qquad (5\text{–}18)$$

Using Eq. (5–13), we determine the a_n coefficients for cosine terms:

$$a_n = \frac{2}{T} \int_{-T/2}^{T/2} f(t) \cos n\omega t \, dt$$

$$= \frac{2}{T} \int_{-T/4}^{T/4} A \cos n\omega t \, dt$$

$$= \frac{2A}{n\omega T} \sin n\omega t \Big|_{-T/4}^{T/4}$$

$$= \frac{2A}{n\pi} \sin\left(\frac{n\pi}{2}\right). \tag{5–19}$$

In obtaining the result in Eq. (5–19), we have made use of the relation $\omega T = 2\pi$. Equation (5–19) tells us that

$$a_n = 0, \qquad \text{if} \qquad n = 2, 4, 6, \ldots,$$

$$a_n = \frac{2A}{n\pi}, \qquad \text{if} \qquad n = 1, 5, 9, \ldots, \tag{5–20}$$

$$a_n = -\frac{2A}{n\pi}, \qquad \text{if} \qquad n = 3, 7, 11, \ldots$$

If $n = 0$, a_n would take a 0/0 indeterminate form which could be readily evaluated. But in this particular case, we can write the average value of the given waveform by inspection:

$$\frac{a_0}{2} = \text{average value of } f(t) = \frac{A}{2}. \tag{5–21}$$

The b_n coefficients for sine terms are found by applying Eq. (5–14):

$$b_n = \frac{2}{T} \int_{-T/2}^{T/2} f(t) \sin n\omega t \, dt$$

$$= \frac{2}{T} \int_{-T/4}^{T/4} A \sin n\omega t \, dt = 0. \tag{5–22}$$

Hence we have

$$f(t) = \frac{A}{2} + \frac{2A}{\pi} \left(\cos \omega t - \tfrac{1}{3} \cos 3\omega t + \tfrac{1}{5} \cos 5\omega t - \cdots\right)$$

$$= \frac{A}{2} + \frac{2A}{\pi} \sum_{k=1}^{\infty} (-1)^{k-1} \frac{1}{(2k-1)} \cos (2k-1)\omega t, \tag{5–23}$$

where the new parameter k has been introduced to take care of the situation in Eq. (5–20). We note from Eq. (5–23) that for this rectangular wave, which has finite discontinuities, the amplitude of the harmonic components

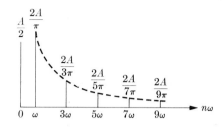

FIG. 5–3. Frequency spectrum of $f(t)$ in Fig. 5–2.

decreases with the same rate as $1/n$. The frequency spectrum of the given rectangular wave is plotted in Fig. 5–3.

The gradual approximation of the given rectangular wave $f(t)$ by taking more and more terms of the Fourier-series expansion in Eq. (5–23) can be visualized from Fig. 5–4(a), where the wave shapes that result from taking the first four and seven terms respectively are shown. However, although the rate of oscillation of the ripples near the discontinuities of $f(t)$ increases with the number of terms we take, the ripples do not disappear when the number of terms becomes very large. This is illustrated in Fig. 5–4(b). As a matter of fact, the amplitude of the ripples near the discontinuities always exceeds that of the given rectangular wave and tends to approach an *overshoot* of 18 percent in both directions relative to the average value $A/2$ even when the number of terms taken approaches infinity. This phenomenon is referred to as the *Gibbs phenomenon* and is due to the fact that the Fourier series expansion of $f(t)$ fails to converge uniformly at discontinuities.* At the discontinuities, the partial sums of Fourier expansions take the mean values. This will be discussed again in Section 5–4.

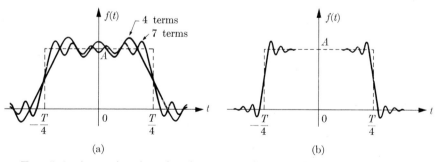

FIG. 5–4. Approximations for the rectangular wave in Fig. 5–2. (a) Partial sum of a few terms of Eq. (5–23). (b) Partial sum of many terms ($k \gg 1$) of Eq. (5–23).

---

* For a very comprehensive discussion of the Gibbs phenomenon, see E. A. Guillemin *The Mathematics of Circuit Analysis*, Chap. VII, Art. 13, John Wiley and Sons, Inc., New York, N. Y.; 1949.

5–3 Symmetry conditions. When a given function is such that it possesses certain symmetry properties, it is possible to tell what terms will be missing in its Fourier expansion by inspection and, moreover, the expressions for the coefficients of those terms that are present can be simplified. We shall discuss the symmetry properties under two categories: (A) cosine or sine terms only, and (B) even or odd harmonics only.*

A. *Cosine or sine terms only*

A function $f_c(\theta)$ is said to be an *even function* if

$$f_c(\theta) = f_c(-\theta). \tag{5–24}$$

Hence the function shown in Fig. 5–5(a) is an even function.

Geometrically, an even function is symmetric not only with respect to the vertical axis passing through the origin ($\theta = 0$) *but also with respect to all vertical lines at* $\theta = n\pi$ ($n = \pm 1, \pm 2, \ldots$), since all periodic functions with period 2π must also satisfy the condition $f(\theta) = f(\theta + 2\pi)$. When $f_c(\theta)$ is even, we have

$$a_n = \frac{1}{\pi} \int_{-\pi}^{\pi} f_c(\theta) \cos n\theta \, d\theta$$

$$= \frac{2}{\pi} \int_0^{\pi} f_c(\theta) \cos n\theta \, d\theta, \qquad (n = 0, 1, 2, \ldots). \tag{5–25}$$

Equation (5–25) is a direct consequence of the fact that the integrand $f_c(\theta) \cos n\theta$, being an even function, integrates into exactly the same value in the intervals $-\pi \le \theta \le 0$ and $0 \le \theta \le \pi$. Also,

$$b_n = \frac{1}{\pi} \int_{-\pi}^{\pi} f_c(\theta) \sin n\theta \, d\theta = 0, \qquad (n = 1, 2, 3, \ldots). \tag{5–26}$$

The integrand in Eq. (5–26), $f_c(\theta) \sin n\theta$, is an odd function (sin $n\theta$ is an odd function, and the product of an even function with an odd function is an odd function); hence the integral over the interval $-\pi \le \theta \le 0$ cancels with the integral over the interval $0 \le \theta \le \pi$. Equation (5–26) tells us that the coefficients of all sine terms are zero. In other words, *the Fourier series expansion of an even periodic function $f_c(\theta)$ contains only cosine terms and the constant term (if $a_0 \ne 0$)*. The coefficients a_n can be determined from Eq. (5–25) for all values of n.

A function $f_s(\theta)$ is said to be an *odd function* if

$$f_s(\theta) = -f_s(-\theta). \tag{5–27}$$

---

* Besides these two categories, there are other less common types of symmetries which also lead to certain simplifications in the Fourier expansion. See problem 5–8.

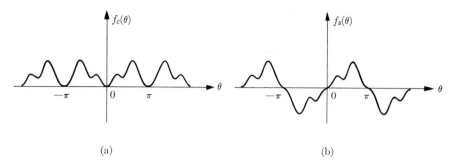

(a)   (b)

FIG. 5–5. (a) An even function. (b) An odd function.

Hence the function shown in Fig. 5–5(b) is an odd function. *Geometrically, an odd function is antisymmetric with respect to all vertical lines at θ = nπ, (n = 0, ±1, ±2, ...).* When $f_s(\theta)$ is odd, we have

$$a_n = \frac{1}{\pi} \int_{-\pi}^{\pi} f_s(\theta) \cos n\theta \, d\theta = 0, \qquad (n = 0, 1, 2, \ldots), \qquad (5\text{–}28)$$

since the integrand $f_s(\theta) \cos n\theta$ is an odd function; and

$$b_n = \frac{1}{\pi} \int_{-\pi}^{\pi} f_s(\theta) \sin n\theta \, d\theta$$

$$= \frac{2}{\pi} \int_{0}^{\pi} f_s(\theta) \sin n\theta \, d\theta, \qquad (n = 1, 2, 3, \ldots). \qquad (5\text{–}29)$$

Hence *the Fourier series expansion of an odd periodic function $f_s(\theta)$ contains*

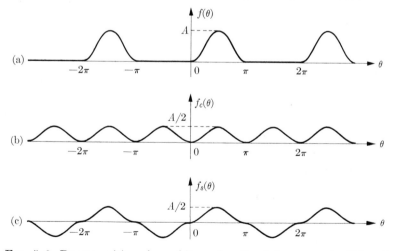

FIG. 5–6. Decomposition of an arbitrary function into even and odd functions. (a) An arbitrary periodic function. (b) An even function with half amplitude. (c) An odd function with half amplitude.

only sine terms. The coefficients b_n for the sine terms can be determined from Eq. (5–29).

In general, any function $f(\theta)$ with no symmetry properties can be decomposed into an even function and an odd function, each with one-half the original amplitude. This is illustrated in Fig. 5–6. The sum of the functions in Figs. 5–6(b) and 5–6(c) gives the original function in Fig. 5–6(a):

$$f(\theta) = f_c(\theta) + f_s(\theta), \tag{5–30}$$

and the given function $f(\theta)$ contains both cosine and sine terms.

B. *Even or odd harmonics only*

A *periodic function* $f_e(\theta)$ *with period* 2π *contains only even harmonics if it satisfies the following condition:*

$$f_e(\theta \pm \pi) = f_e(\theta). \tag{5–31}$$

An example of such a function is shown in Fig. 5–7(a).

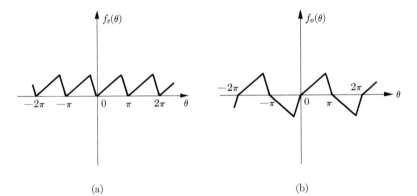

(a)   (b)

Fig. 5–7. A function containing (a) even harmonics only, and (b) odd harmonics only.

Geometrically, it is seen that a function satisfying Eq. (5–31) actually repeats with a period π, that is, it repeats with double frequency if 2π is retained as the fundamental period. It then follows directly that $f_e(\theta)$ contains no odd harmonics ($n = 0, 2, 4, \ldots$ only), and is, in general, quite different from $f_c(\theta)$ (cosine terms only), since the Fourier expansion of $f_e(\theta)$ may contain both sine and cosine terms and the Fourier expansion of $f_c(\theta)$ may contain both even and odd harmonics.

A *periodic function* $f_o(\theta)$ *with period* 2π *contains only odd harmonics if it satisfies the following condition:*

$$f_o(\theta \pm \pi) = -f_o(\theta). \tag{5–32}$$

The above statement can be proved by examining the expressions for a_n and b_n, Eqs. (5–4) and (5–5), in the light of Eq. (5–32):

$$a_n = \frac{1}{\pi} \int_{-\pi}^{\pi} f_o(\theta) \cos n\theta \, d\theta$$

$$= \frac{1}{\pi} \left[\int_{-\pi}^{0} f_o(\theta) \cos n\theta \, d\theta + \int_{0}^{\pi} f_o(\theta) \cos n\theta \, d\theta \right] \cdot \quad (5\text{–}33)$$

A change of variable from θ to $\theta - \pi$ in the first integral gives

$$\int_{-\pi}^{0} f_o(\theta) \cos n\theta \, d\theta = \int_{0}^{\pi} f_o(\theta - \pi) \cos n(\theta - \pi) \, d\theta$$

$$= -\cos n\pi \int_{0}^{\pi} f_o(\theta) \cos n\theta \, d\theta. \quad (5\text{–}34)$$

Substituting Eq. (5–34) in Eq. (5–33), we have

$$a_n = \frac{1}{\pi} (1 - \cos n\pi) \int_{0}^{\pi} f_o(\theta) \cos n\theta \, d\theta$$

$$= \begin{cases} 0, \text{ if } n \text{ is even,} \\ \dfrac{2}{\pi} \displaystyle\int_{0}^{\pi} f_o(\theta) \cos n\theta \, d\theta, \text{ if } n \text{ is odd.} \end{cases} \quad (5\text{–}35)$$

Similarly, we can prove that

$$b_n = \begin{cases} 0, \text{ if } n \text{ is even,} \\ \dfrac{2}{\pi} \displaystyle\int_{0}^{\pi} f_o(\theta) \sin n\theta \, d\theta, \text{ if } n \text{ is odd.} \end{cases} \quad (5\text{–}36)$$

Thus we have proved the statement in connection with Eq. (5–32). An example of $f_o(\theta)$, containing odd harmonics only, is shown in Fig. 5–7(b).

Table 5–1 lists the symmetry conditions and the corresponding characteristic properties of the Fourier expansion of $f(\theta)$. A corresponding list for $f(t)$ is provided in Table 5–2.

In analyzing a given periodic function, it is always advisable to determine first whether simplifying symmetry conditions exist. When they do, some terms are immediately known to vanish according to Tables 5–1 and 5–2; integration over only half a period is necessary for the determination of those Fourier coefficients which do not vanish.

EXAMPLE 5–2. Find the Fourier series expansion of the periodic triangular wave shown in Fig. 5–8.

TABLE 5-1. SYMMETRY CONDITIONS FOR $f(\theta)$

	Symmetry conditions	Characteristic properties	a_n	b_n
$f_c(\theta)$	$f_c(\theta) = f_c(-\theta)$	cos terms only	$\dfrac{2}{\pi} \displaystyle\int_0^{\pi} f_c(\theta) \cos n\theta \, d\theta$	0
$f_s(\theta)$	$f_s(\theta) = -f_s(-\theta)$	sin terms only	0	$\dfrac{2}{\pi} \displaystyle\int_0^{\pi} f_s(\theta) \sin n\theta \, d\theta$
$f_e(\theta)$	$f_e(\theta \pm \pi) = f_e(\theta)$	even n only	$\dfrac{2}{\pi} \displaystyle\int_0^{\pi} f_e(\theta) \cos n\theta \, d\theta$	$\dfrac{2}{\pi} \displaystyle\int_0^{\pi} f_e(\theta) \sin n\theta \, d\theta$
$f_o(\theta)$	$f_o(\theta \pm \pi) = -f_o(\theta)$	odd n only	$\dfrac{2}{\pi} \displaystyle\int_0^{\pi} f_o(\theta) \cos n\theta \, d\theta$	$\dfrac{2}{\pi} \displaystyle\int_0^{\pi} f_o(\theta) \sin n\theta \, d\theta$

TABLE 5-2. SYMMETRY CONDITIONS FOR $f(t)$

	Symmetry conditions	Characteristic properties	a_n	b_n
$f_c(t)$	$f_c(t) = f_c(-t)$	cos terms only	$\dfrac{4}{T} \displaystyle\int_0^{T/2} f_c(t) \cos (2\pi n/T)t \, dt$	0
$f_s(t)$	$f_s(t) = -f_s(-t)$	sin terms only	0	$\dfrac{4}{T} \displaystyle\int_0^{T/2} f_s(t) \sin (2\pi n/T)t \, dt$
$f_e(t)$	$f_e\left(t \pm \dfrac{T}{2}\right) = f_e(t)$	even n only	$\dfrac{4}{T} \displaystyle\int_0^{T/2} f_e(t) \cos (2\pi n/T)t \, dt$	$\dfrac{4}{T} \displaystyle\int_0^{T/2} f_e(t) \sin (2\pi n/T)t \, dt$
$f_o(t)$	$f_o\left(t \pm \dfrac{T}{2}\right) = -f_o(t)$	odd n only	$\dfrac{4}{T} \displaystyle\int_0^{T/2} f_o(t) \cos (2\pi n/T)t \, dt$	$\dfrac{4}{T} \displaystyle\int_0^{T/2} f_o(t) \sin (2\pi n/T)t \, dt$

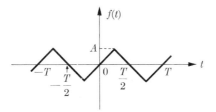

Fig. 5–8. A periodic triangular wave.

Solution. First we examine the given waveform for symmetry conditions, and we see that

$$f(t) = -f(-t) \rightarrow \text{sine terms only } (a_n = 0 \text{ for all } n),$$

$$f\left(t \pm \frac{T}{2}\right) = -f(t) \rightarrow \text{odd harmonics only (odd } n \text{ only).}$$

Hence the Fourier series expansion of the given wave contains only b_n for n odd. When *two* symmetry conditions exist, it is necessary to integrate over only *one quarter period* in the determination of Fourier coefficients. (The proof of this statement is left to the reader.)

$$b_n = \frac{8}{T} \int_0^{T/4} f(t) \sin n \left(\frac{2\pi}{T}\right) t \, dt, \qquad (n \text{ odd}). \qquad (5\text{–}37)$$

Since

$$f(t) = \frac{4A}{T} t, \qquad 0 \le t \le T/4,$$

it follows that

$$b_n = \frac{8}{T} \int_0^{T/4} \left(\frac{4A}{T} t\right) \sin n \left(\frac{2\pi}{T}\right) t \, dt$$

$$= \frac{8A}{n^2\pi^2} \sin \left(\frac{n\pi}{2}\right), \qquad (n \text{ odd}) \qquad (5\text{–}38)$$

$$= \begin{cases} \dfrac{8A}{n^2\pi^2}, & n = 1, 5, 9, \ldots, \\[2mm] -\dfrac{8A}{n^2\pi^2}, & n = 3, 7, 11, \ldots \end{cases}$$

Hence we have

$$f(t) = \frac{8A}{\pi^2} \left(\sin \omega t - \frac{1}{3^2} \sin 3\omega t + \frac{1}{5^2} \sin 5\omega t - \cdots \right)$$

$$= \frac{8A}{\pi^2} \sum_{k=1}^{\infty} (-1)^{k-1} \frac{1}{(2k-1)^2} \sin (2k-1)\omega t, \qquad (5\text{–}39)$$

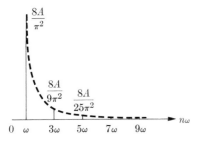

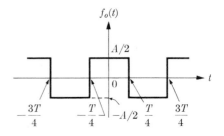

FIG. 5–9. Frequency spectrum of $f(t)$ in Fig. 5–8.

FIG. 5–10. Rectangular wave of Fig. 5–2 shifted down by $A/2$.

where $\omega = 2\pi/T$. We note that for this triangular wave, which is everywhere continuous but has discontinuous derivatives, the amplitude of the harmonic components decreases with the same rate as $1/n^2$, much more rapidly than that for the rectangular wave in Example 5–1. The frequency spectrum of the given triangular wave is plotted in Fig. 5–9.

Sometimes a given waveform may not satisfy any of the symmetry conditions given in this section, but a simple shift of the horizontal axis may put a "concealed" symmetry condition in evidence, thus simplifying the analysis. Take, for example, the rectangular wave in Fig. 5–2. Since $f(t) = f(-t)$, we know it is an even function and contains only cosine terms. As it stands, it does not satisfy the condition $f(t \pm T/2) = -f(t)$, yet its Fourier expansion in Eq. (5–23) contains only odd harmonics and an average term $A/2$. The reason becomes quite clear when we look at the waveform of $f_o(t) = f(t) - A/2$ in Fig. 5–10. Obviously, the new function $f_o(t)$ contains cosine terms of odd harmonics only. *A parallel shift of the horizontal axis changes only the average value of a given function; symmetry conditions with respect to even and odd harmonic components should be examined after the horizontal axis has been shifted to the average level of the given function. A parallel shift in the position of the vertical axis amounts to a shift in t (or θ); it does not change the harmonic content of a periodic wave, but may disturb the sine and cosine structure of the Fourier series expansion.*

In Example 5–1 we observed in connection with Eq. (5–23) that the amplitude of the harmonic components of a rectangular wave, which has finite discontinuities, decreases with the same rate as $1/n$. Similarly, in Example 5–2 we observed in connection with Eq. (5–39) that the amplitude of the harmonic components of a triangular wave, which is everywhere continuous but has discontinuous first derivatives, decreases at a much faster rate, $1/n^2$. Geometrically this is reasonable, since a triangular wave is "smoother" than a rectangular wave, hence its Fourier series converges more rapidly. These observations can be summarized into several rather general statements. These statements can be proved in a general manner,* and are useful in estimating the behavior of the Fourier series expansion of periodic functions. We assume that the Dirichlet conditions are satisfied.

1. The coefficients of the Fourier series expansion of periodic functions having one or more discontinuities decrease with the same rate as $1/n$ when n is sufficiently large.

2. The coefficients of the Fourier series expansion of periodic functions, which are everywhere continuous but have discontinuous first derivatives at one or more points, decrease with the same rate as $1/n^2$ when n is sufficiently large.

3. The coefficients of the Fourier series expansion of periodic functions, which together with its derivatives up to the $(k-1)$th order are everywhere continuous but whose kth derivative is discontinuous at one or more points, decrease with the same rate as $1/n^{k+1}$ when n is sufficiently large.

4. The coefficients of the Fourier series expansion of any periodic function decrease as least as fast as $1/n$ when n is sufficiently large.

5–4 Fourier series as least mean-square-error approximation. In this section we shall take an entirely different approach to the Fourier series expansion of an arbitrary periodic function $f(\theta)$ having a period 2π. We wish to approximate $f(\theta)$ in the interval $-\pi \leq \theta \leq \pi$ by a finite trigonometric series of $2N+1$ terms:

$$S_N(\theta) = \frac{a_0}{2} + \sum_{n=1}^{N} (a_n \cos n\theta + b_n \sin n\theta), \qquad (5\text{–}40)$$

such that the coefficients a_0, a_n, and b_n are chosen to give a least mean-square-error.

The error function of the approximation, $\epsilon_N(\theta)$, is

$$\epsilon_N(\theta) = f(\theta) - S_N(\theta). \qquad (5\text{–}41)$$

Hence the mean-square-error is

$$E_N = \overline{\epsilon_N^2(\theta)} = \frac{1}{2\pi} \int_{-\pi}^{\pi} \epsilon_N^2(\theta)\, d\theta, \qquad (5\text{–}42)$$

where E_N is a function of the coefficients a_0, a_n, and b_n, but is no longer a function of θ. In order to make E_N a minimum, we require

---

* L. A. Pipes, *Applied Mathematics for Engineers and Physicists*, Chap. III, Sec. 5, McGraw-Hill Book Company, New York, N. Y.; 1946.

$$\frac{\partial E_N}{\partial a_n} = 0, \qquad (n = 0, 1, 2, \ldots, N), \tag{5-43}$$

and

$$\frac{\partial E_N}{\partial b_n} = 0, \qquad (n = 1, 2, 3, \ldots, N). \tag{5-44}$$

Equations (5–43) and (5–44) give us a total of $2N + 1$ equations from which the $(N + 1)$ a_n's and N b_n's can be determined. Substituting Eq. (5–42) in Eq. (5–43) and remembering Eqs. (5–40) and (5–41), we obtain

$$\frac{\partial E_N}{\partial a_n} = \frac{1}{\pi} \int_{-\pi}^{\pi} \epsilon_N(\theta) \frac{\partial \epsilon_N(\theta)}{\partial a_n} d\theta$$

$$= \frac{1}{\pi} \int_{-\pi}^{\pi} [f(\theta) - S_N(\theta)](-\cos n\theta) \, d\theta = 0,$$

which requires

$$\int_{-\pi}^{\pi} S_N(\theta) \cos n\theta \, d\theta = \int_{-\pi}^{\pi} f(\theta) \cos n\theta \, d\theta. \tag{5-45}$$

The left side of Eq. (5–45) is

$$\int_{-\pi}^{\pi} S_N(\theta) \cos n\theta \, d\theta = \int_{-\pi}^{\pi} \left[\frac{a_0}{2} + \sum_{n=1}^{N} (a_n \cos n\theta + b_n \sin n\theta) \right] \cos n\theta \, d\theta.$$

In the above integral, all terms except

$$\int_{-\pi}^{\pi} a_n \cos^2 n\theta \, d\theta = \pi a_n \tag{5-46}$$

vanish because of the orthogonal relations in Eqs. (5–7) and (5–8). Combining Eqs. (5–45) and (5–46), we obtain

$$a_n = \frac{1}{\pi} \int_{-\pi}^{\pi} f(\theta) \cos n\theta \, d\theta, \qquad (n = 0, 1, 2, \ldots, N), \tag{5-47}$$

which is exactly the same as the formula for the Fourier coefficients a_n given in Eq. (5–4). Similarly, from Eq. (5–44) we can determine

$$b_n = \frac{1}{\pi} \int_{-\pi}^{\pi} f(\theta) \sin n\theta \, d\theta, \qquad (n = 1, 2, 3, \ldots, N), \tag{5-48}$$

which is identical with Eq. (5–5). Therefore we have proved that *a Fourier series with a finite number of terms represents the best approximation (in the least mean-square-error sense) possible for the given periodic function by any trigonometric series with the same number of terms.* It is important to note that each a_n and b_n is determined directly and separately, and that, once determined, the values of these coefficients are definite and *final; the*

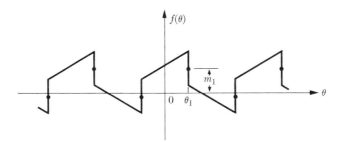

Fig. 5–11. A periodic function with finite discontinuities.

necessity for adding more terms to the series and finding new coefficients with a view to improving the approximation requires no change in the coefficients already calculated.

At the end of Section 5–2 we mentioned that at the finite discontinuities of a given function the partial sums of Fourier series expansion take the values of the arithmetic mean of the values of the given function adjacent to the discontinuities. Thus the Fourier series for the waveform in Fig. 5–11 will converge to the points marked by dots. At the typical point $\theta = \theta_1$, the Fourier series will assume the value

$$m_1 = \tfrac{1}{2}[f(\theta_1-) + f(\theta_1+)]. \tag{5–49}$$

Based upon this mean value, the mean-square-error at $\theta = \theta_1$ becomes

$$E_1 = \tfrac{1}{2}\{[m_1 - f(\theta_1-)]^2 + [m_1 - f_1(\theta_1+)]^2\}. \tag{5–50}$$

This mean-square-error can be minimized by setting

$$\frac{\partial E_1}{\partial m_1} = 0, \tag{5–51}$$

which yields Eq. (5–49) immediately and hence is consistent with the inherent least mean-square-error property of the Fourier series.

5–5 Exponential form of Fourier series. So far we have written Fourier series as trigonometric series consisting of sine and cosine terms; separate formulas are required to determine the coefficients a_n and b_n. We can also write Fourier series in an alternative form consisting of exponential terms. This alternative form is not only mathematically more elegant and easier to manipulate, but it also provides a convenient transition to Fourier integrals and Fourier transforms, as we shall see in a later section.

Let us start with the trigonometric form of the Fourier series expansion of a periodic function $f(t)$ with period T:

$$f(t) = \frac{a_0}{2} + \sum_{n=1}^{\infty} (a_n \cos n\omega t + b_n \sin n\omega t), \qquad (5\text{--}52)$$

where $\omega = 2\pi/T$ and

$$a_n = \frac{2}{T} \int_{-T/2}^{T/2} f(t) \cos n\omega t \, dt, \qquad (n = 0, 1, 2, \ldots), \qquad (5\text{--}53)$$

$$b_n = \frac{2}{T} \int_{-T/2}^{T/2} f(t) \sin n\omega t \, dt, \qquad (n = 1, 2, 3, \ldots). \qquad (5\text{--}54)$$

Both sine and cosine functions can be expressed in terms of exponential functions with imaginary exponents, therefore

$$\sin n\omega t = \frac{1}{2j} (\epsilon^{jn\omega t} - \epsilon^{-jn\omega t}), \qquad (5\text{--}55)$$

$$\cos n\omega t = \tfrac{1}{2}(\epsilon^{jn\omega t} + \epsilon^{-jn\omega t}). \qquad (5\text{--}56)$$

Substituting Eqs. (5–55) and (5–56) in Eq. (5–52), we have

$$j(t) = \frac{a_0}{2} + \sum_{n=1}^{\infty} \left(\frac{a_n - jb_n}{2} \epsilon^{jn\omega t} + \frac{a_n + jb_n}{2} \epsilon^{-jn\omega t} \right)$$

$$= \alpha_0 + \sum_{n=1}^{\infty} (\alpha_n \epsilon^{jn\omega t} + \alpha_{-n} \epsilon^{-jn\omega t}), \qquad (5\text{--}57)$$

where the new coefficients

$$\alpha_n = \tfrac{1}{2}(a_n - jb_n), \qquad (5\text{--}58)$$

$$\alpha_{-n} = \tfrac{1}{2}(a_n + jb_n) \qquad (5\text{--}59)$$

are now complex quantities.

$$\alpha_0 = \frac{a_0}{2} \qquad (5\text{--}60)$$

is real and is consistent with Eqs. (5–58) and (5–59) when n is set equal to zero. The second exponential series in Eq. (5–57) can be written in a slightly different form by substituting n for $-n$ and changing the limits of the summation properly:

$$\sum_{n=1}^{\infty} \alpha_{-n} \epsilon^{-jn\omega t} = \sum_{n=-1}^{-\infty} \alpha_n \epsilon^{jn\omega t}. \qquad (5\text{--}61)$$

Hence the right side of Eq. (5–57) can be very simply combined into a single summation:

$$f(t) = \sum_{n=-\infty}^{\infty} \alpha_n \epsilon^{jn\omega t}, \qquad (5\text{--}62)$$

which is the *exponential form** of *Fourier series*. The complex coefficient α_n is obtained by substituting Eqs. (5–53) and (5–54) for a_n and b_n in Eq. (5–58):

$$\alpha_n = \frac{1}{T} \int_{-T/2}^{T/2} f(t)\,\epsilon^{-jn\omega t}\,dt, \tag{5–63}$$

where n ranges over all integral values from $-\infty$ to $+\infty$. It is clear that the exponential form in Eq. (5–62), which needs only one formula for all its coefficients, is more compact and easier to manipulate than the trigonometric form in Eq. (5–52), which requires two separate formulas for the coefficients a_n and b_n.

The relations between the complex coefficients α_n and α_{-n} and the real coefficients a_n and b_n as expressed in Eqs. (5–58) and (5–59) can be seen more clearly from the phasor representation in Fig. 5–12 (at $t = 0$) for a pair of general terms $\alpha_n \epsilon^{jn\omega t} + \alpha_{-n}\epsilon^{-jn\omega t}$ from Eq. (5–62). Since α_n and α_{-n} are complex conjugates and the phasors representing the pair of terms rotate with equal and opposite angular frequencies, the sum of these two terms will always be real and will vary sinusoidally at the rate of $n\omega$ radians per second. The amplitude of the sinusoidal variation is reached when α_n and α_{-n} are in phase, and is equal to

$$2|\alpha_n| = \sqrt{a_n^2 + b_n^2}. \tag{5–64}$$

Of course, no matter in what form the Fourier series may be expressed, its value at any given t should always be real, since $f(t)$ is a real function of t.

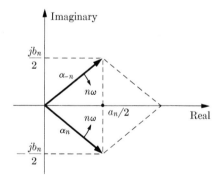

FIG. 5–12. Phasor representation of a pair of terms $(\alpha_n \epsilon^{jn\omega t} + \alpha_{-n}\epsilon^{-jn\omega t})$.

---

* Equation (5–62) is commonly referred to as the *complex form* of Fourier series. However, since it is basically an exponential series which represents a *real function f(t)*, the designation *exponential form* seems to be more appropriate. We do not call the trigonometric expansion in Eq. (5–52) the real form!

In Eq. (5–62), n is seen to take on negative values also; this gives rise to the so-called *negative frequencies*, which have no physical significance. Their appearance is a result of the mathematical manipulation that converts sine and cosine functions into pairs of exponential functions. There must be negative frequency components to combine with corresponding positive frequency components if the resultants are to be real; they are purely a mathematical concept.

EXAMPLE 5–3. Find the exponential form of the Fourier series expansion for the periodic rectangular pulse train shown in Fig. 5–13.

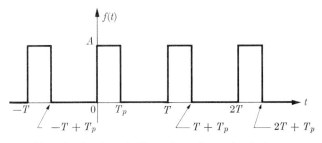

FIG. 5–13. A periodic rectangular pulse train.

Solution. We first write the expressions for $f(t)$ in one period ($-T/2 < t < T/2$):

$$f(t) = \begin{cases} 0, & -T/2 < t < 0, \\ A, & 0 < t < T_p, \\ 0, & T_p < t < T/2. \end{cases} \tag{5–65}$$

From Eq. (5–63) we have

$$\alpha_n = \frac{1}{T} \int_0^{T_p} A\,\epsilon^{-jn(2\pi/T)t}\,dt$$

$$= \frac{A}{-jn2\pi} \left(\epsilon^{-jn2\pi T_p/T} - 1 \right)$$

$$= \frac{A}{jn2\pi}\,\epsilon^{-jn\pi T_p/T} \left(\epsilon^{jn\pi T_p/T} - \epsilon^{-jn\pi T_p/T} \right)$$

$$= \frac{A}{n\pi}\,\epsilon^{-jn\pi T_p/T} \sin\left(n\pi T_p/T \right). \tag{5–66}$$

Hence

$$f(t) = \sum_{n=-\infty}^{\infty} \alpha_n \epsilon^{jn(2\pi/T)t}$$

$$= \frac{A}{\pi} \sum_{n=-\infty}^{\infty} \frac{1}{n} \sin\left(n\pi\,\frac{T_p}{T} \right) \epsilon^{jn(2\pi/T)[t-(T_p/2)]} \tag{5–67}$$

Since the given function in Fig. 5–13 has no symmetry conditions, the trigonometric form of its Fourier series expansion would contain both sine and cosine terms for all values of n.

The frequency spectrum of the given pulse train is a plot of the amplitude

$$2|\alpha_n| = \frac{2A}{n\pi}\left|\sin\left(n\pi\frac{T_p}{T}\right)\right| \tag{5–68}$$

versus $n\omega$ ($=2\pi n/T$). For $n = 0$, Eq. (5–66) becomes indeterminate, but it can be readily evaluated by applying l'Hospital's rule.

$$\begin{aligned}
\alpha_0 &= \lim_{n\to 0} \frac{A}{n\pi}\sin\left(n\pi\frac{T_p}{T}\right) \\
&= \frac{A}{\pi}\lim_{n\to 0}\frac{(d/dn)\sin[n\pi(T_p/T)]}{(d/dn)(n)} \\
&= \frac{A}{\pi}\lim_{n\to 0}\left(\pi\frac{T_p}{T}\right)\cos\left(n\pi\frac{T_p}{T}\right) = A\frac{T_p}{T},
\end{aligned} \tag{5–69}$$

which is seen to be the correct average value of the given function. The frequency spectrum is plotted in Fig. 5–14 for $T_p/T = 1/6$ according to the following table.

n	0	1	2	3	4
α_n	$A/6$	A/π	$\sqrt{3}A/2\pi$	$2A/3\pi$	$-\sqrt{3}A/4\pi$
$2|\alpha_n|$		$0.637A$	$0.551A$	$0.424A$	$0.275A$

n	5	6	7	8	9
α_n	$-A/5\pi$	0	$-A/7\pi$	$-\sqrt{3}A/8\pi$	$-2A/9\pi$
$2|\alpha_n|$	$0.127A$	0	$0.091A$	$0.138A$	$0.142A$

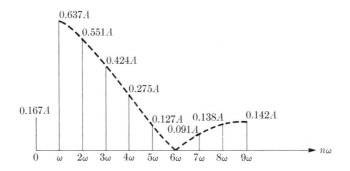

FIG. 5–14. Frequency spectrum of $f(t)$ in Fig. 5–13 for $T_p/T = 1/6$.

5–6 Fourier integrals and Fourier transforms. The frequency spectrum shown in Fig. 5–14 for the periodic pulse train of Fig. 5–13 is a *discrete* (or *line*) *spectrum* because amplitude values have significance only at discrete values of $n\omega$. Every line erected at $n\omega$ for an integral n represents the amplitude of the harmonic component whose frequency is n times that of the fundamental frequency $1/T$ ($=\omega/2\pi$), the spacing between adjacent harmonic components being equal to ω. As period T increases, ω decreases by an inverse proportion, and the lines in the discrete spectrum become closer. In the limit as T increases indefinitely, ω tends to approach zero and the lines will become so dense that the discrete frequency spectrum eventually approaches a smooth curve, resulting in a *continuous spectrum*. We note for the particular example in Fig. 5–13 that as T approaches infinity we are left essentially with a *single* rectangular pulse of duration T_p which never repeats. We are then led to investigate the feasibility of modifying the Fourier series expansion for periodic functions such that it may represent nonperiodic (transient) functions.

We repeat below the exponential form of Fourier series and the formula for its complex coefficients:

$$f(t) = \sum_{n=-\infty}^{\infty} \alpha_n \epsilon^{jn\omega t}, \tag{5–70}$$

$$\alpha_n = \frac{1}{T} \int_{-T/2}^{T/2} f(t) \epsilon^{-jn\omega t} \, dt. \tag{5–71}$$

When $f(t)$ is a transient function, some changes in notation are desirable. First of all, ω approaches zero and n becomes meaningless as T approaches infinity. Second, for a continuous spectrum, the angular frequency ($n\omega$) can assume any value. The following changes in notation are appropriate:

Any angular frequency, $n\omega \rightarrow \omega$,

Spacing between adja-
cent components, $\omega \rightarrow \Delta\omega$, $\tag{5–72}$

Period, $T \rightarrow 2\pi/\Delta\omega$.

From Eqs. (5–70) and (5–71),

$$f(t) = \sum_{\omega=-\infty}^{\infty} \alpha_\omega \epsilon^{j\omega t}, \qquad (\omega = 0, \pm\Delta\omega, \pm 2\,\Delta\omega, \ldots), \tag{5–73}$$

$$\alpha_\omega = \frac{\Delta\omega}{2\pi} \int_{-T/2}^{T/2} f(t) \epsilon^{-j\omega t} \, dt. \tag{5–74}$$

Substituting Eq. (5–74) into Eq. (5–73), we have

$$f(t) = \frac{1}{2\pi} \left[\sum_{\omega=-\infty}^{\infty} \int_{-T/2}^{T/2} f(t)\epsilon^{-j\omega t}\, dt \right] \epsilon^{j\omega t}\, \Delta\omega. \qquad (5\text{–}75)$$

As $T \to \infty$, $\Delta\omega \to d\omega$, and $\sum \to \int$, Eq. (5–75) becomes

$$f(t) = \frac{1}{2\pi} \int_{-\infty}^{\infty} \left[\int_{-\infty}^{\infty} f(t)\epsilon^{-j\omega t}\, dt \right] \epsilon^{j\omega t}\, d\omega. \qquad (5\text{–}76)$$

Equation (5–76) is one form of the *Fourier integral* of $f(t)$. Other forms are possible, but they are all simple variations of Eq. (5–76) and we shall not have to deal with them in this book. Note that we arrived at Eq. (5–76) in a heuristic way. In a rigorous development, fulfillment of the Dirichlet conditions enumerated in Section 5–2 is required, with the slight change that the integral $\int_{-\infty}^{\infty} |f(t)|\, dt$ must now be finite (instead of the original condition No. 3).

The Fourier integral in Eq. (5–76) can be broken into a pair of relations by calling the quantity inside the brackets $g(\omega)$:

$$f(t) = \frac{1}{2\pi} \int_{-\infty}^{\infty} g(\omega)\epsilon^{j\omega t}\, d\omega, \qquad (5\text{–}77)^*$$

$$g(\omega) = \int_{-\infty}^{\infty} f(t)\epsilon^{-j\omega t}\, dt. \qquad (5\text{–}78)$$

Comparison of Eqs. (5–77) and (5–78) with Eqs. (5–73) and (5–74) indicates that we can regard $f(t)$ as being analyzed into an infinite number of frequency components with infinitesimal amplitude $(1/2\pi)g(\omega)\, d\omega$. In general, $g(\omega)$ is a complex function of ω, and the plot of $|g(\omega)|$ versus ω shows the *relative frequency distribution* of $f(t)$.

The expressions for $f(t)$ and $g(\omega)$ given by the pair of relations in Eqs. (5–77) and (5–78) are called a *Fourier transform pair*; $g(\omega)$ is called the *Fourier transform* of $f(t)$ and, conversely, $f(t)$ is called the *inverse Fourier transform* of $g(\omega)$. Their relationship can be written symbolically as

$$f(t) = \mathfrak{F}^{-1}[g(\omega)], \qquad (5\text{–}79)$$

$$g(\omega) = \mathfrak{F}[f(t)]. \qquad (5\text{–}80)$$

* Whether the factor $1/2\pi$ is included in the expression for $f(t)$ [as it is in Eq. (5–77)] or in the expression for $g(\omega)$ is unimportant, so long as the combination yields Eq. (5–76). Some books include a factor $1/\sqrt{2\pi}$ in both expressions to achieve formal symmetry between the expressions for $f(t)$ and $g(\omega)$. We prefer to include the factor $1/2\pi$ in the inverse Fourier transform because it then closely resembles the expression for the inversion integral for the Laplace transform, as we shall see later.

The (direct) Fourier transformation of Eq. (5–80) transforms a function of t (in the *time domain*) into a function of ω (in the *frequency domain*), and the inverse Fourier transformation of Eq. (5–79) does the reverse. $g(\omega)$ is also called the *spectrum function* of $f(t)$.

EXAMPLE 5–4. Determine the relative frequency distribution of a single rectangular pulse of amplitude A and duration T_p.

Solution. The pulse has the form shown in Fig. 5–15. For this non-periodic function $f(t)$, we find the Fourier transform $g(\omega)$ by means of Eq. (5–78):

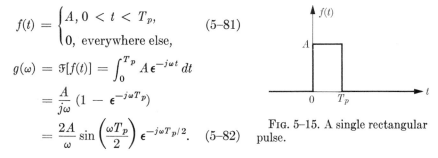

$$f(t) = \begin{cases} A, 0 < t < T_p, \\ 0, \text{ everywhere else,} \end{cases} \quad (5\text{–}81)$$

$$g(\omega) = \mathfrak{F}[f(t)] = \int_0^{T_p} A\epsilon^{-j\omega t}\, dt$$

$$= \frac{A}{j\omega}(1 - \epsilon^{-j\omega T_p})$$

$$= \frac{2A}{\omega}\sin\left(\frac{\omega T_p}{2}\right)\epsilon^{-j\omega T_p/2}. \quad (5\text{–}82)$$

FIG. 5–15. A single rectangular pulse.

The relative frequency distribution is a plot of $|g(\omega)|$ versus ω, where

$$|g(\omega)| = \frac{2A}{\omega}\left|\sin\left(\frac{\omega T_p}{2}\right)\right| = AT_p\left|\frac{\sin(\omega T_p/2)}{\omega T_p/2}\right|. \quad (5\text{–}83)$$

Note that here $\omega T_p \neq 2\pi$ because T_p is not the period of a periodic function and ω can have any value; $|g(\omega)|$ takes the form of the function $|(\sin x)/x|$ and is plotted in Fig. 5–16.

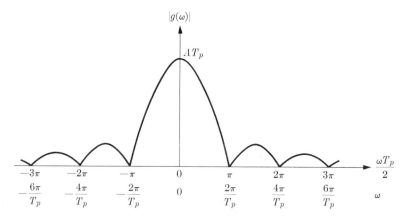

FIG. 5–16. Relative frequency distribution of the single rectangular pulse in Fig. 5–15.

It is seen that the most important part of the relative frequency distribution function, $|g(\omega)|$, of a rectangular pulse lies in the range $0 \leq |\omega| \leq 2\pi/T_p$. This range widens as the duration T_p of the pulse is made shorter. This relationship forms the basis of the fact that *the frequency bandwidth required to faithfully reproduce a rectangular pulse increases with the reciprocal of the pulse duration.*

Analogously to the Fourier series case, it is possible to foretell certain special properties of the Fourier transform of a transient function when the latter possesses symmetry conditions. Let us rewrite Eq. (5–78) in the following form:

$$g(\omega) = \int_{-\infty}^{\infty} f(t)[\cos \omega t - j \sin \omega t]\, dt$$

$$= \left[\int_{-\infty}^{\infty} f(t) \cos \omega t\, dt - j \int_{-\infty}^{\infty} f(t) \sin \omega t\, dt \right]. \qquad (5\text{–}84)$$

Now if $f(t) = f_c(t)$ is an even function of t, then $f_c(t) \sin \omega t$ will be an odd function of t and the second integral in Eq. (5–84) will vanish. We then have

$$g_c(\omega) = \int_{-\infty}^{\infty} f_c(t) \cos \omega t\, dt = 2 \int_{0}^{\infty} f_c(t) \cos \omega t\, dt, \qquad (5\text{–}85)$$

and $g_c(\omega)$, the Fourier transform of $f_c(t)$, will therefore be real and will be an even function of ω. On the other hand, if $f(t) = f_s(t)$ is an odd function of t, then $f_s(t) \cos \omega t$ will be an odd function of t and the first integral in Eq. (5–84) will vanish. We will then have

$$g_s(\omega) = -j \int_{-\infty}^{\infty} f_s(t) \sin \omega t\, dt = -2j \int_{0}^{\infty} f_s(t) \sin \omega t\, dt, \qquad (5\text{–}86)$$

and $g_s(\omega)$, the Fourier transform of $f_s(t)$, will therefore be purely imaginary and will be an odd function of ω.

In general, any arbitrary function $f(t)$ can be decomposed into an even function and an odd function, as illustrated in Fig. 5–6. We write, as in Eq. (5–30),

$$f(t) = f_c(t) + f_s(t). \qquad (5\text{–}87)$$

It follows that

$$g(\omega) = g_c(\omega) + g_s(\omega), \qquad (5\text{–}88)$$

where $g_c(\omega)$ and $g_s(\omega)$ are given by Eqs. (5–85) and (5–86) respectively. Hence the Fourier transform $g(\omega)$ of an arbitrary function will, in general, be a complex function of ω, and

$$|g(\omega)| = \sqrt{g_c^2(\omega) + |g_s(\omega)|^2}. \qquad (5\text{–}89)$$

5–7 Analysis by Fourier methods. The most valuable property of Fourier analysis undoubtedly lies in the fact that arbitrary periodic functions can be expanded into Fourier series of harmonic sinusoidal components, and that transient nonperiodic functions can be expressed as Fourier integrals. This property enables us to transform a function in the time domain into one in the frequency domain. In other words, we can discuss a function of time (provided that it satisfies the rather liberal Dirichlet conditions) in terms of its frequency components, and the response of a linear system to an arbitrary excitation can then be examined in terms of its responses to sinusoidal excitations by superposition. Problems related to harmonic content, frequency response, waveform preservation, and the like are particularly adaptable to solution by Fourier methods. Physical significance can be attached to the mathematical procedure involved in the solution.

Limitations of Fourier methods lie mainly in the difficulty of arriving at a solution in a closed form. For nonsinusoidal periodic excitations, it is a relatively simple matter to determine the solution of a system for each of the frequency components. However, it is rarely possible to find a neat, closed expression for the total response, which is a summation of all component responses. For transient nonperiodic excitations the method of Fourier integrals must be used. In order to determine the system response, it is necessary to find the inverse Fourier transform of the product of the spectrum function of the excitation and the system transfer function, that is, an infinite integration of the type shown in Eq. (5–77) has to be performed. The infinite integral along the ω-axis is, in general, difficult to evaluate. Although extensive tables* of Fourier transform pairs are available, the method of Laplace transforms usually proves to be much more convenient to use and easier to handle in these circumstances, as we

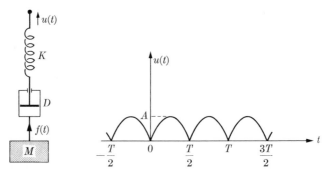

Fig. 5–17. A mechanical system with applied rectified sine-wave velocity.

---

* G. A. Campbell and R. M. Foster, "Fourier Integrals for Practical Applications," Bell System Monograph Series, No. B-584; 1931.

shall see in following chapters.　Moreover, important functions such as the unit step and an undamped sinusoid do not have Fourier transforms.

Some of the applications of Fourier series and Fourier integrals are illustrated in the following examples.

EXAMPLE 5–5.　The upper end of the spring in the mechanical system shown in Fig. 5–17 moves with a periodic velocity in the form of a rectified sine wave.　Determine the steady-state force on the mass M.

Solution.　We shall solve this problem by means of its force-current analogous electrical circuit.　Figure 5–18 is such an analogous circuit, in which all circuit elements have been expressed in terms of their analogous mechanical quantities.　Inspection of the given waveform for $u(t)$ reveals that it contains only an average term and cosine terms of even harmonics.

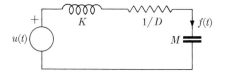

FIG. 5–18.　f-i analogous electrical circuit for the mechanical system in Fig. 5–17.

We have

$$u(t) = \frac{a_0}{2} + \sum_{k=1}^{\infty} a_{2k} \cos 2k\omega t, \qquad (5\text{–}90)$$

where $\omega = 2\pi/T$.　Because two symmetry conditions exist, it is necessary to integrate over only one quarter period in the determination of the Fourier coefficients (see remark in Example 5–2):

$$a_{2k} = \frac{8}{T} \int_0^{T/4} A \sin \omega t \cos 2k\omega t \, dt$$

$$= \frac{4A}{T} \left[\frac{\cos (2k-1)\omega t}{(2k-1)\omega} - \frac{\cos (2k+1)\omega t}{(2k+1)\omega} \right]_0^{T/4}$$

$$= -\frac{4A}{(4k^2 - 1)\pi}, \qquad (k = 0, 1, 2, \ldots). \qquad (5\text{–}91)$$

Hence

$$u(t) = \frac{2A}{\pi} - \frac{4A}{\pi} \left(\frac{1}{3} \cos 2\omega t + \frac{1}{15} \cos 4\omega t + \frac{1}{35} \cos 6\omega t + \cdots \right). \qquad (5\text{–}92)$$

Since we are interested in the steady-state response only, we can conveniently make use of the impedance concept.　An excitation of angular frequency $n\omega$, $u_n(t) = a_n \cos n\omega t$, applied to the circuit in Fig. 5–18 will produce a response $f_n(t)$:

$$f_n(t) = \frac{a_n}{\sqrt{(1/D)^2 + [n\omega K - (1/n\omega M)]^2}} \cos (n\omega t - \Psi_n), \qquad (5\text{-}93)$$

where

$$\Psi_n = \tan^{-1} [n\omega K - (1/n\omega M)]D. \qquad (5\text{-}94)$$

Clearly, the average component $a_0/2$ of the applied $u(t)$ does not contribute to the steady-state response function $f(t)$. (There is no d-c current through the analogous capacitor, which presents an infinite impedance to the zero-frequency component.) By the principle of superposition, we then have for the steady-state force on the mass M

$$f(t) = - \sum_{k=1}^{\infty} \frac{4A}{(4k^2 - 1)\pi |Z_{2k}|} \cos (2k\omega t - \Psi_{2k}), \qquad (5\text{-}95)$$

where

$$|Z_{2k}| = \sqrt{(1/D)^2 + [2k\omega K - (1/2k\omega M)]^2}, \qquad (5\text{-}96)$$

$$\Psi_{2k} = \tan^{-1} [2k\omega K - (1/2k\omega M)]D. \qquad (5\text{-}97)$$

For all values of t, $f(t)$ can be computed from Eqs. (5–95), (5–96), and (5–97). In practice, we expect that the most important contributions to $f(t)$ will come from those components whose angular frequencies $2k\omega = 4k\pi/T$ are equal to or close to the natural (resonant) angular frequency ω_r of the given system, where

$$\omega_r K = \frac{1}{\omega_r M},$$

or

$$\omega_r = \frac{1}{\sqrt{MK}} \qquad (5\text{-}98)$$

and

$$\Psi_r = 0. \qquad (5\text{-}99)$$

Components with angular frequencies much smaller or much larger than ω_r will have very little effect on the steady-state response; only a few terms are needed in the actual numerical computation of $f(t)$.

EXAMPLE 5–6. Determine the response of the circuit in Fig. 5–19 to an applied single half-sine voltage of amplitude A and duration $T/2$ as shown.

Solution. Before we attempt a solution to this particular problem, let us first formulate a formal procedure for solving problems of this type by Fourier transforms. Four steps are involved. First, we must find the spectrum function $g_i(\omega)$ of the input transient excitation function $f_i(t)$. This step determines the relative amplitude and phase of the frequency components of the input function. Next, the steady-state response of the

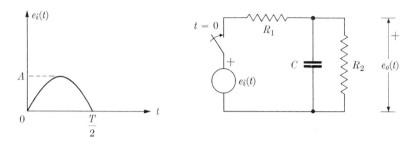

Fig. 5–19. An electrical circuit with a single half-sine voltage applied at $t = 0$.

given system to a general component of angular frequency ω and with unit amplitude and reference phase is determined. This response is left in sinor form (although not written in bold-faced letters) as a complex function of ω. It is the *transfer function*, $H(j\omega)$, of the given system for sinusoidal excitation. With $g_i(\omega)$ as the input spectrum, the output spectrum is then clearly the product $g_i(\omega)H(j\omega)$. Finally, the output response function is the inverse Fourier transform of this product. Thus the steps are as follows:

1. Find the Fourier transform of the input excitation function: $g_i(\omega) = \mathcal{F}[f_i(t)]$.

2. Determine the transfer function, $H(j\omega)$, of the given system.

3. Form the product $g_o(\omega) = g_i(\omega)H(j\omega)$, which is the spectrum function of the desired response at the output.

4. Find the response function, $f_o(t)$, by taking the inverse Fourier transform of the output spectrum function $g_o(\omega)$: $f_o(t) = \mathcal{F}^{-1}[g_o(\omega)]$.

We shall now solve our problem in four steps, as outlined above.

Step 1. The given excitation function is

$$e_i(t) = A \sin (2\pi/T)t$$

$$= A \sin \omega_i t, \qquad (0 \le t \le T/2),$$

(5–100)

where $\omega_i = 2\pi/T$. Note that *we cannot use ω (without a subscript) to represent $2\pi/T$ here*, because ω is used as a variable in Fourier transforms. It can assume *any* value, not just integral multiples of a fundamental angular frequency (as in the Fourier series expansion of a periodic function). On the other hand, $2\pi/T$ is a given, *fixed* quantity. The spectrum function of $e_i(t)$ is

$$g_i(\omega) = \mathcal{F}[e_i(t)]$$

$$= \int_0^{T/2} A \sin \omega_i t \cdot \epsilon^{-j\omega t}\, dt$$

$$= \left(\frac{2A\omega_i}{\omega_i^2 - \omega^2} \cos \omega \frac{T}{4} \right) \epsilon^{-j\omega T/4}.$$

(5–101)

Note again that ω in Eq. (5–101) is a variable and that ωT is *not* equal to 2π except when $\omega = \omega_i$.

Step 2. The transfer function $H(j\omega)$ for the circuit of Fig. 5–19 is

$$H(j\omega) = \frac{\text{Parallel impedance of } C \text{ and } R_2}{R_1 + (\text{parallel impedance of } C \text{ and } R_2)}$$

$$= \frac{R_2/(1 + j\omega C R_2)}{R_1 + [R_2/(1 + j\omega C R_2)]} = \frac{R_2}{R_1 + R_2 + j\omega C R_1 R_2}$$

$$= \frac{R_2}{\sqrt{(R_1 + R_2)^2 + \omega^2 C^2 R_1^2 R_2^2}} \, \epsilon^{-j \tan^{-1} \omega C R_1 R_2/(R_1+R_2)}. \qquad (5\text{–}102)$$

Step 3. The spectrum function of the response function is

$$g_o(\omega) = g_i(\omega)H(j\omega)$$

$$= \frac{2A\omega_i R_2 \cos(\omega T/4)}{(\omega_i^2 - \omega^2)\sqrt{(R_1 + R_2)^2 + \omega^2 C^2 R_1^2 R_2^2}}$$

$$\times \; \epsilon^{-j[\omega T/4 + \tan^{-1} \omega C R_1 R_2/(R_1+R_2)]}. \qquad (5\text{–}103)$$

Step 4. The desired response at the output is

$$e_o(t) = \mathcal{F}^{-1}[g_o(\omega)] = \frac{1}{2\pi} \int_{-\infty}^{\infty} g_o(\omega)\epsilon^{j\omega t} \, d\omega. \qquad (5\text{–}104)$$

It is obvious that the evaluation of the infinite integral in Eq. (5–104) along the ω-axis with $g_o(\omega)$ given by Eq. (5–103) is a formidable task. We shall not attempt it here because problems of this type can be solved very easily by the method of Laplace transforms, which we shall introduce in the following chapter.

Fourier transforms enable us to discuss system response to transient excitations in terms of the steady-state response to sinusoidal excitations. Although Fourier methods are not as readily applicable to the solution of differential equations or to the determination of the response of specific systems as some other methods, they are indispensable in problems involving signal analysis, frequency response, and system bandwidth. Fourier series and Fourier integrals are of such fundamental importance that they also make their appearances in noise studies, communication theory, and boundary-value problems, which are beyond the scope of this book.

PROBLEMS

For each of the periodic waveforms in Figs. 5–20 through 5–23, (a) find the Fourier series expansion, and (b) sketch the frequency spectrum approximately to scale:

5–1.

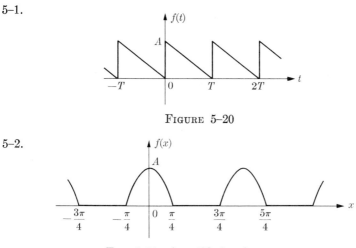

FIGURE 5–20

5–2.

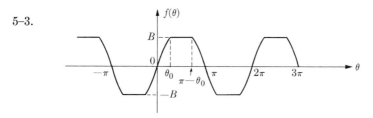

FIG. 5–21. A rectified cosine wave.

5–3.

FIG. 5–22. A truncated sine wave.

5–4.

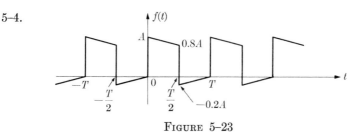

FIGURE 5–23

5–5. Expand $f(t) = A \cos (2\pi/T)t$ in a sine series for the interval $0 < t < T/2$.

5–6. Expand $f(x) = \epsilon^{-x}$ in a cosine series for the interval $0 < x < 2$.

5–7. Expand $f(x) = x^2$ in an exponential series for the interval $0 < x < 1$.

5-8. Determine the effects of each of the following symmetry conditions on the coefficients of the Fourier series expansion for $f(\theta)$, and obtain formulas for those coefficients which do not vanish.

(a) $f(\theta) = f(\pi - \theta)$

(b) $f(\theta) = -f(\pi - \theta)$

(c) $f(\theta) = f\left(\dfrac{\pi}{2} - \theta\right)$

(d) $f(\theta) = f\left(\dfrac{\pi}{2} + \theta\right)$

5-9. A periodic function $f(t)$ is known to have the form given in Fig. 5-24 in the first quarter cycle ($0 \leq t \leq T/4$). Complete the waveform of $f(t)$ for one period (t from 0 to T) for the following cases:

(a) $f(t)$ is an even function of t and contains only even harmonics.
(b) $f(t)$ is an even function of t and contains only odd harmonics.
(c) $f(t)$ is an even function of t and contains both even and odd harmonics.
(d) $f(t)$ is an odd function of t and contains both even and odd harmonics.
(e) $f(t)$ is an odd function of t and contains only odd harmonics.
(f) $f(t)$ is an odd function of t and contains only even harmonics.

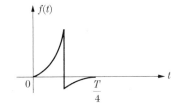

FIG. 5-24. First quarter cycle of a periodic wave.

5-10. (a) Assume that a periodic function $f(x)$ with a period 2π is expressible as a Fourier series as follows:

$$f(x) = \frac{a_0}{2} + \sum_{n=1}^{\infty} (a_n \cos nx + b_n \sin nx).$$

Prove that

$$\frac{1}{2\pi} \int_{-\pi}^{\pi} [f(x)]^2 \, dx = \left(\frac{a_0}{2}\right)^2 + \frac{1}{2} \sum_{n=1}^{\infty} (a_n^2 + b_n^2).$$

This relation, known as *Parseval's theorem*, is useful in computing the *effective value* of the given periodic function $f(x)$, which is by definition the square root of the left side of the above equation.

(b) Compute the energy dissipated per period when a triangular voltage wave of the form shown in Fig. 5-8 is impressed across a one-ohm resistor. Compare this result with the energy that would be dissipated in the resistor per period by assuming that each harmonic component of the given waveform acted separately. Discuss in the light of Parseval's theorem.

Determine the spectrum function for each of the waveforms in Figs. 5-25 through 5-29:

5–11.

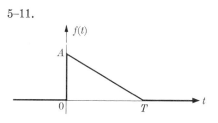

FIGURE 5–25

5–12.

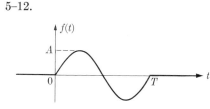

FIG. 5–26. A single cycle of sine wave.

5–13.

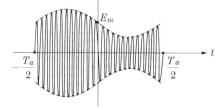

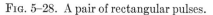

$$e(t) = -E_m(1 - \tfrac{1}{2}\sin\omega_a t)\sin 100\omega_a t,$$
$$\omega_a = 2\pi/T_a.$$

FIG. 5–27. An audio cycle of an amplitude-modulated wave.

5–14.

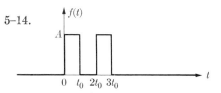

FIG. 5–28. A pair of rectangular pulses.

5–15.

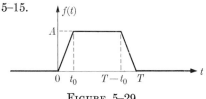

FIGURE 5–29

5–16. If $g(\omega) = \mathcal{F}[f(t)]$, prove:

    (a) $\mathcal{F}[f(t - t_0)] = g(\omega)\epsilon^{-j\omega_0 t_0}$ (b) $\mathcal{F}[f(at)] = \dfrac{1}{a}\, g\left(\dfrac{\omega}{a}\right)$

    (c) $\mathcal{F}\left[\dfrac{d^n}{dt^n}\, f(t)\right] = (j\omega)^n g(\omega)$

5–17. (a) Determine $\mathcal{F}[\epsilon^{-\alpha t}\sin\omega_0 t]$, $t \geq 0$. (b) Sketch the relative frequency distribution.

5–18. For the waveforms in Fig. 5–30, if $\mathcal{F}[f_1(t)] = g_1(\omega)$, what is $\mathcal{F}[f_2(t)]$?

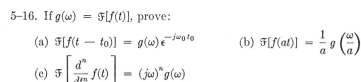

FIGURE 5–30

5–19. A periodic force $f(t)$ following a sawtooth variation as shown in Fig. 5–31 is applied to the platform M_1 in the given mechanical system. Describe quantitatively the steady-state motion of M_1.

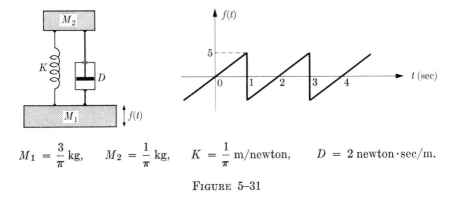

$$M_1 = \frac{3}{\pi} \text{ kg}, \qquad M_2 = \frac{1}{\pi} \text{ kg}, \qquad K = \frac{1}{\pi} \text{ m/newton}, \qquad D = 2 \text{ newton·sec/m.}$$

FIGURE 5–31

5–20. The current function $i(t)$ in Fig. 5–32 consists of periodic appearances of top portions of a cosine wave. This current is to flow through the tank circuit as shown. Determine the amplitude of the steady-state voltage across the tank circuit (a) for the fundamental frequency component, and (b) for the second harmonic component.

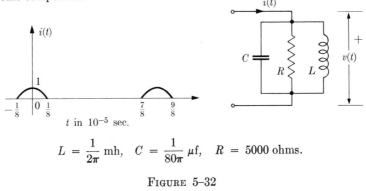

$$L = \frac{1}{2\pi} \text{ mh}, \quad C = \frac{1}{80\pi} \text{ μf}, \quad R = 5000 \text{ ohms.}$$

FIGURE 5–32

5–21. A rectangular pulse of amplitude A and duration t_0 is applied to the input terminals of a hypothetical amplifier which has a uniform amplitude response and a phase lag proportional to frequency within the range $0 < \omega < \omega_c$, where ω_c is the cutoff angular frequency. Find the output.

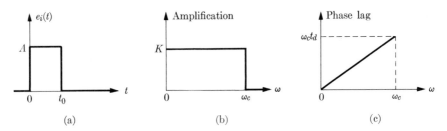

(a)   (b)   (c)

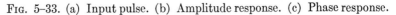

FIG. 5–33. (a) Input pulse. (b) Amplitude response. (c) Phase response.

CHAPTER 6

THE LAPLACE TRANSFORMATION

6–1 Introduction. As we have seen in Chapter 2, the classical method of solving ordinary differential equations can be divided into three major steps, namely: (1) the determination of the complementary function (from the homogeneous equation), (2) the determination of the particular integral (from the nonhomogeneous equation), and (3) the determination of the arbitrary constants (from known values of the dependent variable or its derivatives for specific values of the independent variable). An understanding of the classical approach is important not only because it becomes the method of last resort when transformation methods fail, but also because it contributes to the understanding of the fundamental properties of differential equations and their solutions. For many types of excitation function the classical method of solution yields answers in a simple manner. However, for excitation functions of the more general type, especially when the functions or their derivatives contain discontinuities, the classical method is difficult to apply and the Laplace transform method, which we shall introduce in this chapter and to which the rest of this book will be devoted, proves to be superior.

Pending future detailed expositions to substantiate our claims, we mention at this time the following advantages of the Laplace transform method:

(a) *It simplifies functions.* Laplace transformation transforms the frequently occurring exponential and transcendental functions and their combinations into simple, easier-to-handle algebraic functions. Of particular importance is its ability to transform nonsinusoidal periodic functions and functions with discontinuities or discontinuous derivatives into simple expressions. For example, it is quite difficult to determine the response of a system by the classical approach to excitation functions of the types shown in Fig. 6–1, no matter how simple the system itself is. The reader can easily satisfy himself that this is true by trying to find the solution of a simple first-order differential equation using one of these waveforms as the excitation function. With Laplace transformation, on the other hand, these waveforms present no difficulty; they can all be transformed into simple expressions and add no undue complications to the solution of the system response.

(b) *It simplifies operations.* Laplace transformation transforms differentiation and integration, respectively, into multiplication and division. It essentially transforms integro-differential equations into algebraic

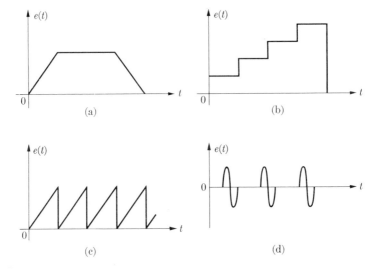

FIG. 6–1. Waveforms not easily handled by classical method.

equations, which are, of course, much easier to handle. In this respect, Laplace transformation is analogous to logarithmic transformation, which transforms multiplication, division, and involution, respectively, into the simpler addition, subtraction, and multiplication operations. Whereas logarithmic operation deals with numbers, Laplace transformation deals with functions.

(c) *It eliminates the necessity for a separate operation to determine the arbitrary constants.* In solving the differential equation for a given system by the method of Laplace transforms, the solution comes out complete; no separate constant-determining operation is necessary. In the classical method the determination of the constants can be a tedious job, especially when there are many (that is, when the order of the differential equation is high).

(d) *It effectively makes use of step and impulse responses.* Briefly, if we know or have determined the response of a system to a step or impulse excitation, then we can determine quite simply the response of the system to any other type of excitation function. This enables us to describe, or "label," systems or networks in terms of their step or impulse responses. The method of Laplace transforms makes effective use of this property.

The above points should all become clear as we gradually expound the techniques of Laplace transformation.

6–2 From Fourier transform to Laplace transform. In Chapter 5 we developed the Fourier integral of a function $f(t)$ that satisfies the Dirichlet conditions into a Fourier transform pair. In most practical applications, the function $f(t)$ is zero before some instant which we can designate as $t = 0$. The Fourier transform pair in Eqs. (5–77) and (5–78) then becomes

$$f(t) = \frac{1}{2\pi} \int_{-\infty}^{\infty} g(\omega) \epsilon^{j\omega t} \, d\omega = \mathcal{F}^{-1}[g(\omega)], \tag{6–1}$$

$$g(\omega) = \int_{0}^{\infty} f(t) \epsilon^{-j\omega t} \, dt = \mathcal{F}[f(t)]. \tag{6–2}$$

Note that the lower limit of the integral in Eq. (6–2) has been changed to zero. Equations (6–1) and (6–2) represent a *unilateral Fourier transform pair*. No difficulty should arise in the assumption that $f(t) = 0$ for $t < 0$, because we can always shift the $t = 0$ instant as far back as we wish, so that it corresponds with the point at which the function $f(t)$ starts to have significance in the problem under consideration. In practice, we need not be concerned about negative values of t because we would be interested in the response of a system only *after* the application of an excitation or a disturbance. What happened before would be included in the initial or boundary condition $f(0)$.

As we pointed out in Chapter 5, $g(\omega)$ is the spectrum function of $f(t)$. Equation (6–2) can be used to find the Fourier transform or spectrum function of an $f(t)$ which starts at $t = 0$ and which satisfies the Dirichlet conditions. Unfortunately, Eq. (6–2) does not yield meaningful results for two of the most important functions in engineering, namely, the unit step function $U(t)$, as shown in Fig. 6–2, and the unit sinusoid $\sin \omega t$, as shown in Fig. 6–3. This is because the integrals $\int_{0}^{\infty} U(t) \, dt$ and $\int_{0}^{\infty} |\sin \omega t| \, dt$ do not exist. Let us examine the case for the unit step function in more detail:

$$U(t) = \begin{cases} 0, & t < 0, \\ 1, & t > 0. \end{cases} \tag{6–3}$$

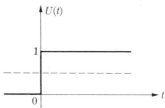

FIG. 6–2. A unit step function.

Substituting into Eq. (6–2), we find

$$g(\omega) = \mathcal{F}[U(t)] = \int_{0}^{\infty} U(t) \epsilon^{-j\omega t} \, dt$$

$$= \int_{0}^{\infty} \epsilon^{-j\omega t} \, dt$$

$$= \frac{1}{-j\omega} \epsilon^{-j\omega t} \Big|_{0}^{\infty}$$

$$= \frac{j}{\omega} (\cos \omega t - j \sin \omega t) \Big|_{0}^{\infty}. \tag{6–4}$$

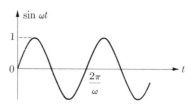

FIG. 6–3. A unit sinusoid.

But Eq. (6–4) cannot be evaluated because cos ∞ and sin ∞ are undefined.

This dilemma can be alleviated by using a limiting process. More specifically, instead of the original $U(t)$, we consider an exponentially decreasing function $\epsilon^{-\sigma t}U(t)$, as shown in Fig. 6–4, which approaches $U(t)$ when we let $\sigma \to 0$:

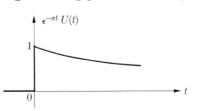

$$\epsilon^{-\sigma t}U(t) = \begin{cases} 0, & t < 0, \\ \epsilon^{-\sigma t}, & t > 0. \end{cases} \quad (6\text{–}5)$$

FIG. 6–4. An exponential decreasing function which approaches the unit step function as $\sigma \to 0$.

With this function, Eq. (6–2) yields

$$g(\omega) = \int_0^\infty \epsilon^{-\sigma t}\epsilon^{-j\omega t}\,dt = \frac{1}{-(\sigma + j\omega)}\,\epsilon^{-(\sigma + j\omega)t}\,\Big|_0^\infty = \frac{1}{\sigma + j\omega}. \quad (6\text{–}6)$$

Now we substitute this $g(\omega)$ into Eq. (6–1):

$$\epsilon^{-\sigma t}U(t) = \frac{1}{2\pi}\int_{-\infty}^\infty \frac{1}{\sigma + j\omega}\,\epsilon^{j\omega t}\,d\omega$$

$$= \frac{1}{2\pi}\int_{-\infty}^\infty \frac{(\sigma\cos\omega t + \omega\sin\omega t) + j(\sigma\sin\omega t - \omega\cos\omega t)}{\sigma^2 + \omega^2}\,d\omega. \quad (6\text{–}7)$$

The last expression in Eq. (6–7) was obtained by expanding $\epsilon^{j\omega t} = \cos\omega t + j\sin\omega t$ and rationalizing the denominator. Since the imaginary part of the integrand,

$$\frac{\sigma\sin\omega t - \omega\cos\omega t}{\sigma^2 + \omega^2},$$

is an odd function of ω, it will vanish upon integration with respect to ω between the limits $-\infty$ and ∞. The integration between the limits $-\infty$ and ∞ of the real part of the integrand, which is an even function of ω, will be equal to twice the integration between limits 0 and ∞. Thus

$$\epsilon^{-\sigma t}U(t) = \frac{1}{\pi}\int_0^\infty \left[\frac{\sigma\cos\omega t}{\sigma^2 + \omega^2} + \frac{\omega\sin\omega t}{\sigma^2 + \omega^2}\right]d\omega. \quad (6\text{–}8)$$

At this point it appears that as we eventually let $\sigma \to 0$ the first term in the integrand in Eq. (6–8) would contribute nothing to the integral. But this hasty opinion is incorrect, because the range of integration includes the region where $\omega \to 0$. In that region, even though $\sigma \to 0$, $\cos\omega t \cong 1$ and the ratio σ/ω^2 is not zero. In the region $0 \leq \omega \leq a$, where a is very small, we have

$$\lim_{\sigma \to 0} \frac{1}{\pi} \int_0^\infty \frac{\sigma \cos \omega t}{\sigma^2 + \omega^2} \, d\omega = \lim_{\sigma \to 0} \frac{1}{\pi} \int_0^a \frac{\sigma \, d\omega}{\sigma^2 + \omega^2}$$

$$= \lim_{\sigma \to 0} \frac{1}{\pi} \left[\tan^{-1} \left(\frac{a}{\sigma} \right) \right] = \frac{1}{2}. \tag{6–9}$$

There is no trouble in letting $\sigma \to 0$ in the second term in the integrand of Eq. (6–8). Hence, as we reduce the exponentially decreasing function in Fig. 6–4 to the unit step function in Fig. 6–2 on letting σ approach zero, we have, from Eq. (6–8),

$$U(t) = \frac{1}{2} + \frac{1}{\pi} \int_0^\infty \frac{\sin \omega t}{\omega} \, d\omega. \tag{6–10}$$

Equation (6–10) tells us that the unit step function $U(t)$ has an average value equal to $1/2$, and superposed upon this average value is an odd function (sine terms only) whose relative frequency distribution is inversely proportional to ω. This is certainly a plausible conclusion, by inspection of Fig. 6–2.

What we have done above was to introduce a factor $\epsilon^{-\sigma t}$ in the integrand of the infinite integral in Eq. (6–2) so that the latter would converge. This factor is called a *convergence factor*. We can extend this technique to $f(t)$, a general function of t. Let us apply the unilateral Fourier transform to the product $\epsilon^{-\sigma t} f(t)$ which is zero for $t < 0$:

$$g(\omega) = \int_0^\infty [\epsilon^{-\sigma t} f(t)] \epsilon^{-j\omega t} \, dt$$

$$= \int_0^\infty f(t) \epsilon^{-(\sigma + j\omega)t} \, dt \tag{6–11}$$

Since ω appears in Eq. (6–11) only in the combination $(\sigma + j\omega)$, it is convenient to replace this combination by a new symbol s, and write $g(\omega)$ as $F(s)$, a function of s. Thus, with

$$s = \sigma + j\omega, \tag{6–12}$$

Eq. (6–11) becomes

$$F(s) = \int_0^\infty f(t) \epsilon^{-st} \, dt. \tag{6–13}$$

From Eq. (6–11) the inverse Fourier transform of $g(\omega)$ is, by definition,

$$\epsilon^{-\sigma t} f(t) = \frac{1}{2\pi} \int_{-\infty}^\infty g(\omega) \epsilon^{j\omega t} \, d\omega. \tag{6–14}$$

Multiplying both sides of Eq. (6–14) by $\epsilon^{\sigma t}$, which is not a function of ω, we have

$$f(t) = \frac{1}{2\pi} \int_{-\infty}^{\infty} g(\omega) \epsilon^{(\sigma+j\omega)t} \, d\omega$$

$$= \frac{1}{2\pi j} \int_{-j\infty}^{j\infty} g(\omega) \epsilon^{(\sigma+j\omega)t} \, d(j\omega). \tag{6-15}$$

Again we make a change of variables as specified in Eq. (6-12) and write $g(\omega)$ as $F(s)$, since ω appears only in the combination $(\sigma + j\omega)$. In so doing, we remember that the limits $-j\infty$ and $j\infty$ for $(j\omega)$ should go over as limits $(\sigma - j\infty)$ and $(\sigma + j\infty)$ for s, respectively. Equation (6-15) then becomes

$$f(t) = \frac{1}{2\pi j} \int_{\sigma-j\infty}^{\sigma+j\infty} F(s) \epsilon^{st} \, ds. \tag{6-16}$$

Equations (6-13) and (6-16) form a *Laplace transform pair*. The integral transformation in Eq. (6-13) is called the (direct) *Laplace transformation* and $F(s)$ is said to be the *Laplace transform* of $f(t)$. For convenience, we represent Eq. (6-13) by the following symbolic notation:

$$F(s) = \mathcal{L}[f(t)], \tag{6-17}$$

where s is a complex variable whose real part σ should be large enough to make the integral in Eq. (6-13) converge.* When this integral exists (converges), we say that the function $f(t)$ is *Laplace transformable*. Both the step function and the sinusoid, which do not have Fourier transforms, are Laplace transformable for $\sigma > 0$. Thus the built-in convergence factor $\epsilon^{-\sigma t}$ makes Laplace transformation a much more versatile tool. Not all functions are Laplace transformable. Functions such as t^t or ϵ^{t^2} are not Laplace transformable no matter how large σ is; but these functions are of no importance in engineering.

The complex integral transformation in Eq. (6-16) is called the *inversion integral* and $f(t)$ is said to be the *inverse Laplace transform* of $F(s)$. For convenience, we represent Eq. (6-16) by the following symbolic notation:

$$f(t) = \mathcal{L}^{-1}[F(s)]. \tag{6-18}$$

In Section 5-7 we indicated that difficulties would be encountered in the evaluation of the infinite integral along the ω-axis when the Fourier

---

* In a strict mathematical sense, we actually require absolute convergence such that

$$\lim_{T \to \infty} \int_0^T |f(t)| \epsilon^{-\sigma t} \, dt < \infty.$$

This is necessary in order to permit us to change the order of certain limit processes in future applications.

transform method is used for determining system response. In the Laplace transform method the evaluation of the inversion integral in Eq. (6–16) is greatly facilitated by the highly developed theory of functions of a complex variable and, in particular, by the theory of residues. In this book we shall show how a great majority of linear systems can be analyzed quite simply by the Laplace transform method without a knowledge of the theory of functions of a complex variable.

There are several remarks we should make about the function $f(t)$ at this time. First of all, the real variable t does not have to represent time; we could just as well use x (representing displacement, for example), or any other symbol suitable for the system under consideration.* Second, $f(t)$ must be a single-valued function in order to have a unique $F(s)$. However, $f(t)$ is allowed to have finite discontinuities. Finally, if there is a discontinuity for $f(t)$ at $t = 0$, then the lower limit of the integral in Eq. (6–13) is to be considered as being approached from the positive side. In other words, the value of $f(t)$ to be used at $t = 0$ is $f(0+)$. For example, if we are dealing with the unit step function $U(t)$ as shown in Fig. 6–2, which is zero for $t < 0$ and unity for $t > 0$ and has a discontinuity (jump) at $t = 0$, then we use $f(0+) = U(0+) = 1$.

The following two theorems are quite obvious.

THEOREM 1. *The Laplace transform of the sum of two functions is equal to the sum of the Laplace transforms of the individual functions:*

$$\mathcal{L}[f_1(t) + f_2(t)] = \int_0^\infty [f_1(t) + f_2(t)]\epsilon^{-st}\,dt$$

$$= \int_0^\infty f_1(t)\epsilon^{-st}\,dt + \int_0^\infty f_2(t)\epsilon^{-st}\,dt$$

$$= \mathcal{L}[f_1(t)] + \mathcal{L}[f_2(t)]. \tag{6–19}$$

THEOREM 2. *The Laplace transform of a constant times a function is equal to the constant times the Laplace transform of the function:*

$$\mathcal{L}[cf(t)] = \int_0^\infty [cf(t)]\epsilon^{-st}\,dt$$

$$= c\int_0^\infty f(t)\epsilon^{-st}\,dt$$

$$= c\mathcal{L}[f(t)]. \tag{6–20}$$

---

* The symbol s for the complex variable, however, is quite standard, since it represents no particular physical quantity (the letter p is sometimes also used).

In general, we have

$$\mathcal{L}[c_1 f_1(t) + c_2 f_2(t)] = c_1 \mathcal{L}[f_1(t)] + c_2 \mathcal{L}[f_2(t)]$$
$$= c_1 F_1(s) + c_2 F_2(s). \tag{6-21}$$

Equation (6–21) assures that $\mathcal{L}$ *is a linear operator.*

6–3 Laplace transforms of some important functions. In engineering problems some functions appear much more frequently than others. In this section we shall develop the Laplace transforms of several important types of functions. With few exceptions, most functions that find applications in engineering are expressible in terms of these basic types. Essentially, we shall deal with only two types of functions: (A) the exponential function of t, and (B) the positive powers of t. Other useful functions, such as the step function, the sinusoidal functions, the hyperbolic sines and cosines, and the damped sinusoidal functions, can all be derived from these two basic types, as we shall see. Eventually we wish to use the method of Laplace transforms to solve differential equations that describe linear systems. But before we can do that, we have to be able to find the Laplace transform of every term in the differential equation, including all possible forms of the excitation function.

A. *An exponential function of t.*

Let us find the Laplace transform of ϵ^{at}, where a is a constant which can be positive or negative real, purely imaginary, or complex.

$$\mathcal{L}[\epsilon^{at}] = \int_0^\infty (\epsilon^{at}) \epsilon^{-st} \, dt = \int_0^\infty \epsilon^{-(s-a)t} \, dt$$

$$= \frac{1}{s-a} \tag{6-22}*$$

Several important special cases follow.

A-1. *The unit step function,* $U(t)$. Setting $a = 0$ in Eq. (6–22), we have directly

$$\mathcal{L}[U(t)] = \frac{1}{s}. \tag{6-23}$$

A-2. *The sine function,* $sin\ \omega t$. Making use of the formula

$$\sin \omega t = \frac{1}{2j} (\epsilon^{j\omega t} - \epsilon^{-j\omega t}),$$

---

* The region of convergence for this direct Laplace transformation is Re $(s) = \sigma >$ Re (a).

we have

$$\mathcal{L}[\sin \omega t] = \frac{1}{2j} \mathcal{L}[\epsilon^{j\omega t} - \epsilon^{-j\omega t}]$$

$$= \frac{1}{2j} \left(\frac{1}{s - j\omega} - \frac{1}{s + j\omega} \right)$$

$$= \frac{\omega}{s^2 + \omega^2}. \tag{6-24}$$

A-3. *The cosine function*, cos ωt. Here,

$$\cos \omega t = \tfrac{1}{2}(\epsilon^{j\omega t} + \epsilon^{-j\omega t}),$$

$$\mathcal{L}[\cos \omega t] = \tfrac{1}{2}\mathcal{L}[\epsilon^{j\omega t} + \epsilon^{-j\omega t}]$$

$$= \frac{1}{2} \left(\frac{1}{s - j\omega} + \frac{1}{s + j\omega} \right)$$

$$= \frac{s}{s^2 + \omega^2}. \tag{6-25}$$

A-4. *The hyperbolic sine and cosine.* By virtue of the relations

$$\sinh bt = \tfrac{1}{2}(\epsilon^{bt} - \epsilon^{-bt})$$

and

$$\cosh bt = \tfrac{1}{2}(\epsilon^{bt} + \epsilon^{-bt}),$$

we can write directly

$$\mathcal{L}[\sinh bt] = \frac{b}{s^2 - b^2}, \tag{6-26}$$

$$\mathcal{L}[\cosh bt] = \frac{s}{s^2 - b^2}. \tag{6-27}$$

A-5. *The damped sine function*, $\epsilon^{-\alpha t} \sin \omega t$. Since

$$\epsilon^{-\alpha t} \sin \omega t = \frac{1}{2j} [\epsilon^{-(\alpha - j\omega)t} - \epsilon^{-(\alpha + j\omega)t}],$$

we have

$$\mathcal{L}[\epsilon^{-\alpha t} \sin \omega t] = \frac{1}{2j} \mathcal{L}[\epsilon^{-(\alpha - j\omega)t} - \epsilon^{-(\alpha + j\omega)t}]$$

$$= \frac{1}{2j} \left[\frac{1}{(s + \alpha) - j\omega} - \frac{1}{(s + \alpha) + j\omega} \right]$$

$$= \frac{\omega}{(s + \alpha)^2 + \omega^2}. \tag{6-28}$$

A-6. *The damped cosine function*, $\epsilon^{-\alpha t} \cos \omega t$. As for Eq. (6–28), we can readily verify that

$$\mathcal{L}[\epsilon^{-\alpha t}\cos\omega t] = \frac{s+\alpha}{(s+\alpha)^2 + \omega^2} \cdot \quad (6\text{-}29)$$

THEOREM. *The Laplace transform of $\epsilon^{-\alpha t}$ times a function is equal to the Laplace transform of that function with s replaced by $(s + \alpha)$.*

Proof:
$$\mathcal{L}[\epsilon^{-\alpha t}f(t)] = \int_0^\infty [\epsilon^{-\alpha t}f(t)]\epsilon^{-st}\, dt$$

$$= \int_0^\infty f(t)\epsilon^{-(s+\alpha)t}\, dt = F(s+\alpha). \quad (6\text{-}30)$$

B. *Positive powers of t.*

Let us now find the Laplace transform of t^n, where n is a positive integer. By definition,

$$\mathcal{L}[t^n] = \int_0^\infty (t^n)\epsilon^{-st}\, dt. \quad (6\text{-}31)$$

We integrate this by parts. Let

$$u = t^n \quad \text{and} \quad dv = \epsilon^{-st}\, dt.$$

Then
$$du = nt^{n-1}\, dt \quad \text{and} \quad v = \int \epsilon^{-st}\, dt = -\frac{1}{s}\epsilon^{-st},$$

$$\int_0^\infty t^n \epsilon^{-st}\, dt = -\frac{t^n}{s}\epsilon^{-st}\Big|_0^\infty + \frac{n}{s}\int_0^\infty t^{n-1}\epsilon^{-st}\, dt$$

$$= \frac{n}{s}\int_0^\infty t^{n-1}\epsilon^{-st}\, dt. \quad (6\text{-}32)$$

Examination of Eq. (6–32) shows that

$$\mathcal{L}[t^n] = \frac{n}{s}\,\mathcal{L}[t^{n-1}]. \quad (6\text{-}33)$$

By extending this process, we have

$$\mathcal{L}[t^n] = \frac{n}{s}\,\mathcal{L}[t^{n-1}] = \frac{n}{s}\cdot\frac{n-1}{s}\,\mathcal{L}[t^{n-2}]$$

$$= \frac{n}{s}\cdot\frac{n-1}{s}\cdot\frac{n-2}{s}\cdots\frac{2}{s}\cdot\frac{1}{s}\,\mathcal{L}[t^0].$$

But t^0 for $t > 0$ is the same as the unit step function $U(t)$, which has a Laplace transform $1/s$. Hence

$$\mathcal{L}[t^n] = \frac{n!}{s^{n+1}} \cdot \quad (6\text{-}34)^*$$

In particular, when $n = 1$,

$$\mathcal{L}[t] = \frac{1}{s^2} \cdot \quad (6\text{-}35)$$

---

* The region of convergence for this direct Laplace transformation is Re $(s) = \sigma > 0$.

TABLE 6–1

SOME IMPORTANT LAPLACE TRANSFORM PAIRS

$f(t)$	$F(s) = \mathcal{L}[f(t)] = \int_0^\infty f(t)\,\epsilon^{-st}\,dt$
(1) $U(t)$	$\dfrac{1}{s}$
(2) ϵ^{at}	$\dfrac{1}{s - a}$
(3) $\sin \omega t$	$\dfrac{\omega}{s^2 + \omega^2}$
(4) $\cos \omega t$	$\dfrac{s}{s^2 + \omega^2}$
(5) $\sinh bt$	$\dfrac{b}{s^2 - b^2}$
(6) $\cosh bt$	$\dfrac{s}{s^2 - b^2}$
(7) $\epsilon^{-at}f(t)$	$F(s + \alpha)$
(8) t^n	$\dfrac{n!}{s^{n+1}}$

We can now tabulate our transform pairs, as in Table 6–1. As has been mentioned above, a great majority of the functions we encounter in engineering can be expressed in terms of these basic types. The reader will do well to remember these basic pairs; he will then rarely find it necessary to refer to an elaborate table of transforms. As is evident in Table 6–1, Laplace transformation transforms functions of the exponential, trigonometric, and power types into algebraic functions which are easier to combine and manipulate.

EXAMPLE 6–1. Find the Laplace transform of the following function:

$$f(t) = A\epsilon^{-at} \sin (\beta t + \theta).$$

Solution. We note that the given function is not one of the types listed in Table 6–1. This problem can be solved in one of the following two ways.

(a) By expressing $f(t)$ in exponential form:

$$f(t) = \frac{A}{2j}\, \epsilon^{-at}\{\epsilon^{j(\beta t+\theta)} - \epsilon^{-j(\beta t+\theta)}\}$$

$$= \frac{A}{2j}\, \{\epsilon^{(-\alpha+j\beta)t}\epsilon^{j\theta} - \epsilon^{(-\alpha-j\beta)t}\epsilon^{-j\theta}\},$$

$\mathcal{L}[f(t)]$

$$= \frac{A}{2j} \{\epsilon^{j\theta}\mathcal{L}[\epsilon^{(-\alpha+j\beta)t}] - \epsilon^{-j\theta}\mathcal{L}[\epsilon^{(-\alpha-j\beta)t}]\}$$

$$= \frac{A}{2j} \left\{\frac{\epsilon^{j\theta}}{s + \alpha - j\beta} - \frac{\epsilon^{-j\theta}}{s + \alpha + j\beta}\right\}$$

$$= \frac{A}{2} \left\{\frac{(s + \alpha + j\beta)(-j\cos\theta + \sin\theta) - (s + \alpha - j\beta)(-j\cos\theta - \sin\theta)}{(s + \alpha)^2 + \beta^2}\right\}$$

$$= A \left\{\frac{(s + \alpha)\sin\theta + \beta\cos\theta}{(s + \alpha)^2 + \beta^2}\right\}.$$

(b) By finding $\mathcal{L}[A\sin(\beta t + \theta)]$ first, and then applying relation (7) in Table 6–1. First we have

$$\mathcal{L}[A\sin(\beta t + \theta)] = A\mathcal{L}[\sin\beta t\cos\theta + \cos\beta t\sin\theta]$$

$$= A\left[\frac{\beta\cos\theta + s\sin\theta}{s^2 + \beta^2}\right],$$

by transform pairs (3) and (4) in Table 6–1. Applying relation (7), we obtain

$$\mathcal{L}[A\epsilon^{-\alpha t}\sin(\beta t + \theta)] = A\left[\frac{(s + \alpha)\sin\theta + \beta\cos\theta}{(s + \alpha)^2 + \beta^2}\right],$$

which is the same as before.

6–4 The Shifting Theorem and its applications. The *Shifting Theorem,** which may be stated as follows, is an important theorem in engineering applications:

THE SHIFTING THEOREM. *If* $\mathcal{L}[f(t)] = F(s)$, *then*

$$\mathcal{L}[f(t - t_0)U(t - t_0)] = \epsilon^{-t_0 s}F(s). \tag{6–36}$$

Proof. By definition,

$$\mathcal{L}[f(t - t_0)U(t - t_0)] = \int_0^\infty [f(t - t_0)U(t - t_0)]\epsilon^{-st}\,dt$$

$$= \int_{t_0}^\infty f(t - t_0)\epsilon^{-st}\,dt, \tag{6–37}$$

---

* This is sometimes referred to as the *Time-Displacement Theorem*, because it involves a shift, or displacement, of the real variable t. Since the real variable does not have to be time, the designation *Shifting Theorem* is preferred.

since the function

$$f(t - t_0)U(t - t_0) = \begin{cases} 0 \text{ for } t < t_0, \\ f(t - t_0) \text{ for } t > t_0. \end{cases}$$

Now let

$$t - t_0 = \tau, \qquad t = \tau + t_0, \qquad dt = d\tau,$$

$$t = t_0 \to \tau = 0.$$

Equation (6–37) then becomes

$$\mathcal{L}[f(t - t_0)U(t - t_0)] = \int_0^\infty f(\tau)\epsilon^{-s(\tau+t_0)}\,d\tau$$

$$= \epsilon^{-t_0 s}\int_0^\infty f(\tau)\epsilon^{-s\tau}\,d\tau$$

$$= \epsilon^{-t_0 s}F(s). \qquad \text{Q.E.D.}$$

Before we demonstrate the usefulness of this Shifting Theorem, we wish to point out that the following four functions are all different:

(a) $f(t - t_0)$,   (b) $f(t - t_0)U(t)$,

(c) $f(t)U(t - t_0)$,   (d) $f(t - t_0)U(t - t_0)$,

and that the Shifting Theorem applies correctly *only* to the last type, (d). Let us assume that

$$f(t) = \sin \omega t.$$

Then

(a) $f(t - t_0) = \sin \omega(t - t_0)$,

(b) $f(t - t_0)U(t) = \sin \omega(t - t_0)U(t)$,

(c) $f(t)U(t - t_0) = \sin \omega t\, U(t - t_0)$,

(d) $f(t - t_0)U(t - t_0) = \sin \omega(t - t_0)U(t - t_0)$.

These functions are shown in parts (a), (b), (c), and (d), respectively, of Fig. 6–5. They are obviously all different.

We shall now find the Laplace transforms of these four different functions. First of all, we note that although $\sin \omega(t - t_0)$ and $\sin \omega(t - t_0)U(t)$ are different functions, they have the same Laplace transform because the range of integration in the definition of the Laplace transformation (6–13) is from $t = 0$ to ∞. Hence, for functions (a) and (b), we have

$$\mathcal{L}[\sin \omega(t - t_0)] = \mathcal{L}[\sin \omega t \cos \omega t_0 - \cos \omega t \sin \omega t_0]$$

$$= \frac{\omega \cos \omega t_0 - s \sin \omega t_0}{s^2 + \omega^2}. \qquad (6\text{–}38)$$

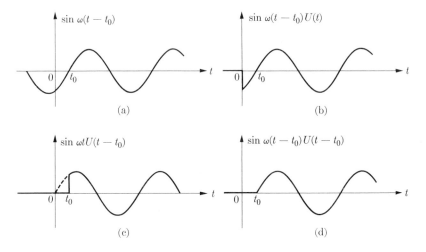

Fig. 6–5. Four different functions.

For the function (c), we have

$$\mathcal{L}[\sin \omega t \; U(t - t_0)] = \int_{t_0}^{\infty} (\sin \omega t) \epsilon^{-st} \, dt$$

$$= \frac{1}{2j} \int_{t_0}^{\infty} \left[\epsilon^{(-s+j\omega)t} - \epsilon^{(-s-j\omega)t} \right] dt$$

$$= \frac{1}{2j} \left[\frac{\epsilon^{(-s+j\omega)t_0}}{s - j\omega} - \frac{\epsilon^{(-s-j\omega)t_0}}{s + j\omega} \right]$$

$$= \epsilon^{-t_0 s} \left[\frac{\omega \cos \omega t_0 + s \sin \omega t_0}{s^2 + \omega^2} \right]. \qquad (6\text{–}39)$$

For the function (d), by applying the Shifting Theorem (6–36), we obtain:

$$\mathcal{L}[\sin \omega(t - t_0) U(t - t_0)] = \epsilon^{-t_0 s} \mathcal{L}[\sin \omega t]$$

$$= \epsilon^{-t_0 s} \left(\frac{\omega}{s^2 + \omega^2} \right). \qquad (6\text{–}40)$$

We see that the expressions in (6–38), (6–39), and (6–40) are quite different from one another.

By making use of the Shifting Theorem (6–36), we are able to find the Laplace transform of a wide variety of waveforms in terms of the basic transform pairs listed in Table 6–1 without actual integration, because waveforms can frequently be constructed from elementary functions with appropriate shifts in the real variable t. This will be made clear by the following examples.

EXAMPLE 6-2. Find the Laplace transform of a function which has the sawtooth form shown in Fig. 6-6.

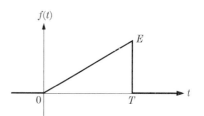

Solution. Before the Laplace transform can be found, it is necessary to write the expression for $f(t)$ which will correctly represent the given sawtooth. The sawtooth can be constructed from the three functions $f_a(t)$, $f_b(t)$, and $f_c(t)$, as shown in parts (a), (b), and (c) of Fig. 6-7:

FIG. 6-6. A sawtooth function

$$f(t) = f_a(t) + f_b(t) + f_c(t),$$

where

$$f_a(t) = \frac{E}{T} tU(t),$$

$$f_b(t) = -EU(t - T),$$

and

$$f_c(t) = -\frac{E}{T}(t - T)U(t - T).$$

Now, by Eq. (6-35),

$$F_a(s) = \mathcal{L}[f_a(t)] = \frac{E}{Ts^2};$$

by Eqs. (6-23) and (6-36),

$$F_b(s) = \mathcal{L}[f_b(t)] = -\frac{E}{s}\epsilon^{-Ts};$$

and, by Eqs. (6-35) and (6-36),

$$F_c(s) = \mathcal{L}[f_c(t)] = -\frac{E}{Ts^2}\epsilon^{-Ts}.$$

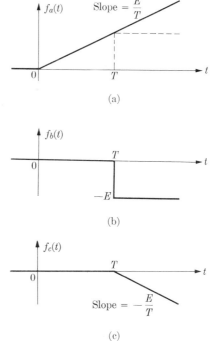

FIG. 6-7. Component functions of the sawtooth in Fig. 6-6.

Hence

$$F(s) = \mathcal{L}[f(t)] = F_a(s) + F_b(s) + F_c(s)$$

$$= \frac{E}{Ts^2}[1 - (Ts + 1)\epsilon^{-Ts}]. \tag{6-41}$$

EXAMPLE 6–3. Find the Laplace transform of the single half-sine cycle shown in Fig. 6–8.

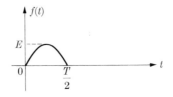

FIG. 6–8. A single half-sine cycle.

Solution. Referring to Fig. 6–9, we find

$$f(t) = f_a(t) + f_b(t)$$

$$= E \sin \frac{2\pi}{T} tU(t) + E \sin \frac{2\pi}{T}\left(t - \frac{T}{2}\right) U\left(t - \frac{T}{2}\right).$$

Note that the period of the sine wave is T (not $T/2$); hence the angular frequency is $2\pi/T$. Since we know the Laplace transform of the sine function, with the aid of the Shifting Theorem we can write

$$F(s) = \mathcal{L}[f(t)] = \mathcal{L}[f_a(t)] + \mathcal{L}[f_b(t)]$$

$$= \frac{E(2\pi/T)}{s^2 + (2\pi/T)^2} + \frac{E(2\pi/T)}{s^2 + (2\pi/T)^2}\, \epsilon^{-Ts/2}$$

$$= \frac{E(2\pi/T)}{s^2 + (2\pi/T)^2}\, (1 + \epsilon^{-Ts/2}). \tag{6–42}$$

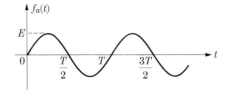

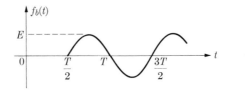

FIG. 6–9. Component functions of the single half-sine cycle in Fig. 6–8.

6–5 The gate function. A function comprising two unit step functions of the form

$$G_{t_0}(T) = U(t - t_0) - U(t - t_0 - T) \qquad (6\text{--}43)$$

represents a rectangular pulse of unit height which starts at $t = t_0$ and lasts for a time T, as depicted in Fig. 6–10. Such a function may be called a *gate function* and is quite useful in the construction of waveforms. Any function multiplied by the gate function $G_{t_0}(T)$ in Eq. (6–43) would have nonzero values only in the duration of the gate, $t_0 < t < t_0 + T$; the value of the function within the gate would be unaffected. The usefulness of gate functions lies in the ease with which the unwanted portions of a continuous function can be "erased" without tedious graphical composition by shifting, and by addition or subtraction. The following examples will serve to illustrate its application.

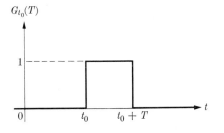

Fig. 6–10. A gate function.

EXAMPLE 6–4. Find the Laplace transform of the sawtooth function in Fig. 6–6 by making use of a gate function.

Solution. This problem was solved in Example 6–2 by resolving the given sawtooth into three component functions properly shifted in t, as shown in Fig. 6–7. We can solve the problem by noting that

$$f(t) = \frac{E}{T} t \cdot G_0(T) = \frac{E}{T} t[U(t) - U(t - T)]. \qquad (6\text{--}44)$$

No graphical composition is then necessary. From Eq. (6–44) we have

$$F(s) = \mathcal{L}[f(t)] = \frac{E}{T} \{\mathcal{L}[tU(t)] - \mathcal{L}[tU(t - T)]\}$$

$$= \frac{E}{T} \left\{ \frac{1}{s^2} - \mathcal{L}[(t - T) + T]U(t - T) \right\}$$

$$= \frac{E}{T} \left\{ \frac{1}{s^2} - \left[\frac{1}{s^2} + \frac{T}{s} \right] \epsilon^{-Ts} \right\}$$

$$= \frac{E}{Ts^2} [1 - (Ts + 1)\epsilon^{-Ts}], \qquad (6\text{--}45)$$

which is the same as Eq. (6–41).

EXAMPLE 6–5. Find the Laplace transform of the single half-sine cycle in Fig. 6–8 by making use of a gate function.

Solution. This problem was solved in Example 6–3 by considering the given half-sine cycle as the sum of two sine waves shifted with respect to each other by half a period, as shown in Fig. 6–9. Using a gate function $G_0(T/2)$, we have

$$f(t) = E\left(\sin\frac{2\pi}{T}t\right)\cdot G_0\left(\frac{T}{2}\right) = E\left(\sin\frac{2\pi}{T}t\right)\left[U(t) - U\left(t - \frac{T}{2}\right)\right].$$

$$(6\text{–}46)$$

Here again the need for graphical composition has been avoided.

$$F(s) = \mathcal{L}[f(t)] = E\mathcal{L}\left[\left(\sin\frac{2\pi}{T}t\right)\cdot U(t)\right]$$

$$- E\mathcal{L}\left[\left(\sin\frac{2\pi}{T}t\right)U\left(t - \frac{T}{2}\right)\right]$$

$$= F_1(s) - F_2(s), \qquad (6\text{–}47)$$

where

$$F_1(s) = E\mathcal{L}\left[\left(\sin\frac{2\pi}{T}t\right)U(t)\right] = \frac{E(2\pi/T)}{s^2 + (2\pi/T)^2} \qquad (6\text{–}48)$$

and

$$F_2(s) = E\mathcal{L}\left\{\left(\sin\frac{2\pi}{T}t\right)U\left(t - \frac{T}{2}\right)\right\}$$

$$= E\mathcal{L}\left\{\sin\frac{2\pi}{T}\left[\left(t - \frac{T}{2}\right) + \frac{T}{2}\right]U\left(t - \frac{T}{2}\right)\right\}$$

$$= E\mathcal{L}\left\{\sin\left[\frac{2\pi}{T}\left(t - \frac{T}{2}\right) + \pi\right]U\left(t - \frac{T}{2}\right)\right\}. \qquad (6\text{–}49)$$

Expanding the sine function, we have

$$F_2(s) = E\mathcal{L}\left\{-\sin\frac{2\pi}{T}\left(t - \frac{T}{2}\right)U\left(t - \frac{T}{2}\right)\right\}$$

$$= -\frac{E(2\pi/T)}{s^2 + (2\pi/T)^2}\,\epsilon^{-Ts/2}, \qquad (6\text{–}50)$$

by virtue of Eqs. (6–24) and (6–36). Substituting Eqs. (6–48) and (6–50) in Eq. (6–47), we find

$$F(s) = \frac{E(2\pi/T)}{s^2 + (2\pi/T)^2}\,(1 + \epsilon^{-Ts/2}), \qquad (6\text{–}51)$$

which is the same as Eq. (6–42).

6–6 Laplace transform of periodic functions. Periodic functions often appear as excitation functions in engineering problems. There exists a method which enables us to write the Laplace transform of a periodic function if the Laplace transform of a single cycle is known. We state this method in the form of a theorem.

THEOREM. *The Laplace transform of a periodic function with period T is equal to $1/(1 - \epsilon^{-Ts})$ times the Laplace transform of the first cycle.*

Proof. Let $f(t)$ be a periodic function with period T, and $f_1(t)$, $f_2(t)$, $f_3(t)$, ... be the functions describing the first cycle, second cycle, third cycle, ... Then

$$f(t) = f_1(t) + f_2(t) + f_3(t) + \cdots$$
$$= f_1(t) + f_1(t - T)U(t - T) + f_1(t - 2T)U(t - 2T) + \cdots,$$

since the second cycle is identical with the first cycle shifted toward the right by a period T, the third cycle is identical with the first cycle shifted toward the right by two periods $(2T)$, and so on. If we call $F_1(s)$ the Laplace transform of the first cycle,

$$\mathcal{L}[f_1(t)] = F_1(s), \tag{6–53}$$

then, by the Shifting Theorem, we have from Eq. (6–52),

$$F(s) = \mathcal{L}[f(t)] = (1 + \epsilon^{-Ts} + \epsilon^{-2Ts} + \cdots)F_1(s)$$

$$= \frac{1}{1 - \epsilon^{-Ts}} F_1(s). \qquad \text{Q.E.D.} \tag{6–54}$$

We shall illustrate the application of this theorem by examples.

EXAMPLE 6–6. Find the Laplace transform of the periodic, rectified half-sine wave shown in Fig. 6–11.

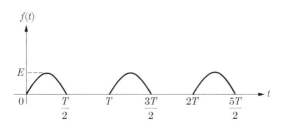

FIG. 6–11. Rectified half-sine wave.

Solution. We refer to Example 6–3, where we found that the Laplace transform of a single half-sine cycle starting at $t = 0$ is [see Eq. (6–42)]

$$F_1(s) = \frac{E(2\pi/T)}{s^2 + (2\pi/T)^2} (1 + \epsilon^{-Ts/2}).$$

Hence, by Eq. (6–54), we have

$$F(s) = \mathcal{L}[f(t)] = \frac{1 + \epsilon^{-Ts/2}}{1 - \epsilon^{-Ts}} \cdot \frac{E(2\pi/T)}{s^2 + (2\pi/T)^2}$$

$$= \frac{1}{1 - \epsilon^{-Ts/2}} \cdot \frac{E(2\pi/T)}{s^2 + (2\pi/T)^2}. \tag{6–55}$$

EXAMPLE 6–7. Find the Laplace transform of the periodic sawtooth wave shown in Fig. 6–12.

FIG. 6–12. A periodic sawtooth wave.

Solution. The first period of the wave is indicated by the heavy line in Fig. 6–12. We solve this problem in three steps.

1. We first determine $f_1(t)$ for the first period (it is simplest to use gate function here):

$$f_1(t) = -\frac{2E}{T}\left(t - \frac{T}{2}\right)[U(t) - U(t - T)]$$

$$= -\frac{2E}{T}\left(t - \frac{T}{2}\right)U(t) + \frac{2E}{T}\left[\left(t - T + \frac{T}{2}\right)\right]U(t - T)$$

$$= -\frac{2E}{T}\left(t - \frac{T}{2}\right)U(t) + \left[\frac{2E}{T}(t - T) + E\right]U(t - T). \tag{6–56}$$

2. We now find $F_1(s)$ from $f_1(t)$:

$$F_1(s) = \mathcal{L}[f_1(t)] = -\frac{2E}{T}\left(\frac{1}{s^2} - \frac{T}{2s}\right) + \left[\frac{2E}{Ts^2} + \frac{E}{s}\right]\epsilon^{-Ts}$$

$$= \frac{2E}{Ts}\left\{\frac{T}{2}(1 + \epsilon^{-Ts}) - \frac{1}{s}(1 - \epsilon^{-Ts})\right\}. \tag{6–57}$$

3. Finally, we obtain $F(s)$ from $F_1(s)$:

$$F(s) = \frac{F_1(s)}{1 - \epsilon^{-Ts}} = \frac{2E}{Ts} \left\{ \frac{T}{2} \left(\frac{1 + \epsilon^{-Ts}}{1 - \epsilon^{-Ts}} \right) - \frac{1}{s} \right\}$$

$$= \frac{2E}{Ts} \left\{ \frac{T}{2} \coth \left(\frac{Ts}{2} \right) - \frac{1}{s} \right\}. \tag{6–58}$$

6–7 Laplace transformation of operations. In applying Laplace transformation to the solution of differential equations, we must be able to transform not only functions but also operations, mainly differentiations and integrations. In this section we shall develop the differentiation and integration theorems.

DIFFERENTIATION THEOREM. *If a function $f(t)$ and its derivative are both Laplace transformable and if $\mathcal{L}[f(t)] = F(s)$, then*

$$\mathcal{L}\left[\frac{df(t)}{dt}\right] = \mathcal{L}[Df(t)] = sF(s) - f(0+). \tag{6–59}$$

Proof. We proceed from the definition

$$F(s) = \mathcal{L}[f(t)] = \int_0^\infty f(t)\epsilon^{-st} \, dt.$$

Integrating by parts, we let

$$u = f(t), \qquad du = \left[\frac{df(t)}{dt}\right] dt,$$

$$dv = \epsilon^{-st} \, dt, \qquad v = -\frac{1}{s}\epsilon^{-st}.$$

Hence

$$F(s) = -\frac{1}{s} f(t)\epsilon^{-st} \Big|_0^\infty + \frac{1}{s}\int_0^\infty \left[\frac{df(t)}{dt}\right] \epsilon^{-st} \, dt$$

$$= \frac{f(0+)}{s} + \frac{1}{s}\mathcal{L}\left[\frac{df(t)}{dt}\right].$$

Rearranging terms, we find

$$\mathcal{L}\left[\frac{df(t)}{dt}\right] = \mathcal{L}[Df(t)] = sF(s) - f(0+). \qquad \text{Q.E.D.}$$

Thus Laplace transformation transforms the differentiating operation into multiplication by s, with $f(0+)$ being subtracted from the product.

Equation (6–59) can be readily extended to cover higher-order derivatives when they are Laplace transformable. Thus the Laplace transform of the second derivative of $f(t)$ is

$$\mathcal{L}[D^2 f(t)] = s\mathcal{L}[Df(t)] - f'(0+)$$
$$= s[sF(s) - f(0+)] - f'(0+)$$
$$= s^2 F(s) - sf(0+) - f'(0+), \tag{6–60}$$

where $f'(0+)$ is the value of the first derivative of $f(t)$ as t approaches zero from the positive side. In general,

$$\mathcal{L}[D^n f(t)]$$

$$= \mathcal{L}\left[\frac{d^n f(t)}{dt^n}\right]$$

$$= s^n F(s) - s^{n-1} f(0+) - s^{n-2} f'(0+) - \cdots - sf^{(n-2)}(0+) - f^{(n-1)}(0+)$$

$$= s^n F(s) - \sum_{k=1}^{n} s^{n-k} f^{(k-1)}(0+), \tag{6–61}$$

with $f^{(0)}(t) = f(t)$. There is no need to remember either Eq. (6–60) or Eq. (6–61), inasmuch as they can be easily derived from Eq. (6–59), which should be remembered.

INTEGRATION THEOREM. *If* $\mathcal{L}[f(t)] = F(s)$, *then*

$$\mathcal{L}\left[\int_0^t f(t)\, dt\right] = \frac{F(s)}{s}, \tag{6–62}$$

Proof. By definition,

$$\mathcal{L}\left[\int_0^t f(t)\, dt\right] = \int_0^\infty \left[\int_0^t f(t)\, dt\right] \epsilon^{-st}\, dt.$$

Integrating by parts, we let

$$u = \int_0^t f(t)\, dt, \qquad du = f(t)\, dt,$$

$$dv = \epsilon^{-st}\, dt, \qquad v = -\frac{1}{s}\epsilon^{-st}.$$

Hence

$$\mathcal{L}\left[\int_0^t f(t)\, dt\right] = -\frac{\epsilon^{-st}}{s}\int_0^t f(t)\, dt\bigg|_0^\infty + \frac{1}{s}\int_0^\infty f(t)\epsilon^{-st}\, dt$$

$$= \frac{1}{s}\mathcal{L}[f(t)] = \frac{F(s)}{s}. \qquad \text{Q.E.D.}$$

Thus, Laplace transformation transforms the integrating operation into simple division by s.

The Laplace transformation of the indefinite integral of a function may be obtained quite easily from Eq. (6–62) because the indefinite integral of a function may be written as

$$\int f(t) \, dt = \int_0^t f(t) \, dt + f^{(-1)}(0+), \tag{6–63}$$

where $f^{(-1)}(0+)$ is the value of the integral of $f(t)$ as t approaches zero from the positive side. Hence

$$\mathcal{L}\left[\int f(t) \, dt\right] = \mathcal{L}\left[\int_0^t f(t) \, dt\right] + \mathcal{L}[f^{(-1)}(0+)]$$

$$= \frac{F(s)}{s} + \frac{f^{(-1)}(0+)}{s}. \tag{6–64}$$

A generalization similar to that leading to the transforms of higher-order derivatives, Eqs. (6–60) and (6–61), permits us to readily extend both Eqs. (6–62) and (6–64) to cover higher-order integrals.

The above theorems involve differentiation and integration of $f(t)$. It is sometimes convenient to consider also the derivatives and integrals of $F(s)$. The following two relations hold:

Multiplication by t: If $\mathcal{L}[f(t)] = F(s)$,   then

$$\mathcal{L}[tf(t)] = -\frac{d}{ds} F(s). \tag{6–65}$$

Division by t: If $\mathcal{L}[f(t)] = F(s)$,   then

$$\mathcal{L}\left[\frac{f(t)}{t}\right] = \int_s^\infty F(s) \, ds. \tag{6–66}$$

The proofs are left as problems at the end of this chapter. We shall illustrate the applications of these two relations in the following examples.

EXAMPLE 6–8. Find $\mathcal{L}[t^2 \sin \omega t]$.

Solution. Repeated application of Eq. (6–65) yields

$$\mathcal{L}[t^2 \sin \omega t] = (-1)^2 \frac{d^2}{ds^2} \{\mathcal{L}[\sin \omega t]\}$$

$$= \frac{d^2}{ds^2} \left\{\frac{\omega}{s^2 + \omega^2}\right\} = \frac{2\omega(3s^2 - \omega^2)}{(s^2 + \omega^2)^3}. \tag{6–67}$$

EXAMPLE 6–9. Find $\mathcal{L}\left[\dfrac{\sin \omega t}{t}\right]$.

Solution. Direct application of Eq. (6–66) yields

$$\mathcal{L}\left[\frac{\sin \omega t}{t}\right] = \int_s^\infty \{\mathcal{L}[\sin \omega t]\} \, ds$$

$$= \int_s^\infty \frac{\omega}{s^2 + \omega^2} \, ds$$

$$= \tan^{-1}\left(\frac{s}{\omega}\right)\Big|_s^\infty = \frac{\pi}{2} - \tan^{-1}\left(\frac{s}{\omega}\right)$$

$$= \tan^{-1}\left(\frac{\omega}{s}\right). \tag{6–68}$$

Table 6–2 collects the Laplace transforms of the various operations for easy reference.

TABLE 6–2

LAPLACE TRANSFORMATION OF OPERATIONS

$f(t)$	$F(s) = \mathcal{L}[f(t)]$
(1) $f(t - t_0) U(t - t_0)$	$\epsilon^{-t_0 s} F(s)$
(2) $\dfrac{d}{dt} f(t)$	$s F(s) - f(0+)$
(3) $\displaystyle\int_0^t f(t) \, dt$	$\dfrac{F(s)}{s}$
(4) $\displaystyle\int f(t) \, dt$	$\dfrac{F(s)}{s} + \dfrac{f^{(-1)}(0+)}{s}$
(5) $t f(t)$	$-\dfrac{d}{ds} F(s)$
(6) $\dfrac{1}{t} f(t)$	$\displaystyle\int_s^\infty F(s) \, ds$

Having learned the ways of performing the Laplace transformation of various types of functions and operations, we are now ready to transform differential and integro-differential equations. The Laplace transform of the unknown dependent variable or variables is first solved in terms of the transforms of known functions and operations. An inverse transformation will yield the desired solution. The procedure and techniques involved will be developed in the next chapter.

Problems

6-1. Find the Laplace transforms of the following functions:

(a) $t^2\epsilon^{-mt}$ (b) $t^3 - 2t + 1$

(c) $3t^2 U(t - 5)$ (d) $t\epsilon^{-\alpha t} U(t - T)$

(e) $(t + 2)^2 U(t - 1)$ (f) $t \cos(\omega t + \theta)$

(g) $\sin 2\omega(t - t_0)$ (h) $\sin 2\omega(t - t_0) \cdot U(t - 2t_0)$

(i) $3\epsilon^{-4t} t \cos 2(t - 1)$ (j) $\epsilon^{-5t} \cosh 3t$

6-2. (a) Show that if $\mathcal{L}[f(t)] = F(s)$, then $\mathcal{L}[f(at)] = (1/a)F(s/a)$.

(b) Find the Laplace transform of $\epsilon^{-t/b} f(t/b)$.

(c) Find the Laplace transform of $\epsilon^{-bt} f(t/b)$.

6-3. (a) Plot the following exponential functions approximately to scale:

(i) $\epsilon^{-2t} U(t)$ (ii) $\epsilon^{-2t} U(t - 1)$

(iii) $\epsilon^{-2(t-1)} U(t)$ (iv) $\epsilon^{-2(t-1)} U(t - 1)$

(b) Find the Laplace transform of each of the above functions.

6-4. Find the Laplace transform of the trapezoidal function in Fig. 6-13.

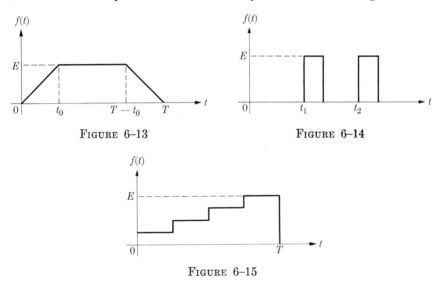

FIGURE 6-13　　　　　　　　FIGURE 6-14

FIGURE 6-15

6-5. Find the Laplace transform of a pair of rectangular pulses, each of duration τ, as shown in Fig. 6-14.

6-6. Find the Laplace transform of the staircase function in Fig. 6-15.

6-7. Find the Laplace transform of an audio cycle of an amplitude-modulated wave. Assume 80% modulation and f_c (carrier frequency) $= 5000 f_a$ (audio-frequency).

6-8. Find the Laplace transform of the periodic triangular wave in Fig. 6-16.

6-9. Find the Laplace transform of the rectangular wave in Fig. 6-17.

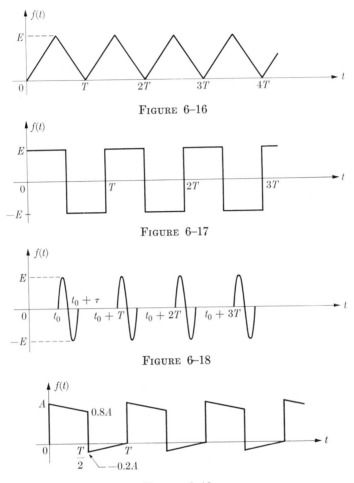

FIGURE 6–16

FIGURE 6–17

FIGURE 6–18

FIGURE 6–19

6–10. Find the Laplace transform of the periodic train of one-cycle sinusoidal pulses shown in Fig. 6–18.

6–11. Find the Laplace transform of the periodic wave shown in Fig. 6–19.

6–12. Derive a general expression for the Laplace transform of the nth-order integral of $f(t)$ in terms of $F(s)$ and initial conditions.

6–13. (a) Prove Eq. (6–65). (b) Find $\mathcal{L}[t \cos^2 3t]$.

6–14. (a) Prove Eq. (6–66). (b) Find $\mathcal{L}[(1/t)(1 - \epsilon^{-\alpha t})]$.

6–15. The function $Si(x)$, "sine integral of x," is defined as follows:

$$Si(x) = \int_0^x \frac{\sin u}{u}\, du.$$

(a) Find $\mathcal{L}[Si(x)]$. (b) Find $\mathcal{L}[Si(ax)]$.

CHAPTER 7

APPLICATIONS OF LAPLACE TRANSFORMATION

7–1 Introduction. Having introduced the definition of Laplace transformation and obtained the Laplace transforms of some fundamental functions and operations in the preceding chapter, we are now ready to demonstrate the applications of Laplace transformation as it is used to determine the response of linear systems under the influence of different types of excitation. Three steps are involved in the analysis of linear systems by the method of Laplace transforms. First, the differential or integro-differential equation or equations describing the system under consideration must be set up from physical laws. Second, Laplace transformation is applied to each and every term, including the excitation functions, in the equation or equations from which the transform of the response function or functions is solved. Finally, an inverse transformation must be performed to obtain the desired response function or functions. We have essentially covered the first step in Chapters 3 and 4, and have provided the techniques for the second step in Chapter 6. In this chapter we shall discuss the inverse transformation and complete the last step for the solution.

7–2 Solution of linear differential equations. In this section we shall demonstrate how the techniques developed in the preceding chapter can be applied to the solution of linear differential equations with constant coefficients. Let us consider a second-order equation:

$$\frac{d^2w}{dt^2} + b_1 \frac{dw}{dt} + b_0w = e(t), \qquad (7\text{–}1)^*$$

where the coefficients b_1 and b_0 are known constants and $e(t)$ is a given Laplace-transformable excitation function. Equation (7–1) can be written in the simpler operator notation

$$(D^2 + b_1 D + b_0)w = e(t). \qquad (7\text{–}2)$$

Let us now apply Laplace transformation to both sides of Eq. (7–2):

$$\mathcal{L}[D^2w] + b_1\mathcal{L}[Dw] + b_0\mathcal{L}[w] = \mathcal{L}[e(t)]. \qquad (7\text{–}3)$$

---

* If the given equation is of the form $a_2(d^2w/dt^2) + a_1(dw/dt) + a_0w = e_1(t)$, it is always possible to reduce it to the form of Eq. (7–1) by dividing it through by a_2. Thus, $b_1 = a_1/a_2$, $b_0 = a_0/a_2$, ond $e(t) = e_1(t)/a_2$.

Since the dependent variable w is as yet an unknown function of t, we shall let

$$\mathcal{L}[w] = W(s). \tag{7-4}$$

Applying the differentiation theorem, Eqs. (6–59) and (6–60), we obtain

$$\mathcal{L}[Dw] = sW(s) - w(0), \tag{7-5}$$

$$\mathcal{L}[D^2w] = s^2W(s) - sw(0) - w'(0), \tag{7-6}$$

where $w(0)$ and $w'(0)$ are understood to assume the values at $t = 0+$ when the functions $w(t)$ and $w'(t)$ have discontinuities at $t = 0$. Substituting Eqs. (7–4), (7–5), and (7–6) back into Eq. (7–3) and calling

$$\mathcal{L}[e(t)] = E(s), \tag{7-7}$$

we obtain

$$[s^2W(s) - sw(0) - w'(0)] + b_1[sW(s) - w(0)] + b_0W(s) = E(s),$$

which can be rearranged as

$$(s^2 + b_1 s + b_0)W(s) = E(s) + (s + b_1)w(0) + w'(0). \tag{7-8}$$

By the use of Laplace transformation, we have succeeded in transforming the given differential equation (7–1) into an algebraic equation (7–8). From Eq. (7–8), we have

$$W(s) = \frac{1}{s^2 + b_1 s + b_0} [E(s) + (s + b_1)w(0) + w'(0)]. \tag{7-9}$$

The desired solution $w(t)$ is then the inverse Laplace transform of $W(s)$:

$$w(t) = \mathcal{L}^{-1}[W(s)] = \mathcal{L}^{-1}\left[\frac{E(s) + (s + b_1)w(0) + w'(0)}{s^2 + b_1 s + b_0} \right]. \tag{7-10}$$

The result of the inverse Laplace transformation in Eq. (7–10) depends, of course, not only on the characteristics of the system, but also on the excitation and the initial conditions. We shall defer discussions on the techniques of finding the inverse transform to a later section.

Let us now examine Eq. (7–9) a little more carefully, since it is typical of the transform solution of all linear differential equations. We can interpret Eq. (7–9) as follows:

$$W(s) = H(s)E_0(s) \tag{7-11}$$

or, in words,

$$\text{(Response transform)} = \text{(Transfer function)}$$
$$\times \text{(Total excitation transform)}, \tag{7-12}$$

where

(a) *Response transform*, $W(s)$, is the $\mathcal{L}$ of the response function sought;

(b) *Transfer function*, $H(s)$, is the ratio of the transform of the output (response) to that of the input (excitation) and is equal to the reciprocal of the characteristic function of the given differential equation obtained by replacing D's by s's in the operator part of Eq. (7–2). It is more general than the transfer function for steady-state sinusoidal excitation, $H(j\omega)$, which we introduced in Section 5–7. $H(j\omega)$ can be obtained from $H(s)$ by setting $s = j\omega$;

(c) *Total excitation transform*, $E_0(s)$, consists of the $\mathcal{L}$ of the applied excitation, $E(s)$, and the contributions due to initial conditions, which do not depend upon the applied excitation.

We see here that the initial conditions of the problem are inserted in the solution automatically in the process of making the Laplace transformation. We need no separate operation for determining arbitrary constants, such as is required in the classical method, because there are no "arbitrary" constants. Changing the excitation function would change only $E(s)$, while different sets of initial conditions would only make $w(0)$ and $w'(0)$ different; the transfer function $H(s)$ remains unaffected.

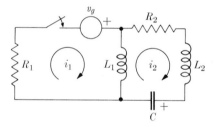

FIG. 7–1. A two-loop network.

Integro-differential equations and simultaneous equations can be dealt with in a similar manner. Let us consider the two-loop network shown in Fig. 7–1. The switch is closed at $t = 0$. Loop analysis yields the following two equations:

$$L_1 \frac{di_1}{dt} + R_1 i_1 - L_1 \frac{di_2}{dt} = v_g, \tag{7–13a}$$

$$-L_1 \frac{di_1}{dt} + (L_1 + L_2) \frac{di_2}{dt} + R_2 i_2 + \frac{1}{C} \int_0^t i_2 \, dt + v_C(0) = 0. \tag{7–13b}$$

We shall now apply Laplace transformation to both the above equations. The Laplace transform for the given excitation function v_g,

$$\mathcal{L}[v_g(t)] = V_g(s), \tag{7–14}$$

can be found by using the techniques of Chapter 6, when $v_g(t)$ is given. We shall call the Laplace transforms of the unknown loop currents i_1 and i_2 (response functions) $I_1(s)$ and $I_2(s)$, respectively:

$$\mathcal{L}[i_1(t)] = I_1(s), \tag{7–15}$$

$$\mathcal{L}[i_2(t)] = I_2(s). \tag{7–16}$$

Thus Laplace transformation of Eqs. (7–13a) and (7–13b) yields

$$L_1[sI_1(s) - i_1(0)] + R_1I_1(s) - L_1[sI_2(s) - i_2(0)] = V_g(s), \tag{7–17a}$$

$$-L_1[sI_1(s) - i_1(0)] + (L_1 + L_2)[sI_2(s) - i_2(0)] + R_2I_2(s)$$
$$+ \frac{I_2(s)}{Cs} + \frac{v_C(0)}{s} = 0. \tag{7–17b}$$

Rearranging terms, we have

$$(L_1s + R_1)I_1(s) - L_1sI_2(s) = V_g(s) + L_1[i_1(0) - i_2(0)] = E_1(s), \tag{7–18a}$$

$$-L_1sI_1(s) + \left[(L_1 + L_2)s + R_2 + \frac{1}{Cs}\right]I_2(s)$$
$$= -\frac{v_C(0)}{s} + L_1[i_2(0) - i_1(0)] + L_2i_2(0)$$
$$= E_2(s). \tag{7–18b}$$

Equations (7–18) are a pair of simultaneous algebraic equations in two unknowns, $I_1(s)$ and $I_2(s)$. For simplicity, we have written $E_1(s)$ and $E_2(s)$ for the total excitation transforms in loops 1 and 2 respectively:

$$E_1(s) = V_g(s) + L_1[i_1(0) - i_2(0)], \tag{7–19a}$$

$$E_2(s) = -\frac{v_C(0)}{s} + L_1[i_2(0) - i_1(0)] + L_2i_2(0). \tag{7–19b}$$

Note that although there is no externally applied excitation source in loop 2 the total excitation transform $E_2(s)$ is not zero unless the circuit is *initially at rest*. Comparing Eqs. (7–18) with Eqs. (7–13), we also note that the left sides of the transformed equations could have been obtained from the given integro-differential equations by replacing differentiation operations with multiplication by s and integration operations with division by s.

Equations (7–18) can be solved for $I_1(s)$ and $I_2(s)$ by algebraic methods. Using Cramer's rule in the method of determinants, we have

$$I_1(s) = \frac{1}{\Delta(s)}\left\{\left[(L_1 + L_2)s + R_2 + \frac{1}{Cs}\right]E_1(s) + L_1sE_2(s)\right\}, \tag{7–20a}$$

$$I_2(s) = \frac{1}{\Delta(s)}\{L_1sE_1(s) + (L_1s + R_1)E_2(s)\}, \tag{7–20b}$$

where

$$\Delta(s) = \begin{vmatrix} L_1 s + R_1 & -L_1 s \\ -L_1 s & (L_1 + L_2)s + R_2 + \dfrac{1}{Cs} \end{vmatrix}$$

$$= (L_1 s + R_1)\left[(L_1 + L_2)s + R_2 + \frac{1}{Cs} \right] - L_1^2 s^2. \quad (7\text{–}21)$$

$\Delta(s)$ is sometimes called the *characteristic function* of the given circuit. The same $\Delta(s)$ appears in the expressions of both $I_1(s)$ and $I_2(s)$. As we shall see later, $\Delta(s)$ determines the form of the transient solution; hence *the response functions in all parts of the system have the same types of transients.* We have already taken note of this fact in Chapter 2, Section 2–6, in connection with a discussion of the classical solutions for simultaneous differential equations. We noted there that the characteristic equations for all dependent variables are the same. As a result, the complementary functions of the solutions for all dependent variables have the *same* numbers and types of terms. The complementary-function part of the solutions of differential equations describing physical systems represents the transient solutions.

The loop currents $i_1(t)$ and $i_2(t)$ in the given two-loop network of Fig. 7–1 are determined by taking, respectively, the inverse Laplace transforms of $I_1(s)$ and $I_2(s)$ in Eqs. (7–20).

7–3 The inverse transformation; Heaviside's expansion theorem. The solution of the differential equation (7–1) is not complete until we have carried out the inverse Laplace transformation as indicated in Eq. (7–10) and have determined $w(t)$. Likewise, the expressions for $I_1(s)$ and $I_2(s)$ in Eqs. (7–20a) and (7–20b) are not useful unless we can find their inverse transforms. In this section we shall discuss the techniques of inverse Laplace transformation. The theory of the functions of a complex variable will not be used here in effecting the inverse transformation. Fortunately, we are able to complete the solutions for most linear systems in engineering with a knowledge of the algebra of partial fractions.

First of all, we note from Eqs. (7–9) and (7–20) that the response transforms are rational algebraic functions of s,* i.e., they are expressible in the form of a ratio of two polynomials in s; no nonintegral powers of s

---

* Not all response transforms are rational functions of s, although most of those we encounter are. In general, a good knowledge of the theory of functions of a complex variable is necessary in order to handle the inverse transformation of irrational functions. The inverse Laplace transforms of some special irrational functions will be obtained in Section 8–6.

are involved. We shall then direct our attention to a general rational algebraic fraction of the following form:

$$F(s) = \frac{A(s)}{B(s)} = \frac{a_m s^m + a_{m-1} s^{m-1} + \cdots + a_1 s + a_0}{b_n s^n + b_{n-1} s^{n-1} + \cdots + b_1 s + b_0}, \quad (7\text{-}22)$$

where the coefficients a and b are all real constants, and m and n are *positive* integers. A further restriction is to be imposed on the relative values of m and n. That is, we require $m < n$. When this condition is satisfied, we say that $A(s)/B(s)$ is a *proper fraction*. If $m \geq n$, then we have an improper fraction, as exemplified by

$$F(s) = \frac{3s^3 - 2s^2 - 7s + 1}{s^2 + s - 1}$$

$$= (3s - 5) + \frac{s - 4}{s^2 + s - 1}.$$

We see that the improper fraction can be expressed as the sum of a polynomial $(3s - 5)$ and a proper fraction $(s - 4)/(s^2 + s - 1)$. Improper fractions can appear as response transforms only under ideal conditions (for example, in dissipationless circuits), and are not important in practice. We shall deal with the inverse transformation of improper fractions in the next chapter.

When $F(s) = A(s)/B(s)$ in Eq. (7–22) is a proper rational fraction, we can distinguish two different situations, namely, when the roots of the equation $B(s) = 0$ are all distinct and when $B(s) = 0$ has multiple roots. We shall discuss these two situations separately.

I. $m < n$, *roots of* $B(s) = 0$ *all distinct*. $B(s)$, being a polynomial in s of the nth degree, can be factored* as

$$B(s) = b_n(s - s_1)(s - s_2) \cdots (s - s_k) \cdots (s - s_n) = b_n \prod_{k=1}^{n} (s - s_k).$$
$$(7\text{-}23)$$

The condition that the roots of the equation $B(s) = 0$ are all distinct means that no two of the roots, $s_1, s_2, \ldots, s_k, \ldots, s_n$, are equal. We have

$$\frac{A(s)}{B(s)} = \frac{A(s)}{b_n(s - s_1)(s - s_2) \cdots (s - s_k) \cdots (s - s_n)}, \quad (7\text{-}24)$$

which can be expanded as a sum of n simple partial fractions, each having one of the factors of $B(s)$ as its denominator:

---

* See Appendix A, *Numerical Solution of Algebraic Equations*.

$$\frac{A(s)}{B(s)} = \frac{1}{b_n} \left[\frac{K_1}{s - s_1} + \frac{K_2}{s - s_2} + \cdots + \frac{K_k}{s - s_k} + \cdots + \frac{K_n}{s - s_n} \right].$$

(7–25)

The numerators of the partial fractions, $K_1, K_2, \ldots, K_k, \ldots, K_n$, are constants which have yet to be determined.

To determine the typical constant K_k, we multiply both sides of Eq. (7–25) by the factor $(s - s_k)$:

$$(s - s_k) \frac{A(s)}{B(s)}$$

$$= \frac{1}{b_n} \left[K_1 \frac{s - s_k}{s - s_1} + K_2 \frac{s - s_k}{s - s_2} + \cdots + K_k + \cdots + K_n \frac{s - s_k}{s - s_n} \right].$$

(7–26)

The left side of Eq. (7–26) is same as the right side of Eq. (7–24) with the factor $(s - s_k)$ in the denominator cancelled. Setting $s = s_k$ in Eq. (7–26) reduces all terms on the right side to zero except the one containing K_k, hence

$$K_k = b_n \left\{ (s - s_k) \frac{A(s)}{B(s)} \right\}_{s=s_k}.$$

(7–27)

An alternative formula for K_k is obtained by noting that as s is set to equal s_k, the product $(s - s_k)[A(s)/B(s)]$ tends to be an indeterminate of the form 0/0, since both $(s - s_k)$ and $B(s)$ vanish at $s = s_k$. An indeterminate of the 0/0 type can be evaluated by differentiating the numerator and the denominator separately with respect to s and then letting s approach s_k (l'Hospital's rule). Thus, from Eq. (7–27),

$$K_k = b_n \left\{ \lim_{s \to s_k} \left[\frac{(s - s_k) A(s)}{B(s)} \right] \right\} = b_n \left\{ \lim_{s \to s_k} \frac{(d/ds)[(s - s_k) A(s)]}{(d/ds)[B(s)]} \right\},$$

or,

$$K_k = b_n \left\{ \frac{A(s)}{B'(s)} \right\}_{s=s_k}.$$

(7–28)

Of course, there is nothing wrong with evaluating K_k by Eq. (7–27), because the factor $(s - s_k)$ would have cancelled with the one in $B(s)$ before we set $s = s_k$, hence no indeterminacy would actually result. Equations (7–27) and (7–28) are two equivalent formulas for K_k.

Now that we have determined the values of the K's in the partial-fraction expansion of $A(s)/B(s)$, we are ready to write its inverse transform. In view of relation (2) in Table 6–1, we have

$$\mathcal{L}^{-1} \left[\frac{K_k}{s - s_k} \right] = K_k \epsilon^{s_k t}.$$

Hence, from Eqs. (7–25) and (7–27),

$$\mathcal{L}^{-1}\left[\frac{A(s)}{B(s)}\right] = \mathcal{L}^{-1}\left\{\frac{1}{b_n}\sum_{k=1}^{n}\frac{K_k}{s-s_k}\right\}$$

$$= \sum_{k=1}^{n}\left\{(s-s_k)\frac{A(s)}{B(s)}\right\}_{s=s_k}\epsilon^{s_k t}, \qquad t \geq 0, \qquad (7\text{–}29)$$

or, from Eqs. (7–25) and (7–28),

$$\mathcal{L}^{-1}\left[\frac{A(s)}{B(s)}\right] = \sum_{k=1}^{n}\left\{\frac{A(s_k)}{B'(s_k)}\right\}\epsilon^{s_k t}, \qquad t \geq 0. \qquad (7\text{–}30)$$

Equations (7–29) and (7–30) are two forms of *Heaviside's Expansion Theorem*, which finds the inverse Laplace transform of rational algebraic fractions as a summation of exponential terms. In view of the definition of the Laplace transformation, Eq. (6–13), the inverse transformations above have no meaning in the region $t < 0$. It can be proved that the inverse Laplace transformation is unique; that is, $f(t) = \mathcal{L}^{-1}[F(s)]$, once determined from $F(s)$ by whatever method, is the only possible $f(t)$, except at discontinuities. The proof is beyond the scope of this book.

EXAMPLE 7–1. Find $\mathcal{L}^{-1}\left[\dfrac{s+4}{2s^2+5s+3}\right]$.

Solution. We shall solve this problem in systematic steps. After we are more experienced in this type of problem, we train ourselves to follow a mental procedure. The actual numbering of the steps is, of course, not essential to the solution.

Before we proceed with the solution, we first make sure that the given $F(s)$ is a rational algebraic fraction in s and that the denominator is of a *higher* degree than the numerator, since we are now equipped to handle only those types of $F(s)$ which satisfy the stated restrictions.

Step 1. First we factor the denominator [i.e., determine the roots of the equation $B(s) = 0$].

$$B(s) = 2s^2 + 5s + 3 = 2(s^2 + \tfrac{5}{2}s + \tfrac{3}{2})$$

$$= 2(s+1)(s+\tfrac{3}{2}), \qquad b_n = 2,$$

and we find two distinct roots for $B(s) = 0$:

$$s_1 = -1 \qquad \text{and} \qquad s_2 = -\tfrac{3}{2}.$$

Step 2. Next we expand the given function of s into partial fractions and determine the constants.

$$\frac{s+4}{2s^2+5s+3} = \frac{1}{2}\left[\frac{K_1}{s+1} + \frac{K_2}{s+\frac{3}{2}}\right].$$

By Eq. (7-27),

$$K_1 = 2\left\{(s+1)\frac{s+4}{2(s+1)(s+\frac{3}{2})}\right\}_{s=-1} = \left\{\frac{s+4}{s+\frac{3}{2}}\right\}_{s=-1} = 6,$$

$$K_2 = 2\left\{(s+\frac{3}{2})\frac{s+4}{2(s+1)(s+\frac{3}{2})}\right\}_{s=-\frac{3}{2}} = \left\{\frac{s+4}{s+1}\right\}_{s=-\frac{3}{2}} = -5.$$

Hence

$$\frac{s+4}{2s^2+5s+3} = \frac{1}{2}\left[\frac{6}{s+1} - \frac{5}{s+\frac{3}{2}}\right].$$

Step 3. Finally, we find the inverse transform:

$$\mathcal{L}^{-1}\left[\frac{s+4}{2s^2+5s+3}\right] = \frac{1}{2}\mathcal{L}^{-1}\left[\frac{6}{s+1} - \frac{5}{s+\frac{3}{2}}\right]$$

$$= \frac{1}{2}(6\epsilon^{-t} - 5\epsilon^{-3t/2}).$$

EXAMPLE 7-2. Find $\mathcal{L}^{-1}\left[\dfrac{s\epsilon^{-2s}}{s^2+2s+5}\right]$.

Solution. The given function is *not* a rational function of s, hence we *cannot* expand the whole function in partial fractions in the following fashion:

$$\frac{s\epsilon^{-2s}}{s^2+2s+5} = \frac{C_1}{s-s_1} + \frac{C_2}{s-s_2}.$$

However, from Chapter 6 [relation (1), Table 6–2] we know that multiplication by a factor ϵ^{-2s} simply amounts to a shift in the independent variable from t to $(t-2)$. We can therefore first perform

$$\mathcal{L}^{-1}\left[\frac{s}{s^2+2s+5}\right]$$

and then apply the shift in t.

Step 1. First we factor the denominator [i.e., determine the roots of the equation $B(s) = 0$].

$$B(s) = s^2 + 2s + 5 = 0, \qquad b_n = 1,$$

$$s = \tfrac{1}{2}(-2 \pm \sqrt{4-20}) = -1 \pm j2.$$

The two distinct roots are

$$s_1 = -1 + j2 \quad \text{and} \quad s_2 = -1 - j2.$$

Here we have a pair of complex roots, but they are distinct; normal procedures apply.

Step 2. Next we expand into partial fractions. Let us use Eq. (7–28) in this problem.

$$B'(s) = 2s + 2 = 2(s + 1),$$

$$K_1 = \left\{\frac{s}{2(s + 1)}\right\}_{s=s_1=-1+j2} = \frac{-1 + j2}{j4} = \frac{1}{4}(2 + j1),$$

$$K_2 = \left\{\frac{s}{2(s + 1)}\right\}_{s=s_2=-1-j2} = \frac{-1 - j2}{-j4} = \frac{1}{4}(2 - j1).$$

Actually, knowing that $s_2 = s_1^*$ (complex roots appear in conjugate pairs), we also know that $K_2 = K_1^*$. Hence once K_1 has been evaluated, there is no need to evaluate K_2 by formulas. We have

$$\frac{s}{s^2 + 2s + 5} = \frac{1}{4}\left[\frac{2 + j1}{s + 1 - j2} + \frac{2 - j1}{s + 1 + j2}\right].$$

Step 3. Now we find

$$\mathcal{L}^{-1}\left[\frac{s}{s^2 + 2s + 5}\right] = \frac{1}{4}\mathcal{L}^{-1}\left[\frac{2 + j1}{s + 1 - j2} + \frac{2 - j1}{s + 1 + j2}\right]$$

$$= \frac{1}{4}[(2 + j1)\epsilon^{(-1+j2)t} + (2 - j1)\epsilon^{(-1-j2)t}]$$

$$= \frac{1}{2}\epsilon^{-t}(2 \cos 2t - \sin 2t).$$

Step 4. Finally, we find $\mathcal{L}^{-1}\left[\dfrac{s\epsilon^{-2s}}{s^2 + 2s + 5}\right].$

By the Shifting Theorem [Eq. (6–36), or relation (1) in Table 6–2], we have

$$\mathcal{L}^{-1}\left[\frac{s\epsilon^{-2s}}{s^2 + 2s + 5}\right] = \frac{1}{2}\epsilon^{-(t-2)}[2 \cos 2(t - 2) - \sin 2(t - 2)]U(t - 2).$$

This function is zero in the region $t < 2$.

At this time we shall demonstrate a short cut for finding the inverse transform of a function whose denominator is a polynomial in s of the second degree. We see from the above example that when $B(s)$ has complex roots the procedure of expanding into partial fractions, determining the constants, and recombining the exponential terms becomes quite tedious. The trick here is to complete a square in the denominator and to consider a pair of complex roots as a unit, instead of treating them separately. Thus, for the fraction in Example 7–2, we have

$$\frac{s}{s^2 + 2s + 5} = \frac{s}{(s^2 + 2s + 1) + 4} = \frac{s}{(s + 1)^2 + 2^2},$$

$$\mathcal{L}^{-1}\left[\frac{s}{s^2 + 2s + 5}\right] = \mathcal{L}^{-1}\left[\frac{(s + 1)}{(s + 1)^2 + 2^2} - \frac{1}{(s + 1)^2 + 2^2}\right]$$

$$= \epsilon^{-t} \cos 2t - \tfrac{1}{2}\epsilon^{-t} \sin 2t$$

$$= \tfrac{1}{2}\epsilon^{-t}(2 \cos 2t - \sin 2t),$$

by relations (3), (4), and (7) in Table 6–1. The result is the same as before but the work involved is much less.

This method of completing a square when the denominator $B(s)$ is in a quadratic form is not limited to cases where $B(s) = 0$ has complex roots, although the advantage is most obvious in such cases. Take the problem in Example 7–1 for illustration:

$$\frac{s + 4}{2s^2 + 5s + 3} = \frac{s + 4}{2(s^2 + \tfrac{5}{2}s + \tfrac{3}{2})}$$

$$= \frac{s + 4}{2[(s^2 + \tfrac{5}{2}s + \tfrac{25}{16}) + (\tfrac{3}{2} - \tfrac{25}{16})]} = \frac{s + 4}{2[(s + \tfrac{5}{4})^2 - (\tfrac{1}{4})^2]}$$

$$= \frac{1}{2}\left[\frac{s + \tfrac{5}{4}}{(s + \tfrac{5}{4})^2 - (\tfrac{1}{4})^2} + \frac{\tfrac{11}{4}}{(s + \tfrac{5}{4})^2 - (\tfrac{1}{4})^2}\right].$$

The inverse Laplace transforms of the above expressions can be readily written if we recall relations (5), (6), and (7) in Table 6–1. Hence

$$\mathcal{L}^{-1}\left[\frac{s + 4}{2s^2 + 5s + 3}\right] = \tfrac{1}{2}[\epsilon^{-5t/4} \cosh (t/4) + 11\epsilon^{-5t/4} \sinh (t/4)]$$

$$= \tfrac{1}{2}(6\epsilon^{-t} - 5\epsilon^{-3t/2}),$$

as before. Doing the problem this way, we have even avoided the necessity for partial fraction expansion and the associated steps for evaluating the constants. However, this method is not useful when $B(s)$ is of a degree higher than 2.

II. $m < n$, $B(s) = 0$ *has multiple roots.* Let us assume now that the denominator $B(s)$ in Eq. (7–22) has a root s_1 of multiplicity p; that is, the factor $(s - s_1)$ repeats p times:

$$B(s) = b_n(s - s_1)^p(s - s_{p+1})(s - s_{p+2}) \cdots (s - s_n). \qquad (7\text{–}31)$$

It is quite obvious that

$$\frac{A(s)}{B(s)} \neq \frac{1}{b_n} \left[\frac{K_1}{s - s_1} + \frac{K_2}{s - s_1} + \cdots + \frac{K_p}{s - s_1} \right.$$

$$\left. + \frac{K_{p+1}}{s - s_{p+1}} + \frac{K_{p+2}}{s - s_{p+2}} + \cdots + \frac{K_n}{s - s_n} \right], \quad (7\text{–}32)$$

because the first p terms on the right side can be combined and are in reality a single fraction:

$$\frac{K_1}{s - s_1} + \frac{K_2}{s - s_1} + \cdots + \frac{K_p}{s - s_1} = \frac{K}{s - s_1},$$

where

$$K = K_1 + K_2 + \cdots + K_p.$$

Hence the right side of (7–32) cannot restore the $(s - s_1)^p$ factor in $B(s)$. We must modify the expansion of $A(s)/B(s)$ as follows:

$$\frac{A(s)}{B(s)} = \frac{1}{b_n} \left[\frac{K_{1p}}{(s - s_1)^p} + \frac{K_{1(p-1)}}{(s - s_1)^{p-1}} + \cdots + \frac{K_{12}}{(s - s_1)^2} + \frac{K_{11}}{s - s_1} \right.$$

$$\left. + \frac{K_{p+1}}{s - s_{p+1}} + \frac{K_{p+2}}{s - s_{p+2}} + \cdots + \frac{K_n}{s - s_n} \right], \quad (7\text{–}33)$$

where the first p terms are the partial-fraction expansion corresponding to the root s_1 of multiplicity p. Either Eq. (7–27) or Eq. (7–28) can be used to determine the constants K_{p+1}, K_{p+2}, $\ldots$, K_n corresponding to the distinct roots s_{p+1}, s_{p+2}, $\ldots$, s_n. We shall now see how the first p constants K_{1p}, $K_{1(p-1)}$, $\ldots$, K_{12}, K_{11}, can be found.

To determine K_{1p}, we multiply both sides of Eq. (7–33) by $(s - s_1)^p$:

$$(s - s_1)^p \frac{A(s)}{B(s)} = \frac{1}{b_n} [K_{1p} + K_{1(p-1)}(s - s_1) + \cdots$$

$$+ K_{12}(s - s_1)^{p-2} + K_{11}(s - s_1)^{p-1}]$$

$$+ \frac{(s - s_1)^p}{b_n} \left[\frac{K_{p+1}}{s - s_{p+1}} + \frac{K_{p+2}}{s - s_{p+2}} + \cdots + \frac{K_n}{s - s_n} \right]. \quad (7\text{–}34)$$

Setting $s = s_1$ in Eq. (7–34), we obtain directly

$$K_{1p} = b_n \left[(s - s)^p \frac{A(s)}{B(s)} \right]_{s=s_1}. \quad (7\text{–}35)$$

If we now differentiate both sides of Eq. (7–34) with respect to s, we have

$$\frac{d}{ds}\left[(s - s_1)^p \frac{A(s)}{B(s)}\right] = \frac{1}{b_n}\left[K_{1(p-1)} + \cdots\right.$$

$$+ K_{12}(p - 2)(s - s_1)^{p-3} + K_{11}(p - 1)(s - s_1)^{p-2}]$$

$$+ \frac{1}{b_n}\frac{d}{ds}\left\{(s - s_1)^p \left[\frac{K_{p+1}}{s - s_{p+1}} + \frac{K_{p+2}}{s - s_{p+2}} + \cdots + \frac{K_n}{s - s_n}\right]\right\}. \quad (7\text{–}36)$$

Setting $s = s_1$ in Eq. (7–36), we find

$$K_{1(p-1)} = b_n \left\{\frac{d}{ds}\left[(s - s_1)^p \frac{A(s)}{B(s)}\right]\right\}_{s=s_1}. \quad (7\text{–}37)$$

By extending this process, the following general formula is obtained:

$$K_{1k} = \frac{b_n}{(p - k)!}\left\{\frac{d^{(p-k)}}{ds^{(p-k)}}\left[(s - s_1)^p \frac{A(s)}{B(s)}\right]\right\}_{s=s_1}. \quad (7\text{–}38)$$

Once the coefficient K_{1k} is determined, we can find the inverse transform of the typical term by relations (7) and (8) in Table 6–1:

$$\mathcal{L}^{-1}\left[\frac{K_{1k}}{(s - s_1)^k}\right] = \frac{K_{1k}}{(k - 1)!} t^{k-1}\epsilon^{s_1 t}. \quad (7\text{–}39)$$

Therefore, from Eq. (7–33), we have

$$\mathcal{L}^{-1}\left[\frac{A(s)}{B(s)}\right] = \frac{1}{b_n}\left[\frac{K_{1p}}{(p - 1)!} t^{p-1} + \frac{K_{1(p-1)}}{(p - 2)!} t^{p-2} + \cdots\right.$$

$$\left. + \frac{K_{12}}{1!} t + K_{11}\right]\epsilon^{s_1 t}$$

$$+ \frac{1}{b_n}\sum_{q=p+1}^{n} K_q\epsilon^{s_q t}, \quad t \geq 0. \quad (7\text{–}40)$$

EXAMPLE 7–3. Find $\mathcal{L}^{-1}\left[\dfrac{s + 2}{s(s + 1)^2(s + 3)}\right]$.

Solution. The denominator $B(s)$ has four roots.

$$\text{2 simple roots:} \quad s = 0 \quad \text{and} \quad s = -3,$$
$$\text{1 double root:} \quad s = -1.$$

Partial-fraction expansion ($b_n = 1$):

$$\frac{A(s)}{B(s)} = \frac{s + 2}{s(s + 1)^2(s + 3)} = \frac{K_1}{s} + \left[\frac{K_{22}}{(s + 1)^2} + \frac{K_{21}}{s + 1}\right] + \frac{K_3}{s + 3},$$

$$K_1 = \left\{ s\, \frac{A(s)}{B(s)} \right\}_{s=0} = \left\{ \frac{s+2}{(s+1)^2(s+3)} \right\}_{s=0} = \frac{2}{3},$$

$$K_{22} = \left\{ (s+1)^2\, \frac{A(s)}{B(s)} \right\}_{s=-1} = \left\{ \frac{s+2}{s(s+3)} \right\}_{s=-1} = -\frac{1}{2},$$

$$K_{21} = \left\{ \frac{d}{ds} \left[\frac{s+2}{s(s+3)} \right] \right\}_{s=-1} = \left\{ \frac{s(s+3) - (s+2)(2s+3)}{s^2(s+3)^2} \right\}_{s=-1}$$

$$= -\frac{3}{4},$$

$$K_3 = \left\{ (s+3)\, \frac{A(s)}{B(s)} \right\}_{s=-3} = \left\{ \frac{s+2}{s(s+1)^2} \right\}_{s=-3} = \frac{1}{12}.$$

Hence

$$\frac{s+2}{s(s+1)^2(s+3)} = \frac{2}{3s} - \frac{1}{2(s+1)^2} - \frac{3}{4(s+1)} + \frac{1}{12(s+3)}$$

and

$$\mathcal{L}^{-1}\left[\frac{s+2}{s(s+1)^2(s+3)} \right] = \frac{2}{3} - \frac{1}{2}\left[t + \frac{3}{2} \right]\epsilon^{-t} + \frac{1}{12}\,\epsilon^{-3t}, \qquad t \geq 0.$$

EXAMPLE 7–4. Find $\mathcal{L}^{-1}\left[\dfrac{1}{3s^2(s^2+4)} \right]$.

Solution. The denominator $B(s)$ has four roots.

$$\begin{aligned} &\text{1 double root:} && s = 0, \\ &\text{2 conjugate roots:} && s = +j2 \quad \text{and} \quad s = -j2. \end{aligned}$$

If we follow the normal procedure, we would write the partial-fraction expansion as ($b_n = 3$):

$$\frac{1}{3s^2(s^2+4)} = \frac{1}{3}\left[\frac{K_{12}}{s^2} + \frac{K_{11}}{s} + \frac{K_2}{s-j2} + \frac{K_3}{s+j2} \right], \qquad (7\text{-}41)$$

where the constants K_2 and K_3 could be determined with the aid of Eq. (7–27) or Eq. (7–28), and K_{12} and K_{11} could be determined using Eq. (7–38). But in view of Example 7–2, we expect that the procedure will be simplified if we keep the pair of conjugate roots together as a unit. We shall demonstrate here how this problem can be solved simply without having to deal with complex numbers or to remember any formulas. We write the partial-fraction expansion as follows:

$$\frac{1}{3s^2(s^2+4)} = \frac{1}{3}\left[\frac{K_{12}}{s^2} + \frac{K_{11}}{s} + \frac{C_1 s + C_2}{s^2+4} \right]. \qquad (7\text{-}42)$$

Note that the numerator of the last term in Eq. (7–42) is written as $C_1s + C_2$, instead of just a single constant, because it should represent the last two terms of Eq. (7–41):

$$\frac{K_2}{s - j2} + \frac{K_3}{s + j2} = \frac{(K_2 + K_3)s + j2(K_2 - K_3)}{s^2 + 4}.$$

We have simply written C_1 for $(K_2 + K_3)$ and C_2 for $j2(K_2 - K_3)$. It is necessary to determine *four* constants with a *fourth-degree* $B(s)$. To evaluate the four constants K_{12}, K_{11}, C_1, and C_2 without using the formulas, we multiply both sides of Eq. (7–42) by $B(s) = 3s^2(s^2 + 4)$:

$$1 = K_{12}(s^2 + 4) + K_{11}s(s^2 + 4) + (C_1s + C_2)s^2$$
$$= (K_{11} + C_1)s^3 + (K_{12} + C_2)s^2 + 4K_{11}s + 4K_{12}.$$

Since the above equation has to hold for all values of s, the coefficients for like powers of s on both sides of the equation must be equal. Thus,

$$K_{11} + C_1 = 0, \qquad K_{12} + C_2 = 0,$$
$$4K_{11} = 0, \qquad 4K_{12} = 1.$$

From these four equations, the four constants can be easily determined:

$$K_{12} = \tfrac{1}{4}, \qquad K_{11} = 0, \qquad C_1 = 0, \qquad C_2 = -\tfrac{1}{4}.$$

Hence, from Eq. (7–42),

$$\frac{1}{3s^2(s^2 + 4)} = \frac{1}{3}\left[\frac{1}{4s^2} - \frac{1}{4(s^2 + 4)}\right]. \tag{7–43}$$

From relations (3) and (8) in Table 6–1, we find

$$\mathcal{L}^{-1}\left[\frac{1}{3s^2(s^2 + 4)}\right] = \mathcal{L}^{-1}\frac{1}{12}\left[\frac{1}{s^2} - \frac{1}{s^2 + 4}\right]$$
$$= \frac{1}{12}\left[t - \frac{1}{2}\sin 2t\right], \qquad t \geq 0.$$

We have, in fact, used the method of undetermined coefficients to evaluate the constants in this example. This method is straightforward and simple when the number of constants is not more than, say, three or four; there is no need for remembering special formulas. However, this method becomes laborious when the number of constants to be determined is large.

It is not possible to formulate general rules that will indicate which method is simplest to use for a given problem, and there is actually no such need so long as we solve the problem correctly. By and large, only experience and alertness will enable obtaining the solution with the least labor. For instance, we might have noticed that the given function in Example 7–4 is a function of s^2; hence we would not expect s to appear alone in the partial-fraction expansion. As a matter of fact, we could have let $s^2 = S$ and expanded the given function as follows:

$$\frac{1}{3s^2(s^2+4)} = \frac{1}{3S(S+4)} = \frac{1}{3}\left[\frac{A_1}{S} + \frac{A_2}{S+4}\right], \qquad (7\text{–}44)$$

where the constants A_1 and A_2 can be found very easily by the standard formula (7–27) to be $\frac{1}{4}$ and $-\frac{1}{4}$ respectively. Therefore we obtain from Eq. (7–44) directly

$$\frac{1}{3s^2(s^2+4)} = \frac{1}{3}\left[\frac{1}{4S} - \frac{1}{4(S+4)}\right] = \frac{1}{3}\left[\frac{1}{4s^2} - \frac{1}{4(s^2+4)}\right],$$

which is the same as Eq. (7–43).

7–4 Analysis of system response. As was pointed out at the beginning of this chapter, three major steps are involved in the analysis of linear systems by the method of Laplace transforms, namely, the setting up of the differential or integro-differential equations that describe the system, the application of Laplace transformation to all terms of the equations, and the inverse transformation of the response transform. The response transform is equal to the product of the system transfer function and the total excitation transform. The transfer function, which we call $H(s)$, is the reciprocal of the characteristic function of the differential or integro-differential equation obtained by replacing D's by s's in its operator part; it is slightly more complicated for simultaneous equations (see Section 2–6). The total excitation transform, which we call $E_0(s)$, consists of the Laplace transform of the applied excitation function and the contributions due to initial conditions. We shall illustrate the procedure with a number of examples.

EXAMPLE 7–5. A rectangular voltage pulse of unit height and duration T is applied to a series R-C combination at $t = 0$. Determine the voltage across the capacitance C as a function of time.

Solution. The problem is illustrated in Fig. 7–2. First, let us write the expression for the input rectangular voltage pulse:

$$v_i(t) = U(t) - U(t - T). \qquad (7\text{–}45)$$

Next we write the differential equation of this simple one-loop circuit, using Kirchhoff's voltage law. Since the desired response is $v_C = q/C$,

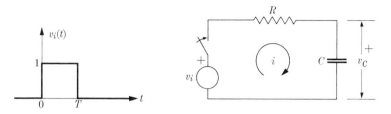

FIG. 7–2. A rectangular pulse applied to an R-C circuit.

and $i = dq/dt = C(dv_C/dt)$, we write the equation with v_C as the dependent variable:

$$RC \frac{dv_C}{dt} + v_C = v_i(t) = U(t) - U(t - T). \tag{7-46}$$

We now apply Laplace transformation to Eq. (7–46). Calling

$$\mathcal{L}[v_C(t)] = V_C(s), \tag{7-47}$$

we have

$$RC[sV_C(s) - v_C(0+)] + V_C(s) = \frac{1}{s}(1 - \epsilon^{-Ts}).$$

With an initially uncharged C, $v_C(0+) = 0$. Hence

$$(RCs + 1)V_C(s) = \frac{1}{s}(1 - \epsilon^{-Ts})$$

or

$$V_C(s) = \frac{1 - \epsilon^{-Ts}}{s(RCs + 1)}. \tag{7-48}$$

Equation (7–48) can also be obtained by noting that the response transform $V_C(s)$ is equal to $H(s)E_0(s)$:

$$V_C(s) = H(s)E_0(s), \tag{7-49}$$

where the transfer function

$$H(s) = \frac{1/sC}{R + 1/sC} = \frac{1}{RCs + 1} \tag{7-50}$$

is the ratio of the output voltage transform and the input voltage transform, and the total excitation transform $E_0(s)$ is equal to the input voltage transform in the absence of initial conditions:

$$E_0(s) = V_i(s) = \frac{1}{s}(1 - \epsilon^{-Ts}). \tag{7-51}$$

Substitution of Eqs. (7–50) and (7–51) in Eq. (7–49) yields Eq. (7–48) directly.

Now we must find the inverse transform of $V_C(s)$ from Eq. (7–48). Remembering that the factor ϵ^{-Ts} amounts to a simple shift in t in the inverse transformation, we proceed as follows:

$$\frac{1}{s(RCs + 1)} = \frac{1}{RCs(s + 1/RC)} = \frac{1}{RC}\left[\frac{K_1}{s} + \frac{K_2}{s + 1/RC}\right],$$

where

$$K_1 = \left[\frac{1}{s + 1/RC}\right]_{s=0} = RC,$$

$$K_2 = \left[\frac{1}{s}\right]_{s=-1/RC} = -RC.$$

Hence

$$\mathcal{L}^{-1}\left[\frac{1}{s(RCs + 1)}\right] = \mathcal{L}^{-1}\left[\frac{1}{s} - \frac{1}{s + 1/RC}\right] = 1 - \epsilon^{-t/RC}. \quad (7\text{–}52)$$

This would be the solution for v_C if the input voltage was a unit step function. From Eqs. (7–48) and (7–52), we have finally

$$v_C(t) = \mathcal{L}^{-1}\left[\frac{1 - \epsilon^{-Ts}}{s(RCs + 1)}\right]$$

$$= (1 - \epsilon^{-t/RC})U(t) - (1 - \epsilon^{-(t-T/RC)})U(t - T). \quad (7\text{–}53)$$

A plot of $v_C(t)$ given by Eq. (7–53) is roughly shown in Fig. 7–3, the exact shape of $v_C(t)$ being dependent upon the relative values of T and RC.

EXAMPLE 7–6. The circuit in Fig. 7–4 is initially under steady-state conditions. The switch is opened at $t = 0$. Find the voltage across the inductance L as a function of t.

Solution. Let us first determine the steady-state conditions just before the switch is opened. As shown in Fig. 7–5, the battery will be sending a

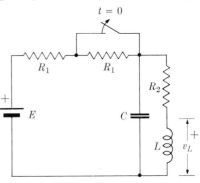

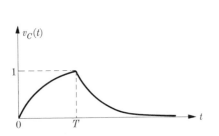

FIG. 7–3. Plot of $v_C(t)$ vs. t for problem shown in Fig. 7–2.

FIG. 7–4.   Circuit initially under steady state.   (Switch is opened at $t = 0$.)

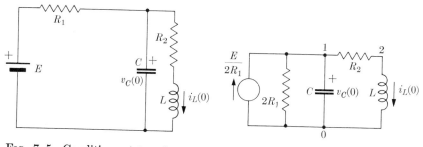

FIG. 7–5. Conditions at $t = 0-$.   FIG. 7–6. Conditions at $t = 0+$.

direct current through the path containing R_1, R_2, and L, and the capacitor C will have been charged to a voltage equal to that appearing across R_2, the inductance L being effectively a short circuit to direct current. We have

$$v_C(0-) = \frac{R_2 E}{R_1 + R_2} = v_C(0+), \tag{7–54}$$

$$i_L(0-) = \frac{E}{R_1 + R_2} = i_L(0+). \tag{7–55}$$

After the switch is opened, an additional R_1 is put in the circuit. Since the voltage across the inductance L is required, it is more direct to use node analysis. We redraw the circuit as in Fig. 7–6, where the combination of the voltage source E in series with the resistance $2R_1$ has been changed to a current source $E/2R_1$ in parallel with $2R_1$. Two independent nodes, 1 and 2, are in evidence. Calling the voltages between these nodes and the datum node, 0, respectively v_1 and v_2, we write two node equations, using Kirchhoff's current law, as follows:

Node 1: $$\left(\frac{1}{2R_1} + \frac{1}{R_2}\right) v_1 + C \frac{dv_1}{dt} - \frac{1}{R_2} v_2 = \frac{E}{2R_1}. \tag{7–56}$$

Node 2: $$-\frac{1}{R_2} v_1 + \frac{1}{R_2} v_2 + \frac{1}{L} \int_0^t v_2 \, dt + i_L(0+) = 0. \tag{7–57}$$

If we let

$$\mathcal{L}[v_1(t)] = V_1(s) \qquad \text{and} \qquad \mathcal{L}[v_2(t)] = V_2(s),$$

then Laplace transformation of Eqs. (7–56) and (7–57) yields

$$\left(\frac{1}{2R_1} + \frac{1}{R_2}\right) V_1(s) + C[sV_1(s) - v_1(0+)] - \frac{1}{R_2} V_2(s) = \frac{E}{2R_1 s}$$

or

$$\left(\frac{1}{2R_1} + \frac{1}{R_2} + Cs\right) V_1(s) - \frac{1}{R_2} V_2(s) = \frac{E}{2R_1 s} + C \frac{R_2 E}{R_1 + R_2}, \tag{7–58}$$

and

$$-\frac{1}{R_2} V_1(s) + \left(\frac{1}{R_2} + \frac{1}{Ls}\right) V_2(s) = -\frac{E}{(R_1 + R_2)s}. \qquad (7\text{--}59)$$

We can now solve for $V_2(s)$ from Eqs. (7–58) and (7–59) algebraically. By Cramer's rule, we have

$$V_2(s) = \frac{\begin{vmatrix} \dfrac{1}{2R_1} + \dfrac{1}{R_2} + Cs & \dfrac{E}{2R_1 s} + \dfrac{CR_2E}{R_1 + R_2} \\[2mm] -\dfrac{1}{R_2} & -\dfrac{E}{(R_1 + R_2)s} \end{vmatrix}}{\begin{vmatrix} \dfrac{1}{2R_1} + \dfrac{1}{R_2} + Cs & -\dfrac{1}{R_2} \\[2mm] -\dfrac{1}{R_2} & \dfrac{1}{R_2} + \dfrac{1}{Ls} \end{vmatrix}}$$

$$= \frac{-\dfrac{E}{2R_2(R_1 + R_2)s}}{\dfrac{C}{R_2}\left[s + \left(\dfrac{1}{2CR_1} + \dfrac{R_2}{L}\right) + \dfrac{1}{LCs}\left(1 + \dfrac{R_2}{2R_1}\right)\right]}$$

$$= -\frac{E}{2C(R_1 + R_2)\left[s^2 + \left(\dfrac{1}{2CR_1} + \dfrac{R_2}{L}\right)s + \dfrac{1}{LC}\left(1 + \dfrac{R_2}{2R_1}\right)\right]}$$

$$= -\frac{E}{2C(R_1 + R_2)}\left[\frac{1}{(s - s_1)(s - s_2)}\right], \qquad (7\text{--}60)$$

where

$$s_1 = -\frac{1}{2}\left(\frac{1}{2CR_1} + \frac{R_2}{L}\right) + \sqrt{\frac{1}{4}\left(\frac{1}{2CR_1} - \frac{R_2}{L}\right)^2 - \frac{1}{LC}}, \qquad (7\text{--}61)$$

$$s_2 = -\frac{1}{2}\left(\frac{1}{2CR_1} + \frac{R_2}{L}\right) - \sqrt{\frac{1}{4}\left(\frac{1}{2CR_1} - \frac{R_2}{L}\right)^2 - \frac{1}{LC}}. \qquad (7\text{--}62)$$

In order to find the inverse transform of $V_2(s)$, we expand Eq. (7–60) in partial fractions:

$$V_2(s) = -\frac{E}{2C(R_1 + R_2)}\left[\frac{K_1}{s - s_1} + \frac{K_2}{s - s_2}\right],$$

$$K_1 = \left[\frac{1}{s - s_2}\right]_{s=s_1} = \frac{1}{s_1 - s_2} = \frac{1}{\sqrt{\left(\dfrac{1}{2CR_1} - \dfrac{R_2}{L}\right)^2 - \dfrac{4}{LC}}}, \qquad (7\text{--}63)$$

$$K_2 = \left[\frac{1}{s - s_1}\right]_{s=s_2} = \frac{1}{s_2 - s_1} = -K_1. \qquad (7\text{--}64)$$

Hence

$$V_2(s) = -\frac{E}{2C(R_1 + R_2)(s_1 - s_2)} \left[\frac{1}{s - s_1} - \frac{1}{s - s_2}\right]. \qquad (7\text{–}65)$$

Finally, we have

$$v_2(t) = v_L(t) = \mathcal{L}^{-1}[V_2(s)] = -\frac{E}{2C(R_1 + R_2)(s_1 - s_2)}(\epsilon^{s_1 t} - \epsilon^{s_2 t}),$$

$$t \geq 0, \qquad (7\text{–}66)$$

where the values of s_1 and s_2 are given in Eqs. (7–61) and (7–62) respectively.

EXAMPLE 7–7. A certain cushioned package is to be dropped from a height $h = 10$ ft. The package can be represented by the schematic diagram of Fig. 7–7. Determine the motion of the mass M. Assume that the package falls onto the ground with no rebound.

Solution. The v-f analogous electrical circuit for this mechanical system can be easily drawn as in Fig. 7–8, where the original symbols for the mechanical elements have been retained. Note that the only externally applied force on the system is the static weight Mg of the mass M.

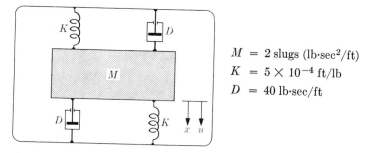

$M = 2$ slugs (lb·sec^2/ft)
$K = 5 \times 10^{-4}$ ft/lb
$D = 40$ lb·sec/ft

FIG. 7–7. Schematic diagram of a cushioned package.

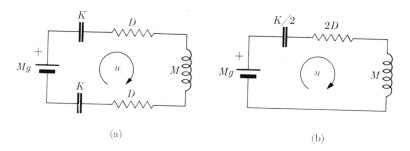

(a)   (b)

FIG. 7–8. v-f analogous electrical circuits for the mechanical system in Fig. 7–7. (a) Electrical circuit analogous to the mechanical system in Fig. 7–7. (b) Simplified analogous electrical circuit.

We write the equation of motion in terms of velocity u as follows:

$$M \frac{du}{dt} + (2D)u + \frac{2}{K}\left[\int_0^t u\,dt + x(0)\right] = Mg. \qquad (7\text{-}67)$$

Equation (7–67) can be solved for u, and displacement and acceleration can then be derived from it. Alternatively, we. can write the equation of motion in terms of displacement x:

$$M \frac{d^2 x}{dt^2} + 2D \frac{dx}{dt} + \frac{2}{K} x = Mg. \qquad (7\text{-}68)$$

Now the initial conditions at $t = 0$ (the instant at which the package first touches the ground) are

$$x(0) = 0, \qquad x'(0) = u(0) = \sqrt{2gh}. \qquad (7\text{-}69)$$

Laplace transformation of Eq. (7–68), taking due account of the initial conditions, yields

$$\left(Ms^2 + 2Ds + \frac{2}{K}\right) X(s) = M\sqrt{2gh} + \frac{Mg}{s} \qquad (7\text{-}70)$$

or

$$X(s) = \frac{\sqrt{2gh}}{s^2 + 2(D/M)s + (2/MK)} + \frac{g}{s[s^2 + 2(D/M)s + (2/MK)]}. \qquad (7\text{-}71)$$

Inserting the given numerical values in Eq. (7–71), we have

$$X(s) = \frac{\sqrt{2 \times 32.2 \times 10}}{s^2 + 40s + 2000} + \frac{32.2}{2000}\left[\frac{1}{s} - \frac{s + 40}{s^2 + 40s + 2000}\right]$$

$$= \frac{25.4}{(s + 20)^2 + 40^2} + 0.016\left[\frac{1}{s} - \frac{s + 40}{(s + 20)^2 + 40^2}\right]. \qquad (7\text{-}72)$$

Inverse transformation of Eq. (7–72) gives the displacement function, $x(t)$, in feet:

$$x(t) = \mathcal{L}^{-1}[X(s)]$$

$$= \frac{25.4}{40}\,\epsilon^{-20t}\sin 40t + 0.016[1 - \epsilon^{-20t}(\cos 40t + \tfrac{1}{2}\sin 40t)]$$

$$= 0.016 + \epsilon^{-20t}(0.627 \sin 40t - 0.016 \cos 40t)$$

$$= 0.016 + 0.627\epsilon^{-20t}\sin(40t - 1.46°)\ \text{ft}, \qquad t \geq 0. \qquad (7\text{-}73)$$

Both velocity and acceleration functions can be obtained from $x(t)$ in Eq. (7–73) by differentiation. If we had neglected the static deflection due to the weight of the mass M, we would have been left with only the first term in Eqs. (7–71) and (7–72), the inverse transformation of which gives

$$x(t) \simeq 0.635\epsilon^{-20t} \sin 40t \text{ ft}, \qquad t \geq 0. \tag{7–74}$$

Equation (7–74) is, in fact, a good approximation of Eq. (7–73).

7–5 Initial and final values. Frequently it is desirable to determine the initial and/or final values of the response function before completing the solution of a problem. Even when these values can be determined by inspection, the possibility of checking them from the response transform is of great value, because the process of carrying out the inverse transformation of the response transform is, in general, the most tedious part of a problem. Hence it is always a good idea to check the correctness of the response transform against known initial or final values before embarking on the task of performing the inverse transformation. We can do this with the aid of the initial-value and final-value theorems.

INITIAL-VALUE THEOREM. *If $f(t)$ and its first derivative are Laplace transformable, then the initial value of $f(t)$ is*

$$f(0+) = \lim_{t \to 0+} f(t) = \lim_{s \to \infty} sF(s). \tag{7–75}$$

Proof. To prove this theorem, we start with the transform of the first derivative of $f(t)$. From Eq. (6–59), we have

$$\int_0^\infty \left[\frac{df(t)}{dt} \right] \epsilon^{-st} \, dt = sF(s) - f(0+).$$

Now let s approach ∞:

$$\lim_{s \to \infty} \int_0^\infty \left[\frac{df(t)}{dt} \right] \epsilon^{-st} \, dt = \lim_{s \to \infty} [sF(s) - f(0+)]. \tag{7–76}$$

By the hypothesis of $df(t)/dt$ being Laplace transformable, the integral on the left side of Eq. (7–76) exists. Moreover, since s is not a function of t, it is allowable to let $s \to \infty$ before integrating. The left side of Eq. (7–76) then vanishes, and

$$0 = \lim_{s \to \infty} [sF(s) - f(0+)]$$

or

$$f(0+) = \lim_{t \to 0+} f(t) = \lim_{s \to \infty} sF(s). \qquad \text{Q.E.D.}$$

Let us apply this theorem to the response transform $V_2(s)$ of Example 7–6, as given in Eq. (7–60):

$$V_2(s) = -\frac{E}{2C(R_1 + R_2)} \left[\frac{1}{(s - s_1)(s - s_2)}\right].$$

We have

$$v_2(0+) = \lim_{s \to \infty} s V_2(s) = \lim_{s \to \infty} \left[-\frac{Es}{2C(R_1 + R_2)(s - s_1)(s - s_2)}\right] = 0,$$

which is known to be correct, since the voltage across the inductance is initially 0. If we could not get a correct check at this point, we certainly would not want to wait until the complete, but wrong, solution is obtained before we go back and try to correct the response transform!

The Initial-Value Theorem is useful in determining not only the initial value of $f(t)$ but also that of its derivatives. This is illustrated in the following example.

EXAMPLE 7–8. Given

$$F(s) = A \frac{(s + \alpha) \sin \theta + \beta \cos \theta}{(s + \alpha)^2 + \beta^2}, \tag{7–77}$$

find the values of $f(t)$ and its slope at $t = 0+$.

Solution. The initial value of $f(t)$ is

$$f(0+) = \lim_{s \to \infty} sF(s) = \lim_{s \to \infty} A \left[\frac{s(s + \alpha) \sin \theta + s\beta \cos \theta}{(s + \alpha)^2 + \beta^2}\right]$$

$$= \lim_{s \to \infty} A \left[\frac{(1 + \alpha/s) \sin \theta + (\beta/s) \cos \theta}{(1 + \alpha/s)^2 + (\beta/s)^2}\right] = A \sin \theta. \tag{7–78}$$

The initial value of $df(t)/dt$ is

$$\left[\frac{df(t)}{dt}\right]_{t=0+} = \lim_{s \to \infty} s\mathcal{L}\left[\frac{df(t)}{dt}\right] = \lim_{s \to \infty} s[sF(s) - f(0+)]$$

$$= \lim_{s \to \infty} As \left[\frac{(s + \alpha)s \sin \theta + \beta s \cos \theta}{(s + \alpha)^2 + \beta^2} - \sin \theta\right]$$

$$= \lim_{s \to \infty} A \left[\frac{\beta s^2 \cos \theta - \alpha s^2 \sin \theta - \alpha^2 s \sin \theta - \beta^2 s \sin \theta}{(s + \alpha)^2 + \beta^2}\right]$$

$$= A(\beta \cos \theta - \alpha \sin \theta). \tag{7–79}$$

The correctness of the initial values of both $f(t)$ and its first derivative can be verified by recalling (see Example 6–1 in Section 6–3) that

$$f(t) = \mathcal{L}^{-1}\left[A \frac{(s + \alpha) \sin \theta + \beta \cos \theta}{(s + \alpha)^2 + \beta^2}\right] = A\epsilon^{-\alpha t} \sin(\beta t + \theta) \tag{7–80}$$

and

$$\frac{df(t)}{dt} = A\epsilon^{-\alpha t}[\beta \cos(\beta t + \theta) - \alpha \sin(\beta t + \theta)]. \tag{7–81}$$

Substitution of $t = 0$ in Eqs. (7–80) and (7–81) yields the values given in Eqs. (7–78) and (7–79) directly.

FINAL-VALUE THEOREM. *If $f(t)$ and its first derivatives are Laplace transformable, then the final value of $f(t)$ is*

$$\lim_{t \to \infty} f(t) = \lim_{s \to 0} sF(s). \tag{7–82}$$

Proof. To prove this theorem, we start, as in the case for the Initial-Value Theorem, with the transform of the first derivative of $f(t)$:

$$\int_0^\infty \left[\frac{df(t)}{dt} \right] \epsilon^{-st} \, dt = sF(s) - f(0+).$$

Now let s approach 0:

$$\lim_{s \to 0} \int_{0.}^\infty \left[\frac{df(t)}{dt} \right] \epsilon^{-st} \, dt = \lim_{s \to 0} [sF(s) - f(0+)]. \tag{7–83}$$

Since s is not a function of t, we let $s \to 0$ before integrating:

$$\lim_{s \to 0} \int_0^\infty \left[\frac{df(t)}{dt} \right] \epsilon^{-st} \, dt = \int_0^\infty \left[\frac{df(t)}{dt} \right] dt = \lim_{t \to \infty} \int_0^t \left[\frac{df(t)}{dt} \right] dt$$

$$= \lim_{t \to \infty} [f(t) - f(0+)]. \tag{7–84}$$

Substituting Eq. (7–84) back into Eq. (7–83), we have

$$\lim_{t \to \infty} f(t) = \lim_{s \to 0} sF(s). \qquad \text{Q.E.D.}$$

In applying this theorem to the response transform $V_2(s)$ of Example 7–6 as given in Eq. (7–60), we find

$$\lim_{t \to \infty} v_2(t) = \lim_{s \to 0} sV_2(s)$$

$$= \lim_{s \to 0} \left[-\frac{Es}{2C(R_1 + R_2)(s - s_1)(s - s_2)} \right] = 0,$$

which is also known to be correct by inspection of the given circuit in Fig. 7–4.

We note here that the Final-Value Theorem does not apply when $f(t)$ is a periodic function. In particular, it cannot be used to determine the final value of a sinusoidal function such as $\sin (\omega t + \theta)$ because $\sin \infty$ does not have a definite value.*

---

* In the language of the theory of the functions of a complex variable, the Final-Value Theorem requires that the function $sF(s)$ be analytic in the right half-plane as well as on the imaginary axis.

7-6 Response to periodic sinusoidal excitations; the impedance concept.
For a number of reasons, one of the most important types of excitation
functions in engineering applications is the periodic sinusoidal function.
First, periodic sinusoidal functions are easy to generate. Second, even
when the actual excitation function is not sinusoidal, it can always be
decomposed into its Fourier components, and hence the response of a
linear system to nonsinusoidal excitations can be expressed as a superposition of its responses to the sinusoidal components. Last but not least
is the fact that periodic sinusoidal functions are easy to handle analytically; specifically, the *impedance concept* applies. We shall examine this
last aspect from the point of view of Laplace transformation.

Let us assume that the following integro-differential equation describes
a physical system:

$$L \frac{di}{dt} + Ri + \frac{1}{C} \left[\int_0^t i\, dt + q(0+) \right] = E \epsilon^{j\omega_0 t}, \qquad (7\text{-}85)$$

where L, R, and C are constant coefficients, i is the response function,
and $E\epsilon^{j\omega_0 t}$ is the excitation function. It is simplest to use this exponential form because we know that if we can determine i as the response to
$E\epsilon^{j\omega_0 t}$, then, for a linear system, the response to the excitation function
$E \sin \omega_0 t = \text{Im}\,(E\epsilon^{j\omega_0 t})$ is simply $\text{Im}\,(i)$, and that to the excitation function $E \cos \omega_0 t = \text{Re}\,(E\epsilon^{j\omega_0 t})$ is $\text{Re}\,(i)$. Laplace transformation of Eq.
(7-85) will yield the following expression for the response transform:

$$I(s) = \frac{1}{Ls + R + \dfrac{1}{Cs}} \left[\frac{E}{s - j\omega_0} + Li(0+) - \frac{q(0+)}{Cs} \right]. \qquad (7\text{-}86)$$

The transfer function $1/(Ls + R + 1/Cs)$ can be written in the form
$s/L(s - s_1)(s - s_2)$, where s_1 and s_2 are related to the coefficients L, R,
and C by the conventional formulas for the roots of the quadratic equation, $s^2 + (R/L)s + 1/LC = 0$. We have

$$I(s) = \frac{Es/L + (s - j\omega_0)[si(0+) - q(0+)/LC]}{(s - s_1)(s - s_2)(s - j\omega_0)}$$

$$= \frac{K_1}{s - s_1} + \frac{K_2}{s - s_2} + \frac{K_3}{s - j\omega_0}, \qquad (7\text{-}87)$$

where

$$K_1 = \frac{Es_1/L + (s_1 - j\omega_0)[s_1 i(0+) - q(0+)/LC]}{(s_1 - s_2)(s_1 - j\omega_0)}, \qquad (7\text{-}88)$$

$$K_2 = \frac{Es_2/L + (s_2 - j\omega_0)[s_2 i(0+) - q(0+)/LC]}{(s_2 - s_1)(s_2 - j\omega_0)}, \qquad (7\text{-}89)$$

$$K_3 = \frac{Ej\omega_0/L}{(j\omega_0 - s_1)(j\omega_0 - s_2)} = \frac{E}{j\omega_0 L + R + 1/j\omega_0 C}. \quad (7\text{--}90)$$

Equation (7–90) may be understood more easily from inspection of the following:

$$K_3 = \frac{Es}{L(s - s_1)(s - s_2)}\bigg|_{s=j\omega_0} = \frac{E}{Ls + R + 1/Cs}\bigg|_{s=j\omega_0}$$

$$= \frac{E}{j\omega_0 L + R + 1/j\omega_0 C}.$$

The complete solution of the response function is then

$$i(t) = \mathcal{L}^{-1}[I(s)] = K_1 \epsilon^{s_1 t} + K_2 \epsilon^{s_2 t} + K_3 \epsilon^{j\omega_0 t}. \quad (7\text{--}91)$$

Up to now, the procedure has been quite routine. However, we wish to examine the response function in Eq. (7–91) more carefully.

First, we note that the two roots of the characteristic equation of the system, s_1 and s_2, can be either negative real, or complex conjugates with negative real parts. They cannot have positive real parts for a physical system because they would then make $i(t)$ increase indefinitely as t increases without limit, which is a physical impossibility; nor can they be purely imaginary (real part = 0) because some form of attenuation or damping (electrical resistance, mechanical friction, etc.) is always present in a physical system. When a system is self-oscillatory because its loss is replenished by an external supply of energy, it then becomes an oscillator or a source, and we do not talk about the response of a source. Because of the negative real parts of s_1 and s_2, the first two terms in the response function $i(t)$ as given in Eq. (7–91) will die out exponentially as t increases. They are the *transient* terms and correspond to the *complementary-function* part of the solution for the given differential equation (7–85).

The rates of decay of these transient terms depend on the magnitude of the negative real parts of the roots s_1 and s_2, which in turn *are fixed by the system parameters L, R, and C and do not depend on either the external excitation or the initial conditions. The initial magnitudes of these transient terms*, however, *are dependent upon both the external excitation and the initial conditions*, as evidenced by the expressions for K_1 and K_2 in Eqs. (7–88) and (7–89).

The last term in the response function, $K_3 \epsilon^{j\omega_0 t}$, has the same form as the original external excitation function, $E\epsilon^{j\omega_0 t}$; its amplitude does not decay with time. Hence it is the *steady-state* term and corresponds to the *particular-integral* part of the solution for the given differential equation (7–85). From Eq. (7–90), we also see that the amplitude K_3 of the steady-state response is equal to the amplitude E of the external excitation function divided by a complex quantity

$$Z = j\omega_0 L + R + \frac{1}{j\omega_0 C} = |Z|\epsilon^{j\theta_z}, \qquad (7\text{-}92)$$

where

$$|Z| = \sqrt{R^2 + \left(\omega_0 L - \frac{1}{\omega_0 C}\right)^2} \qquad (7\text{-}93)$$

and

$$\theta_z = \tan^{-1} \frac{\omega_0 L - 1/\omega_0 C}{R}. \qquad (7\text{-}94)$$

If the excitation function is a voltage and the response function a current, then Z in Eq. (7–92) corresponds to the familiar complex *impedance* of the system at the angular frequency ω_0; it can be obtained by replacing s in the characteristic function of the system, $Ls + R + (1/Cs)$, by $(j\omega_0)$. Hence we can write

$$\frac{\text{Complex amplitude of}}{\text{steady-state current response}} = \frac{\text{Amplitude of periodic sinusoidal voltage}}{\text{Complex impedance at excitation frequency}}$$
$$(7\text{-}95)$$

or

$$K_3 = \frac{E}{j\omega_0 L + R + 1/j\omega_0 C} = \frac{E}{|Z|} \epsilon^{-j\theta_z}. \qquad (7\text{-}96)$$

We note that *the steady-state solution depends only on the system parameters and the amplitude and frequency of the periodic sinusoidal excitation function, and is independent of the initial conditions.*

7–7 Response to periodic nonsinusoidal excitations. When the excitation function is periodic but not sinusoidal, we cannot conveniently make use of the impedance concept because a periodic nonsinusoidal function consists of many (very often, an infinite number of) frequency components. It would then be extremely laborious, if not impossible, to find the amplitudes of all sinusoidal Fourier components of the excitation, determine the responses to these components, and then sum up the many component responses with different frequencies, amplitudes, and phases to yield a nice, compact, steady-state solution. In this section, we shall examine what is involved in determining the response of a system to a periodic nonsinusoidal excitation and, in particular, the technique of finding the steady-state response. Let us first study the general problem with an example.

EXAMPLE 7–9. Find the voltage $v_R(t)$ appearing across the resistance R in the following circuit when the input voltage $v_i(t)$ is a periodic sawtooth wave as shown in Fig. 7–9, and there is an initial voltage $v_C(0+) = E/2$ on the capacitor. Determine also the steady-state expression for v_R.

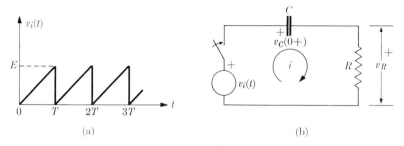

(a)   (b)

Fig. 7–9. A series R-C circuit with an applied periodic sawtooth voltage wave.

Solution. First we write the equation describing the problem:

$$Ri + \frac{1}{C}\int_0^t i\,dt + v_C(0+) = v_i(t). \tag{7–97}$$

In terms of the desired unknown function $v_R = iR$, we have

$$v_R + \frac{1}{CR}\int_0^t v_R\,dt + \frac{E}{2} = v_i(t). \tag{7–98}$$

Laplace transformation of Eq. (7–98) yields

$$\left(1 + \frac{1}{CRs}\right) V_R(s) = V_i(s) - \frac{E}{2s}. \tag{7–99}$$

The Laplace transform $V_i(s)$ of the input voltage $v_i(t)$ can be written from Eq. (6–41) with the aid of the theorem in Section 6–6 regarding the Laplace transform of periodic functions. Thus,

$$\begin{aligned}
V_i(s) &= \frac{E}{Ts^2(1 - \epsilon^{-Ts})}\left[1 - (Ts + 1)\epsilon^{-Ts}\right] \\
&= \frac{E}{T}\left[\frac{(1 + Ts)(1 - \epsilon^{-Ts}) - Ts}{s^2(1 - \epsilon^{-Ts})}\right] \\
&= \frac{E}{T}\left[\frac{1 + Ts}{s^2} - \frac{T}{s(1 - \epsilon^{-Ts})}\right]. \tag{7–100}
\end{aligned}$$

Substituting Eq. (7–100) in Eq. (7–99), we have

$$\left(1 + \frac{1}{CRs}\right) V_R(s) = \frac{E}{T}\left[\frac{2 + Ts}{2s^2} - \frac{T}{s(1 - \epsilon^{-Ts})}\right], \tag{7–101}$$

$$V_R(s) = \frac{E}{T}\left[\frac{2 + Ts}{2s(s + 1/CR)} - \frac{T}{(s + 1/CR)(1 - \epsilon^{-Ts})}\right], \tag{7–102}$$

and

$$v_R(t) = \mathcal{L}^{-1}[V_R(s)]$$

$$= \mathcal{L}^{-1} \frac{E}{2T} \left[\frac{2 + Ts}{s(s + 1/CR)} \right] - \mathcal{L}^{-1} \left[\frac{E}{(s + 1/CR)(1 - \epsilon^{-Ts})} \right].$$

$$(7\text{--}103)$$

The first term on the right side of Eq. (7–103) is easily evaluated:

$$\mathcal{L}^{-1} \frac{E}{2T} \left[\frac{2 + Ts}{s(s + 1/CR)} \right] = \frac{ECR}{T} \mathcal{L}^{-1} \left[\frac{1}{s} - \frac{1 - T/2CR}{s + 1/CR} \right]$$

$$= \frac{ECR}{T} \left[1 - \left(1 - \frac{T}{2CR} \right) \epsilon^{-t/CR} \right],$$

$$t \geq 0. \qquad (7\text{--}104)$$

The second term on the right side of Eq. (7–103) can be evaluated by expanding it in a series, finding the inverse transform of each term, and then summing the results for any given t. Fortunately, the series is a geometric progression and is therefore easy to sum. Thus,

$$\mathcal{L}^{-1} \frac{E}{(s + 1/CR)(1 - \epsilon^{-Ts})} = \mathcal{L}^{-1} \frac{E}{s + 1/CR} [1 + \epsilon^{-Ts} + \epsilon^{-2Ts} + \cdots]$$

$$= E[\epsilon^{-t/CR} U(t) + \epsilon^{-(t-T)/CR} U(t - T) + \epsilon^{-(t-2T)/CR} U(t - 2T) + \cdots].$$

$$(7\text{--}105)$$

It is important to note that the right side of Eq. (7–105) is not an infinite series because in the interval $0 < t < T$ only the first term exists and the other terms vanish because of the shifted unit-step functions, in the interval $T < t < 2T$ only the first two terms exist, and so on. Hence, in the interval of the nth tooth, $(n - 1)T < t < nT$,

$$\mathcal{L}^{-1} \frac{E}{(s + 1/CR)(1 - \epsilon^{-Ts})}$$

$$= E\epsilon^{-t/CR}[1 + \epsilon^{T/CR} + \epsilon^{2T/CR} + \cdots + \epsilon^{(n-1)T/CR}]$$

$$= E\epsilon^{-t/CR} \left(\frac{\epsilon^{nT/CR} - 1}{\epsilon^{T/CR} - 1} \right). \qquad (7\text{--}106)^*$$

Combining the results in Eqs. (7–104) and (7–106), we have the complete solution for $v_R(t)$:

---

* By virtue of the relation for a geometric progression:

$$1 + x + x^2 + \cdots + x^{n-1} = \sum_{k=0}^{n-1} x^k = \frac{x^n - 1}{x - 1}.$$

$$v_R(t) = E\left\{\frac{CR}{T} - \left[\left(\frac{CR}{T} - \frac{1}{2}\right) + \left(\frac{\epsilon^{nT/CR} - 1}{\epsilon^{T/CR} - 1}\right)\right]\epsilon^{-t/CR}\right\},$$

$$(n-1)T < t < nT. \tag{7-107}$$

It is not obvious from Eq. (7–107) what the steady-state solution for v_R is. Certainly we cannot simply assume t to be very large and drop all terms which are multiplied by the factor $\epsilon^{-t/CR}$, because we would then be left with v_R (t very large) $= ECR/T$, which we know is not the correct steady-state response. Let us rearrange Eq. (7–107) as follows:

$$v_R(t) = -E\left[\left(\frac{CR}{T} - \frac{1}{2}\right) - \frac{1}{\epsilon^{T/CR} - 1}\right]\epsilon^{-t/CR} + E\left[\frac{CR}{T} - \frac{\epsilon^{-(t-nT)/CR}}{\epsilon^{T/CR} - 1}\right],$$

$$(n-1)T < t < nT. \tag{7-108}$$

It is now apparent that

$$(v_R)_{\text{tr}} = -E\left[\left(\frac{CR}{T} - \frac{1}{2}\right) - \frac{1}{\epsilon^{T/CR} - 1}\right]\epsilon^{-t/CR}, \qquad t \geq 0, \tag{7-109}$$

is the *transient response*, which dies down as t increases indefinitely, and that

$$(v_R)_{\text{ss}} = E\left[\frac{CR}{T} - \frac{\epsilon^{-(t-nT)/CR}}{\epsilon^{T/CR} - 1}\right], \qquad (n-1)T < t < nT, \tag{7-110}$$

is the *steady-state response*, which is independent of the interval in which t appears, since no matter what n is, $(t - nT)$ always ranges from $-T$ to 0. We note that as t becomes very large, n becomes very large at the same time, and the difference $(t - nT)$ remains finite. Since the steady-state portion of the response is periodic and does not change from one period to another, it is convenient to refer it to the first period. From Eq. (7–110), by setting $n = 1$, we obtain

$$(v_R)_{\text{ss}} = E\left[\frac{CR}{T} - \frac{\epsilon^{-(t-T)/CR}}{\epsilon^{T/CR} - 1}\right], \qquad 0 < t < T, \tag{7-111}$$

which gives the same curve in the first period as Eq. (7–110) does in the nth period.

In a general solution for the system response to a periodic nonsinusoidal excitation such as Eq. (7–107) or (7–108) for the nth period, the constant terms plus the terms containing the combination $(t - nT)$ constitute the steady-state response, and the transient response is described by those terms which decay continually with time. The total response at any time or in any period is the sum of the transient and the steady-state responses.

If only the steady-state response of a system under the influence of a periodic nonsinusoidal excitation is desired, a simpler method can be used which does not require a series expansion and the subsequent summation of the geometric progression as indicated in Eq. (7–106). This method is based upon the fact that under steady-state conditions the responses at the beginning of two consecutive periods are the same, since the excitation is periodic. In particular, the steady-state values at $t = 0+$ and at $t = T+$ are the same. The inclusion of $+$ signs after 0 and T is important when the desired response has discontinuous changes (jumps) as it goes from one period to another (in particular, at $t = T$). We shall illustrate this method by examples.

EXAMPLE 7–10. The periodic rectangular current wave shown in Fig. 7–10(a) is applied to the parallel R-C circuit shown in Fig. 7–10(b). Determine the steady-state expression for the voltage $v(t)$.

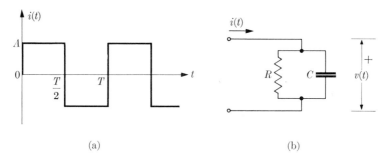

(a)   (b)

FIG. 7–10. A parallel R-C circuit with an applied periodic rectangular current wave.

Solution. As a consequence of the symmetry of the given current waveform, we conclude that under steady-state conditions the response in the second half-cycle will be exactly like the response in the first half-cycle except for a *reversal in sign* and a shift of $T/2$ in t. It is therefore necessary to examine only the steady-state expression for $v(t)$, v_{ss}, in the first half-cycle, $0 < t < T/2$. Within this interval the applied current is constant and is equal to A, and the differential equation for v_{ss} is

$$C \frac{dv_{ss}}{dt} + \frac{1}{R} v_{ss} = A. \qquad (7\text{--}112)$$

Laplace transformation of Eq. (7–112) yields

$$C[s V_{ss}(s) - v_{ss}(0+)] + \frac{1}{R} V_{ss}(s) = \frac{A}{s}$$

or $$V_{ss}(s) = \frac{A}{Cs(s + 1/CR)} + \frac{v_{ss}(0+)}{s + 1/CR}. \qquad (7\text{--}113)$$

The inverse transform of $V_{ss}(s)$ in Eq. (7–113) is

$$v_{ss} = AR(1 - \epsilon^{-t/CR}) + v_{ss}(0)\epsilon^{-t/CR}, \qquad 0 \leq t \leq T/2, \qquad (7\text{–}114)$$

where the $+$ sign in $v_{ss}(0+)$ has been dropped because the voltage across a capacitor cannot change instantaneously and therefore has no discontinuities.

To determine $v_{ss}(0)$, we apply the condition that at steady-state

$$v_{ss}(T/2) = -v_{ss}(0). \qquad (7\text{–}115)$$

From Eq. (7–114), we have

$$-v_{ss}(0) = AR(1 - \epsilon^{-T/2CR}) + v_{ss}(0)\epsilon^{-T/2CR}$$

or

$$v_{ss}(0) = -AR \frac{\epsilon^{T/2CR} - 1}{\epsilon^{T/2CR} + 1}. \qquad (7\text{–}116)$$

The desired steady-state expression, v_{ss}, for $v(t)$ is obtained by substituting $v_{ss}(0)$ as found in Eq. (7–116) into Eq. (7–114):

$$v_{ss} = AR \left[1 - \frac{2\epsilon^{T/2CR}}{\epsilon^{T/2CR} + 1} \epsilon^{-t/CR} \right], \qquad 0 \leq t \leq T/2. \qquad (7\text{–}117)$$

The general shape of v_{ss} represented by Eq. (7–117) is shown in Fig. 7–11. The exact shape depends upon the numerical values of T, C, and R.

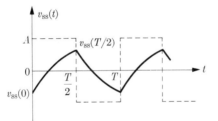

Fig. 7–11. Steady-state response, v_{ss}, for circuit in Fig. 7–10.

EXAMPLE 7–11.   In this example we shall determine the steady-state expression for v_R in Example 7–9 by using the method outlined above, without finding the general solution.

Solution.   Refer to Fig. 7–9. There are two important differences between this problem and the problem in Example 7–10. First, in the present system there exists an initial voltage $v_C(0+)$ across the capacitor C, while the system in the previous problem was initially at rest. In Chapter 3 we learned that a capacitor with an initial voltage across it is equivalent to the uncharged capacitor in series with a d-c voltage source having a magnitude equal to the initial voltage. Since a d-c voltage source in series with the present R-C circuit does not result in a steady current in R, we conclude that the initial voltage $v_C(0+)$ contributes nothing to the steady-state expression for v_R, and that so far as the present problem is concerned we could consider that $v_C(0+)$ did not exist. Second, the discontinuities in the input voltage $v_i(t)$ will appear in v_R because the voltage across a capacitor cannot change instantaneously. Hence, under steady-state conditions,*

$$v_R(0+) = v_R(T+) = v_R(T-) - E, \qquad (7\text{--}118)$$

where the $+$ and $-$ signs are essential and cannot be left out without introducing ambiguity or leading to wrong answers.

Let us first write the integral equation in v_R for the interval $0 < t < T$:

$$v_R + \frac{1}{CR} \int_0^t v_R \, dt = \frac{E}{T} t + v_R(0+), \qquad 0 < t < T. \quad (7\text{--}119)$$

Two things are notable in Eq. (7–119): (1) the omission of the initial voltage $v_C(0+)$ across the capacitor C for the reason explained above; and (2) the introduction of $v_R(0+)$, because under steady-state conditions the voltage across the resistor R is $v_R(0+)$ at the beginning of each cycle of the input voltage v_i; the value of $v_R(0+)$ has yet to be determined.

Laplace transformation of Eq. (7–119) yields

$$\left(1 + \frac{1}{CRs}\right) V_R(s) = \frac{E}{Ts^2} + \frac{v_R(0+)}{s},$$

which, after some rearrangement, becomes

$$V_R(s) = \frac{ECR}{T}\left(\frac{1}{s} - \frac{1}{s + 1/CR}\right) + \frac{v_R(0+)}{s + 1/CR}. \qquad (7\text{--}120)$$

The inverse transformation of $V_R(s)$ in Eq. (7–120) is

$$v_R = \frac{ECR}{T}(1 - \epsilon^{-t/CR}) + v_R(0+)\epsilon^{-t/CR}, \qquad 0 < t < T. \quad (7\text{--}121)$$

From Eq. (7–118), we have

$$v_R(T-) = v_R(0+) + E. \qquad (7\text{--}122)$$

Putting $t = T$ in Eq. (7–121) and equating it to $v_R(0+) + E$, we get

$$\frac{ECR}{T}(1 - \epsilon^{-T/CR}) + v_R(0+)\epsilon^{-T/CR} = v_R(0+) + E,$$

---

* All v_R's in this example are steady-state quantities; the double subscripts ss are omitted for simplicity.

from which $v_R(0+)$ can be solved:

$$v_R(0+) = \frac{ECR}{T} - \frac{E}{1 - \epsilon^{-T/CR}}. \qquad (7\text{–}123)$$

The desired steady-state expression, $(v_R)_{ss}$, for $v_R(t)$ is obtained by substituting $v_R(0+)$ as found in Eq. (7–123) into Eq. (7–121):

$$(v_R)_{ss} = E\left[\frac{CR}{T} - \frac{\epsilon^{-(t-T)/CR}}{\epsilon^{T/CR} - 1}\right], \qquad 0 < t < T, \qquad (7\text{–}124)$$

which is the same as Eq. (7–111) obtained in Example 7–9.

This problem could have proceeded from a differential equation obtained by differentiating Eq. (7–98) with respect to t, instead of from the integral equation (7–119):

$$\frac{dv_R}{dt} + \frac{1}{CR}v_R = \frac{dv_i}{dt} = \frac{E}{T}, \qquad 0 < t < T. \qquad (7\text{–}125)$$

Laplace transformation of Eq. (7–125) yields

$$[sV_R(s) - v_R(0+)] + \frac{1}{CR}V_R(s) = \frac{E}{Ts},$$

which gives the same expression for $V_R(s)$ as in Eq. (7–120).

PROBLEMS

Find the inverse Laplace transform of the functions given in Problems 7–1 through 7–9.

7–1. $F(s) = \dfrac{s+2}{s(s^2-1)}$

7–2. $F(s) = \dfrac{\epsilon^{-s}}{4s(s^2+1)}$

7–3. $F(s) = \dfrac{(s+3)\epsilon^{-s/2}}{s^2+4s+9}$

7–4. $F(s) = \dfrac{s^2+5}{s^3+2s^2+4s}$

7–5. $F(s) = \dfrac{s}{s^4+5s^2+4}$

7–6. $F(s) = \dfrac{3s+1}{5s^3(s-2)^2}$

7–7. $F(s) = \dfrac{3\epsilon^{-s/3}}{s^2(s^2+2)^2}$

7–8. $F(s) = \dfrac{1-\epsilon^{-4s}}{3s^3+2s^2}$

7–9. $F(s) = \ln\left(\dfrac{s}{s^2+9}\right)$

7–10. The switch in the circuit of Fig. 7–12 is closed at $t = 0$. Find v_L as a function of time.

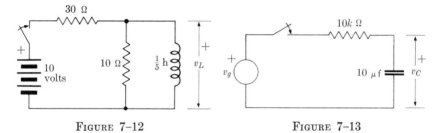

FIGURE 7–12   FIGURE 7–13

7–11. A voltage source $v_g = 20 \sin (10t + \alpha)$ volts is applied to the R-C circuit in Fig. 7–13 at $t = 0$. The capacitor is initially charged to a voltage $v_C(0) = -5$ volts. Determine (a) the voltage v_C, and (b) the phase angle α for which there will be no transient upon closing the switch.

7–12. A voltage source $v_g = 10 \cos 100t$ volts is applied to the primary circuit in Fig. 7–14 at $t = 0$. Determine i_2.

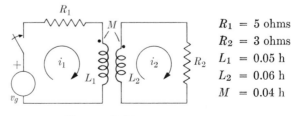

$R_1 = 5$ ohms
$R_2 = 3$ ohms
$L_1 = 0.05$ h
$L_2 = 0.06$ h
$M = 0.04$ h

FIGURE 7–14

7-13. In Fig. 7-15 is shown the equivalent circuit of an audio-amplifier stage. The switch is opened at $t = 0$. The capacitor C is initially uncharged. Determine the output voltage $v_0(t)$ (a) if $i(t) = 20 \sin 10^4 t$ ma, and (b) if $i(t)$ is a rectangular current pulse of amplitude 20 ma and duration of 1 msec.

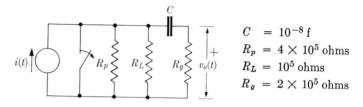

$$C = 10^{-8} \text{ f}$$
$$R_p = 4 \times 10^5 \text{ ohms}$$
$$R_L = 10^5 \text{ ohms}$$
$$R_g = 2 \times 10^5 \text{ ohms}$$

FIGURE 7-15

7-14. Two cycles of a sinusoidal voltage are applied to an R-C circuit at $t = 0$, as shown in Fig. 7-16. The capacitor initially has a voltage of $\frac{1}{2}$ volt on it. Determine $v_C(t)$.

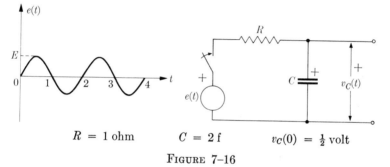

$$R = 1 \text{ ohm} \qquad C = 2 \text{ f} \qquad v_C(0) = \frac{1}{2} \text{ volt}$$

FIGURE 7-16

7-15. A triangular force is applied to the mass M_2 in the system shown in Fig. 7-17. Describe the motion of mass M_1.

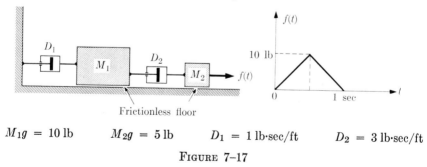

$$M_1 g = 10 \text{ lb} \qquad M_2 g = 5 \text{ lb} \qquad D_1 = 1 \text{ lb·sec/ft} \qquad D_2 = 3 \text{ lb·sec/ft}$$

FIGURE 7-17

7-16. Wheel A, which is positively coupled to wheel B in Fig. 7-18, is given an initial angular displacement θ_0 and then released at $t = 0$. Obtain the expression for the displacement of wheel B as a function of time.

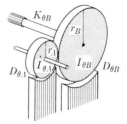

FIGURE 7–18

7–17. Solve the following simultaneous differential equations for y and z by using Laplace transformation. At $x = 0$, $y = 0$ and $z = -1$.

$$\frac{dy}{dx} = z - 1, \qquad \frac{dz}{dx} = y + x.$$

7–18. Solve the following simultaneous differential equations for v and w by using Laplace transformation. At $x = 0$, $v = 0$ and $w = -1$.

$$\frac{dv}{dx} + w = \epsilon^x, \qquad \frac{dw}{dx} - v = x.$$

7–19. Determine the initial and final values of $f(t)$'s corresponding to the $F(s)$'s in problems 7–4, 7–5, and 7–6.

7–20. Solve the following linear differential equation by using Laplace transformation. At $x = 0$, $w = 0$ and $dw/dx = -2$.

$$2 \frac{d^2 w}{dx^2} + x \frac{dw}{dx} - w = 0.$$

7–21. A periodic rectangular wave is applied to a series L-R circuit at $t = 0$, as shown in Fig. 7–19. (a) Find $v_R(t)$. (b) Determine the steady-state expression for $v_R(t)$.

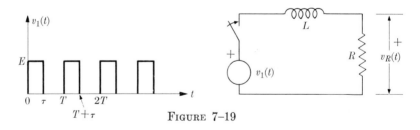

FIGURE 7–19

7–22. A periodic exponential voltage wave, which is $5\epsilon^{-10t}$ volts in any one period, is applied to a series R-C circuit at $t = 0$, as shown in Fig. 7–20. Find $v_C(t)$ for $0.3 \leq t \leq 0.4$ sec.

7–23. A full-wave rectified voltage is applied to an R-C filter as shown in Fig. 7–21. Determine the steady-state expression for $v_0(t)$.

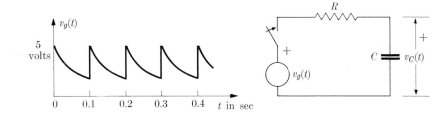

$$R = \tfrac{1}{2} \text{ ohm} \quad C = 1 \text{ f}$$

FIGURE 7–20

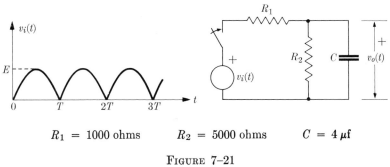

$$R_1 = 1000 \text{ ohms} \qquad R_2 = 5000 \text{ ohms} \qquad C = 4 \,\mu\text{f}$$

FIGURE 7–21

CHAPTER 8

ADDITIONAL CONCEPTS AND THEOREMS

8-1 Introduction. In Chapters 6 and 7 we laid the foundation for the solution of problems in linear systems by the method of Laplace transforms. The method involved the establishment of the integro-differential equations describing the system under consideration, transformation of these equations, solution for the desired response transform, and inverse transformation of the response transform. In general, this procedure leads to the desired solution in a straightforward manner, but in certain circumstances difficulties may be encountered. First, if the transform method is to be used we must be able to find the Laplace transform of the excitation function. In some cases the excitation function may be known only graphically (from experimental data); it may not be representable by a function of a simple analytical form and hence does not lead to an excitation transform. Problems involving such excitation functions cannot be handled by the classical method. The question then arises: Is the Laplace transform method useful in determining the response of linear systems to excitation functions that are known only graphically? The answer is in the affirmative, and there are two ways to do it. Briefly, we can derive the desired response to an arbitrary excitation function from the response to a unit impulse function or from the response to a unit step function. The theorems and techniques required to do this will be developed in this chapter, and a detailed discussion of the impulse function will also be given.

A second difficulty in the application of the transform method to linear systems analysis lies in the possibility that the transform of the desired response may be an irrational function of s, that is, a function containing nonintegral powers of s. This can occur when the system under consideration has distributed (nonlumped) parameters. In such an eventuality, Heaviside's expansion theorem developed in Chapter 7 is no longer applicable. In this chapter we shall present methods which can be used to handle certain irrational functions of s that will be important in the study of systems with distributed parameters, to be taken up in Chapter 11.

Such system concepts as the relation between responses to unit step and unit impulse functions, the response of cascaded stages, and so on, will also be discussed in this chapter.

8–2 The impulse function. Let us consider the single rectangular pulse of height $1/a$ and width a shown in Fig. 8–1(a). The expression for this pulse can be written in terms of step functions as

$$f_a(t) = \frac{1}{a} [U(t) - U(t - a)], \tag{8–1}$$

where we have used the subscript a on $f_a(t)$ to indicate explicitly that it is a function of the pulse width. The Laplace transform of $f_a(t)$ in Eq. (8–1) is

$$F_a(s) = \mathcal{L}[f_a(t)] = \frac{1}{as} (1 - \epsilon^{-as}). \tag{8–2}$$

If we now allow the pulse width a to decrease, the pulse height will increase correspondingly while the area under the pulse remains unity, as shown by the dashed lines in Fig. 8–1(a). In the limit as a approaches zero, a pulse of zero width, infinite amplitude, and unit area will result. This limiting pulse will be called a *unit impulse*. It is customary to represent a unit impulse by a *unit impulse function* $\delta(t)$. Thus,

$$\delta(t) = \lim_{a \to 0} \frac{1}{a} [U(t) - U(t - a)], \tag{8–3}$$

If we had started with a pulse of width a and height k/a (instead of $1/a$ as before), an impulse of area k (instead of unity as before) will result as a approaches zero; it is representable by the function $k\delta(t)$. In this case we say that the impulse has a *strength* k.* An impulse function is

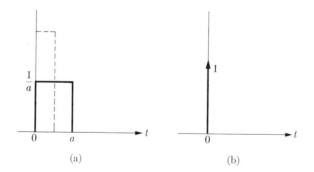

(a)   (b)

Fig. 8–1. Representation for the unit impulse, $\delta(t)$, as $a \to 0$.

---

* Since the coefficient of an impulse function represents the strength or the area of the impulse, an impulsive current function $i(t) = 3\delta(t)$ amperes then represents a current impulse of infinite magnitude at $t = 0$ and zero everywhere else, which transfers a *charge of 3 coulombs* instantaneously. In general, impulsive *current* functions will be written in the form $Q\delta(t)$, where Q represents the amount of instantaneous *charge* transfer.

represented graphically by a vertical arrow, with its strength designated alongside the arrow, as shown in Fig. 8–1(b).

The Laplace transform of the unit impulse function can be found from Eq. (8–2) by letting a approach zero:

$$\mathcal{L}[\delta(t)] = \lim_{a \to 0} \frac{1}{as} (1 - \epsilon^{-as})$$

$$= \lim_{a \to 0} \frac{1}{as} [1 - (1 - as + \tfrac{1}{2}a^2 s^2 - \cdots)]$$

$$= 1. \tag{8–4}*$$

Note that Eq. (8–4) is a particularly simple relationship. In general, *the Laplace transform of an impulse function at $t = 0$ is numerically equal to its strength:*

$$\mathcal{L}[k\,\delta(t)] = k. \tag{8–5}$$

Equation (8–5) enables us to take the inverse Laplace transform of a constant, which we have been unable to do so far.

Impulse functions have occasioned a great deal of controversy among mathematicians and engineers. From the way an impulse function is defined (infinite amplitude, zero duration, and finite area), we suspect whether it is really justifiable on the basis of conventional mathematics. As a matter of fact, it is not! However, a new mathematical entity known as a *distribution*, introduced recently by the French mathematician Laurent Schwartz, provides justification for some of the properties of impulse functions on a rigorous basis.† The theory of distributions is

---

* Although the derivation of Eq. (8–4) seems simple and straightforward, there is a "concealed" step that cannot be justified by conventional mathematics. This step can be "exposed" by noting from Eq. (8–3) that $\delta(t)$ itself involves a limiting process. Therefore, in finding the Laplace transform of $\delta(t)$, we must perform the following integration:

$$\mathcal{L}[\delta(t)] = \int_0^\infty \lim_{a \to 0} \frac{1}{a} [U(t) - U(t - a)]\epsilon^{-st}\, dt.$$

In Eq. (8–4), we used

$$\lim_{a \to 0} \frac{1}{as} (1 - \epsilon^{-as}) = \lim_{a \to 0} \int_0^\infty \frac{1}{a} [U(t) - U(t - a)]\epsilon^{-st}\, dt.$$

The right members of the above two equations are equal (that is, the interchange of the order of integration and the limiting process is permissible) only when the function in the integrand is uniformly convergent, which is not the case here.

† An account of the theory of distributions can be found in B. Friedman, *Principles and Techniques of Applied Mathematics*, pp. 135–141, John Wiley and Sons, Inc., New York; 1956.

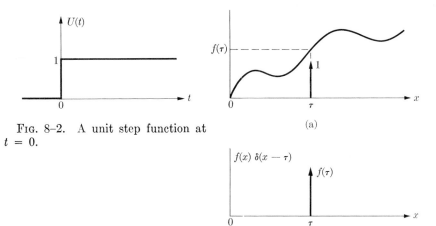

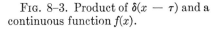

FIG. 8–2. A unit step function at $t = 0$.

(a)

FIG. 8–3. Product of $\delta(x - \tau)$ and a continuous function $f(x)$.

(b)

much too involved to be discussed here. For our purposes it is a comfort to know that justification is possible (though not in terms of conventional mathematics) and that proper use of impulse functions yields the right answers.

A unit impulse function is often considered as the derivative of a unit step function. It is clear that the derivative of a unit step function, $U(t)$, in Fig. 8–2, is zero everywhere except at $t = 0$, where it becomes infinite because of the discontinuous jump there. If we look back at Eq. (8–3), we see that $\delta(t)$ as defined is formally the derivative of $U(t)$ except that the limit in Eq. (8–3) does not really exist. Conceptually, it is more satisfying to define the unit impulse function $\delta(t)$ as a function whose Laplace transform is unity, although we must rely on the theory of distributions for the justification of many of its properties. The following inverse relationship will be considered correct:

$$U(t) = \int_{-\infty}^{t} \delta(x)\, dx, \qquad (8\text{–}6)$$

where x is a dummy variable. Here we might also argue that the definite integral on the right side of Eq. (8–6) should be a continuous function of its upper limit t, while $U(t)$ is obviously discontinuous.

A shifted unit impulse that occurs at $x = \tau$ can be represented by the function $\delta(x - \tau)$. This is the vertical arrow in Fig. 8–3(a). As in Eq. (8–6), we can integrate the shifted unit impulse function, leaving the upper limit variable:

$$\int_{0}^{t} \delta(x - \tau)\, dx = U(t - \tau), \qquad (8\text{–}7)$$

which is equal to 0 for $t < \tau$, and to 1 for $t > \tau$. If $f(x)$ is a continuous function of x, then the product $f(x)\,\delta(x - \tau)$ will be zero everywhere except at $x = \tau$, where there will be an impulse function having a strength equal to $f(\tau)$, the value of $f(x)$ at $x = \tau$. This is illustrated in Fig. 8–3(b). In view of Eq. (8–7), we have

$$\int_0^t f(x)\,\delta(x - \tau)\,dx = f(\tau)U(t - \tau), \qquad (8\text{--}8)$$

which is equal to 0 for $t < \tau$, and to $f(\tau)$ for $t > \tau$. In particular,

$$\int_0^\infty f(x)\,\delta(x - \tau)\,dx = f(\tau). \qquad (8\text{--}8a)$$

Equation (8–8a) expresses the so-called *sifting* or *sampling property* of an impulse function.

It is interesting to determine the Fourier spectrum of a unit impulse $\delta(t)$. From Eq. (5–78) we have, by definition,

$$g_\delta(\omega) = \int_{-\infty}^\infty \delta(t)\epsilon^{-j\omega t}\,dt. \qquad (8\text{--}9)$$

Equation (8–9) is similar to Eq. (8–8a) with $\tau = 0$ except that we are now using t, instead of x, as the variable of integration. The function to be sampled is $\epsilon^{-j\omega t}$. The infinite limits for the integral in Eq. (8–9) are unimportant so long as the range of integration covers the region near the origin where $\delta(t)$ occurs. Hence, according to Eq. (8–8a),

$$g_\delta(\omega) = \epsilon^{-j\omega t}\big|_{t=0} = 1. \qquad (8\text{--}10)$$

The spectrum function of an impulse is therefore a constant. In other words, *the relative frequency distribution of $\delta(t)$ is flat, and the energy contained in an impulse function is uniformly distributed over all frequencies.*

The unit impulse function $\delta(t)$ is a very useful device in linear systems analysis. In the next section we shall show how the response of a system to an arbitrary excitation function can be derived from the response to $\delta(t)$. Before we do so, we shall first show by an example how the function $\delta(t)$ may arise in problems involving linear systems.

EXAMPLE 8–1. Two capacitors ($C_1 = 1$ farad, $C_2 = 2$ farads) and a resistor ($R = 3$ ohms) are arranged in a circuit as shown in Fig. 8–4. C_1 is initially charged to a voltage E volts with polarity as given. The switch is closed at $t = 0$. Determine the current in C_1 as a function of time.

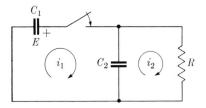

FIG. 8–4. Circuit for Example 8–1.

Solution. Let us use loop analysis. Two independent loops are in evidence. The equations are

$$\frac{1}{C_1}\int_0^t i_1\,dt - E + \frac{1}{C_2}\int_0^t (i_1 - i_2)\,dt = 0, \tag{8–11}$$

$$\frac{1}{C_2}\int_0^t (i_2 - i_1)\,dt + i_2 R = 0. \tag{8–12}$$

Inserting the given numerical values for C_1, C_2, and R in Eqs. (8–11) and (8–12), we have

$$\frac{3}{2}\int_0^t i_1\,dt - \frac{1}{2}\int_0^t i_2\,dt = E, \tag{8–13}$$

$$-\frac{1}{2}\int_0^t i_1\,dt + \frac{1}{2}\int_0^t i_2\,dt + 3i_2 = 0. \tag{8–14}$$

Application of Laplace transformation to Eqs. (8–13) and (8–14) yields

$$\frac{3}{2s} I_1(s) - \frac{1}{2s} I_2(s) = \frac{E}{s}, \tag{8–15}$$

$$-\frac{1}{2s} I_1(s) + \left(\frac{1}{2s} + 3\right) I_2(s) = 0. \tag{8–16}$$

The desired response transform $I_1(s)$ can be solved from Eqs. (8–15) and (8–16) without difficulty:

$$I_1(s) = \left(\frac{6s + 1}{9s + 1}\right) E. \tag{8–17}$$

It is clear that the right side of Eq. (8–17) is *not* a proper fraction, because the numerator and the denominator are polynomials in s of equal degree. Thus Heaviside's expansion theorem cannot be applied directly. To find the inverse transform of $I_1(s)$, we divide the improper fraction out, as follows:

$$I_1(s) = \frac{2E}{3}\left[\frac{s + 1/6}{s + 1/9}\right] = \frac{2E}{3}\left[1 + \frac{1/18}{s + 1/9}\right]. \tag{8–18}$$

Inverse transformation of Eq. (8–18) gives the desired answer:

$$i_1(t) = \mathcal{L}^{-1}[I_1(s)] = \frac{2E}{3}\left[\delta(t) + \frac{1}{18}\,\epsilon^{-t/9}\right]. \qquad (8\text{–}19)$$

Here we encounter the impulse function in a circuit problem for the first time. So far as the problem in Example 8–1 is concerned, it has been completely solved, and $i_1(t)$ in Eq. (8–19) is the desired answer. However, curiosity prompts us to inquire in what respect the circuit given in Fig. 8–4 differs from what we have seen previously. Specifically, why does $\delta(t)$ appear in this problem? An explanation from a physical viewpoint can be given in the following way. We have agreed previously that the voltage across a capacitance cannot change instantaneously. But when the switch in Fig. 8–4 is closed the voltages across C_1 and C_2 must adjust instantly because C_1 and C_2 are effectively in parallel with the switch closed; they must then have equal voltages. This demand of instantaneous equalization of the voltages across C_1 and C_2 is met by an impulsive current $(2E/3)\,\delta(t)$ [first term on the right side of Eq. (8–19)] which represents an instantaneous charge transfer of $2E/3$ coulombs. This can be worked out as follows:

Charge on C_1 at $t = 0-$: $Q_0 = C_1E$. After the switch is closed, voltage equilibrium demands

$$\frac{Q_1}{C_1} = \frac{Q_2}{C_2}. \qquad (8\text{–}20)$$

At $t = 0+$:

$$Q_1 + Q_2 = Q_0 = C_1E. \qquad (8\text{–}21)$$

Solving Eqs. (8–20) and (8–21) for Q_2, we obtain

$$Q_2 = \frac{C_1C_2}{C_1 + C_2}\,E = \frac{2}{3}\,E, \qquad (8\text{–}22)$$

which checks with the strength of the impulsive current. Here, then, is a case where the voltages across the capacitances must change instantaneously (through the mechanism of an impulsive current) in order to achieve voltage equilibrium when two incompatibly charged capacitors are suddenly connected in parallel (as a closed loop).

In general, *impulsive currents may appear in those loops of a network which trace through capacitances or capacitances and voltage sources only.* The loop consisting of C_1 and C_2 in which current i_1 exists in the circuit i Fig. 8–4 is a case in point. Of course, we must realize that this is an idealized circuit, since in a practical circuit there is always some resistance associated with a capacitor and its connecting wires. *By the prin-ciple of duality we can also state that impulsive node-pair voltages may appear at those nodes of a network which join inductances or inductances and current sources only.* Under these idealized circumstances, Eqs. (3–15) through (3–18) in Chapter 3 do not hold, and the values immediately after a disturbance of a circuit must be evaluated from the corresponding values immediately before the disturbance, by the laws of conservation of charge and flux linkage.

8–3 The convolution integral. There is a very important theorem concerning the product of transforms in the theory of Laplace transformation. It not only provides a formula for the inverse transformation of the product of two functions of s in terms of the inverse transforms of the individual functions, but also enables us to evaluate the response of a linear system to an arbitrary excitation in terms of its response to a unit impulse. This theorem is called the *convolution theorem*. We shall devote this section to the derivation, interpretation, and application of this theorem.

CONVOLUTION THEOREM. *If $\mathcal{L}[f_1(t)] = F_1(s)$ and $\mathcal{L}[f_2(t)] = F_2(s)$,* then

$$\mathcal{L}\left[\int_0^t f_1(t-\tau)f_2(\tau)\,d\tau\right] = \mathcal{L}\left[\int_0^t f_1(\tau)f_2(t-\tau)\,d\tau\right]$$

$$= F_1(s)F_2(s). \qquad (8\text{--}23)$$

Proof. By definition,

$$\mathcal{L}\left[\int_0^t f_1(t-\tau)f_2(\tau)\,d\tau\right] = \int_0^\infty \left[\int_0^t f_1(t-\tau)f_2(\tau)\,d\tau\right]\epsilon^{-st}\,dt. \quad (8\text{--}24)$$

To prove Eq. (8–23), we have to be able to convert the right side of Eq. (8–24) into a product of two integrals, both integrating from 0 to ∞. With this in mind, we introduce a unit step function $U(t-\tau)$ inside the brackets. Inasmuch as this function is unity for $\tau < t$ and is zero for $\tau > t$, the range of integration for the integral inside the brackets can be extended to ∞ after the insertion of $U(t-\tau)$ in the integrand. Hence

$$\mathcal{L}\left[\int_0^t f_1(t-\tau)f_2(\tau)\,d\tau\right] = \int_0^\infty \left[\int_0^\infty f_1(t-\tau)f_2(\tau)U(t-\tau)\,d\tau\right]\epsilon^{-st}\,dt.$$

$$(8\text{--}25)$$

Since both $f_1(t)$ and $f_2(t)$ are Laplace transformable, both of the integrals in Eq. (8–25) are absolutely convergent, and the order of performing the integrations with respect to τ and t can be reversed. From Eq. (8–25), we write

$$\mathcal{L}\left[\int_0^t f_1(t - \tau)f_2(\tau)\,d\tau\right] = \int_0^\infty f_2(\tau)\left[\int_0^\infty f_1(t - \tau)U(t - \tau)\epsilon^{-st}\,dt\right]d\tau.$$
(8–26)

The integration within the brackets on the right side of Eq. (8–26) is now with respect to t, all functions of t having being grouped in its integrand. Since $U(t - \tau)$ is zero for $t < \tau$, the integration with respect to t in the range $0 < t < \tau$ contributes nothing to the value of the integral, and the effective range of integration is really from $t = \tau$ to $t = \infty$. Therefore

$$\mathcal{L}\left[\int_0^t f_1(t - \tau)f_2(\tau)\,d\tau\right] = \int_0^\infty f_2(\tau)\left[\int_\tau^\infty f_1(t - \tau)\epsilon^{-st}\,dt\right]d\tau. \quad (8–27)$$

For the inner integral in Eq. (8–27), we let

$$t - \tau = x, \qquad dt = dx.$$

Hence

$$\int_\tau^\infty f_1(t - \tau)\epsilon^{-st}\,dt = \int_0^\infty f_1(x)\epsilon^{-s(x+\tau)}\,dx,$$

and Eq. (8–27) becomes

$$\mathcal{L}\left[\int_0^t f_1(t - \tau)f_2(\tau)\,d\tau\right] = \int_0^\infty f_2(\tau)\left[\int_0^\infty f_1(x)\epsilon^{-sx}\,dx\epsilon^{-s\tau}\right]d\tau$$

$$= \left[\int_0^\infty f_1(x)\epsilon^{-sx}\,dx\right]\left[\int_0^\infty f_2(\tau)\epsilon^{-s\tau}\,d\tau\right]$$

$$= F_1(s)F_2(s), \qquad\qquad (8–28)$$

which is what we started out to prove. Since $F_1(s)F_2(s) = F_2(s)F_1(s)$, f_1 and f_2 on the left side of Eq. (8–28) can be interchanged, giving the second form of the convolution theorem in Eq. (8–23).

The *convolution integral*,† or simply, the *convolution*, of two functions $f_1(t)$ and $f_2(t)$:

$$\int_0^t f_1(t - \tau)f_2(\tau)\,d\tau$$

is usually written conveniently as $f_1(t) * f_2(t)$. Thus the convolution theorem in Eq. (8–23) simplifies to

$$\mathcal{L}[f_1(t) * f_2(t)] = \mathcal{L}[f_2(t) * f_1(t)] = F_1(s)F_2(s), \qquad (8–29)$$

---

† The convolution integral is also known as the *Faltung integral*, which is a German term.

For more general classes of functions which do not vanish for $t < 0$ the lower limit in the convolution integral should be $-\infty$ instead of 0. However, for the cases we consider in this book, the integrand vanishes for negative time.

and we see that *Laplace transformation transforms convolution into multi-plication.* Here is another important instance where Laplace transformation simplifies operations. It is not difficult to extend the convolution theorem to more than two functions.

Before we attempt a graphical interpretation of the convolution integral, we shall first show how it can be used to find the inverse Laplace transform of certain functions which cannot be handled conveniently by partial-fraction expansion.

EXAMPLE 8–2. Find $\mathcal{L}^{-1}\left[\dfrac{1}{(s^2 + a^2)^2}\right]$.

Solution. The given function has two pairs of double roots at $s = +ja$ and $s = -ja$. If partial-fraction expansion is used, we would have

$$\mathcal{L}^{-1}\left[\frac{1}{(s^2 + a^2)^2}\right] = \mathcal{L}^{-1}\left[\frac{1}{(s + ja)^2(s - ja)^2}\right]$$

$$= \mathcal{L}^{-1}\left[\frac{K_{12}}{(s + ja)^2} + \frac{K_{11}}{(s + ja)} + \frac{K_{22}}{(s - ja)^2} + \frac{K_{21}}{(s - ja)}\right],$$

and it would be very laborious to determine the four constants K_{12}, K_{11}, K_{22}, and K_{21}, find the inverse transform of each term, and then properly combine the four functions into a neat, final result.

Instead, let us use convolution theorem in this problem. We have

$$F_1(s) = F_2(s) = \frac{1}{s^2 + a^2}. \tag{8–30}$$

Hence

$$f_1(t) = f_2(t) = \mathcal{L}^{-1}\left[\frac{1}{s^2 + a^2}\right] = \frac{1}{a}\sin at, \tag{8–31}$$

and

$$\mathcal{L}^{-1}\left[\frac{1}{(s^2 + a^2)^2}\right] = f_1 * f_2 = \int_0^t f_1(t - \tau)f_2(\tau)\, d\tau$$

$$= \int_0^t \left[\frac{1}{a}\sin a(t - \tau)\right]\left[\frac{1}{a}\sin a\tau\right] d\tau$$

$$= \frac{1}{a^2}\left[\sin at \int_0^t \cos a\tau \sin a\tau \, d\tau\right.$$

$$\left. - \cos at \int_0^t \sin^2 a\tau \, d\tau\right]$$

$$= \frac{1}{2a^3}(\sin at - at\cos at), \tag{8–32}$$

which is indeed a very neat result, obtained by a simple procedure.

We shall now give a graphical interpretation of the convolution integral of two functions f_1 and f_2. For purposes of illustration, let us take two specific functions:

$$f_1(\tau) = \epsilon^{-a\tau} U(\tau), \tag{8-33}$$

$$f_2(\tau) = \tau U(\tau). \tag{8-34}$$

Note that these two functions are zero for negative values of their arguments. The convolution integral of these two functions is

$$f_1 * f_2 = \int_0^t f_1(t - \tau) f_2(\tau) \, d\tau$$

$$= \int_0^t \epsilon^{-a(t-\tau)} U(t - \tau) \cdot \tau U(\tau) \, d\tau. \tag{8-35}$$

Parts (a), (b), and (c) of Fig. 8–5 are the graphs of the functions $f_1(\tau)$, $f_1(\tau - t_1)$, and $f_1(t_1 - \tau)$, where t_1 is an arbitrary value of t. $f_1(\tau - t_1)$ in (b) is simply a shift of $f_1(\tau)$ in (a) in the positive τ direction (toward the right) by an amount t_1. $f_1(t_1 - \tau) = f_1[-(\tau - t_1)]$ in (c) is a mirror image of $f_1(\tau - t_1)$ in (b) about the $\tau = t_1$ line. This is in general true for any function f_1, and is a very helpful hint in drawing the graph of a function when its argument is changed. $f_2(\tau)$ is shown in Fig. 8–5(d), and the product $f_1(t_1 - \tau) f_2(\tau)$ in Fig. 8–5(e). If we set $t = t_1$ in the convolution integral in Eq. (8–35), then the value of the definite integral

$$\int_0^{t_1} [\epsilon^{-a(t_1-\tau)} U(t_1 - \tau)][\tau U(\tau)] \, d\tau = \epsilon^{-at_1} \int_0^{t_1} \tau \epsilon^{a\tau} \, d\tau \tag{8-36}$$

is the area under the product curve in Fig. 8–5(e). Note that there is really no need to carry the unit step functions under the integral sign in Eq. (8–36), since the product $U(t_1 - \tau) U(\tau)$ is a gate function $G_0(t_1)$ whose value is unity in the interval $0 < \tau < t_1$ and zero everywhere else. The area under the product curve in Fig. 8–5(e) corresponds to the point at $t = t_1$ on the curve in Fig. 8–5(f), which represents the convolution of the two functions, $f_1 * f_2$.

Another graphical interpretation of the convolution of the two functions f_1 and f_2 in Eqs. (8–33) and (8–34), based upon the following alternative formula, is given in Fig. 8–6.

$$f_1 * f_2 = \int_0^t f_1(\tau) f_2(t - \tau) \, d\tau$$

$$= \int_0^t [\epsilon^{-a\tau} U(\tau)][(t - \tau) U(t - \tau)] \, d\tau. \tag{8-37}$$

Here the function f_2, instead of f_1, is shifted, folded, multiplied with f_1,

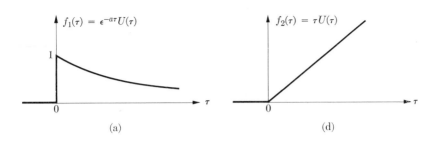

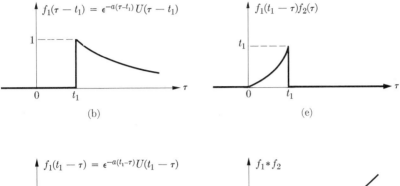

FIG. 8–5. Graphical interpretation of convolution.

and then integrated. The graphs are self-explanatory. The important thing to note here is that the area under the product curve in Fig. 8–6(e) is equal to that in Fig. 8–5(e); hence the convolution curve in part (f) remains unchanged.

The purpose of Figs. 8–5 and 8–6 is to show that the convolution of two functions can be evaluated *graphically* or *numerically* even when the functions themselves are so complicated that no formula is available for the analytical integration of the convolution integral, or when f_1 and f_2 are experimental curves which cannot be represented by analytical functions.

We now wish to explore the application of the convolution theorem in the analysis of linear systems. In Section 7–2 we saw that for a linear system the response transform, $W(s)$, is equal to the product of the transfer function, $H(s)$, of the system and the total excitation transform,

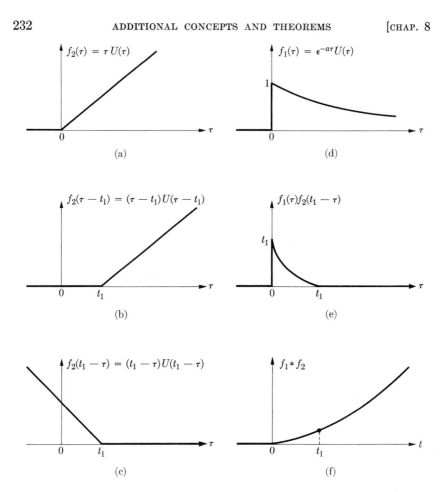

FIG. 8-6. Graphical interpretation of convolution.

$E_0(s)$. We shall assume that the system is initially relaxed (no initial conditions). Hence $E_0(s) = E(s) =$ Laplace transform of the externally applied excitation, and

$$W(s) = H(s)E(s). \tag{8-38}$$

The product form on the right side of Eq. (8-38) immediately suggests the applicability of the convolution integral here. Comparison of Eqs. (8-23) and (8-38) enables us to write the desired response $w(t)$ as

$$w(t) = \mathcal{L}^{-1}[W(s)] = \mathcal{L}^{-1}[H(s)E(s)]$$

$$= \int_0^t h(t - \tau)e(\tau)\, d\tau \tag{8-39}$$

or, alternatively,

$$w(t) = \int_0^t h(\tau)e(t - \tau) \, d\tau. \tag{8–40}$$

In both Eq. (8–39) and Eq. (8–40),

$$e(t) = \mathcal{L}^{-1}[E(s)] \tag{8–41}$$

and

$$h(t) = \mathcal{L}^{-1}[H(s)], \tag{8–42}$$

where $e(t)$ is the externally applied excitation function and $h(t)$ is simply given as the inverse Laplace transform of the transfer function $H(s)$ of the system. But does the function $h(t)$ have physical meaning? To answer this question, we refer back to Eq. (8–38). Suppose the excitation function is a unit impulse:

$$e(t) = \delta(t),$$

$$E(s) = \mathcal{L}[\delta(t)] = 1.$$

From Eq. (8–38), we have

$$W_\delta(s) = H(s) \tag{8–43}$$

and hence

$$w_\delta(t) = h(t). \tag{8–44}$$

Simple as it is, Eq. (8–44) is a very important result. It tells us that *the response of an initially relaxed system to a unit impulse is equal to the inverse Laplace transform of the transfer function.* $h(t)$, the response to a unit impulse, is referred to simply as the *impulse response*.* From Eqs. (8–39) and (8–40), which are consequences of the convolution theorem, we can make the highly important statement that *the response of a linear system to an arbitrary excitation* (which is Laplace transformable) *is the convolution of its impulse response and the excitation function.* This statement is sometimes referred to as *Borel's theorem.* Since the convolution integral can be evaluated graphically or numerically even when there is no analytical expression for the excitation, Borel's theorem provides a way of determining the response to an arbitrary excitation from the impulse response. *The impulse response, like the transfer function, characterizes the input-output relation of a system.*

---

* The term *Green's function* is sometimes used, but the term *impulse response* is preferred because it is more meaningful and because it goes better with the term *step response* which we shall introduce in Section 8–4. It is also referred to as the *weighting function* because, for any given t, the function $h(t - \tau)$ effectively "weighs" the excitation function $e(\tau)$ in the integrand of the convolution integral (8–39).

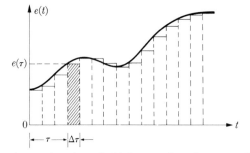

FIG. 8–7. Approximation of $e(t)$ by a series of rectangular pulses.

The above conclusion can also be reached by a purely physical approach. Let the excitation function $e(t)$ be represented by the curve in Fig. 8–7. This curve can be approximated by a series of rectangular pulses, each of duration $\Delta\tau$ as shown, the approximation being better for smaller $\Delta\tau$. As $\Delta\tau$ approaches zero, each of these pulses approaches an impulse with a strength equal to its area (height times $\Delta\tau$). For the typical pulse at $t = \tau$ (shaded in the figure), the impulse as $\Delta\tau \to 0$ is

$$e(\tau)\, \Delta\tau\, \delta(t - \tau).$$

If the response to a unit impulse $\delta(t)$ which occurs at $t = 0$ is $h(t)$, the response to an impulse of strength $e(\tau)\, \Delta\tau$ occurring at $t = \tau$ must be $e(\tau)\, \Delta\tau\, h(t - \tau)$. By the principle of superposition, the response to $e(t)$ up to time t is the sum of the responses to all the impulses up to that time. Hence

$$w(t) = \lim_{\Delta\tau \to 0} \sum_{\tau=0}^{t} e(\tau)\, \Delta\tau\, h(t - \tau)$$

$$= \int_{0}^{t} h(t - \tau)e(\tau)\, d\tau,$$

which is the same as Eq. (8–39). Thus the convolution integral is in reality a superposition integral.

If the given system is initially not relaxed, we *cannot* use the same procedure by simply changing $e(\tau)$ under the convolution integral to $e_0(\tau)$ to mean "total excitation function," although we are tempted to do so. A little reflection will reveal that although an initial voltage on a capacitor can be replaced by a series d-c voltage source and an initial current in an inductance by a parallel d-c current source, these equivalent sources may appear in locations entirely different from where the externally applied excitation is located in the system to be analyzed. Since the transfer function depends critically upon the location of the input, we have no way of combining the effects of the externally applied excitation with those of the equivalent sources due to initial conditions. In other words, "total excitation function" has little meaning for sources in different locations which we cannot combine. For a system with initial conditions, we must find the response due to the externally applied excitation and the responses due to initial conditions separately and superpose the results. The conventional transform method of Chapter 7, which takes care of the initial conditions implicitly, is always simpler to use than the convolution integral approach whenever the conventional method can be applied. On the other hand, the convolution integral formulation can be used even when the applied excitation is not expressible in a simple analytical form. The concept of impulse response also gives us a neat way of describing system performance.

The following example will serve to illustrate the procedure of determining the impulse response and, from it, also the response to a given excitation. It does not imply that this procedure would normally be used for a problem of this type.

EXAMPLE 8–3. The switch in the circuit of Fig. 8–8 is closed at $t = 0$. Find the current i_2 in R_2 for $v_g(t) = V_g \epsilon^{-\alpha t}$ by using the convolution integral.

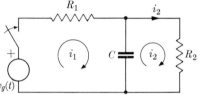

FIG. 8–8. Circuit for Example 8–3.

Solution. In order to use the convolution integral, we must first determine the impulse response $h(t)$, which is the current in R_2 when the applied voltage source is $\delta(t)$. We shall do this in a formal manner first; then we will demonstrate a more direct method which is applicable for a relatively simple problem like this one.

The formal way for finding $h(t)$ is to write the loop equations for the given circuit with $\delta(t)$ as the excitation function:

$$R_1 i_1 + \frac{1}{C} \int_0^t (i_1 - h)\, dt = \delta(t), \tag{8–45}$$

$$\frac{1}{C} \int_0^t (h - i_1)\, dt + R_2 h = 0, \tag{8–46}$$

where h has been written for i_2 as the response to $\delta(t)$. The transformed equations are

$$\left(R_1 + \frac{1}{Cs}\right) I_1(s) - \frac{1}{Cs} H(s) = 1, \tag{8–47}$$

$$-\frac{1}{Cs} I_1(s) + \left(R_2 + \frac{1}{Cs}\right) H(s) = 0. \tag{8–48}$$

The transfer function $H(s)$ can be solved from Eqs. (8–47) and (8–48):

$$H(s) = \frac{\begin{vmatrix} R_1 + \dfrac{1}{Cs} & 1 \\[2mm] -\dfrac{1}{Cs} & 0 \end{vmatrix}}{\begin{vmatrix} R_1 + \dfrac{1}{Cs} & -\dfrac{1}{Cs} \\[2mm] -\dfrac{1}{Cs} & R_2 + \dfrac{1}{Cs} \end{vmatrix}} = \frac{1}{(R_1 + R_2) + R_1 R_2 Cs}$$

$$= \frac{1}{R_1 R_2 C}\left[\frac{1}{s + (R_1 + R_2)/R_1 R_2 C}\right]. \tag{8–49}$$

Hence

$$h(t) = \mathcal{L}^{-1}[H(s)] = \frac{1}{R_1 R_2 C}\, \epsilon^{-\alpha' t}, \tag{8–50}$$

where

$$\alpha' = \frac{R_1 + R_2}{R_1 R_2 C}. \tag{8–51}$$

The transfer function $H(s)$ can be determined much more directly from the given circuit by remembering that $H(s)$ is the ratio of the response transform to the excitation transform:

$$H(s) = \frac{I_2(s)}{V_g(s)} = \frac{1}{R_1 + 1/(Cs + 1/R_2)} \cdot \frac{1/Cs}{R_2 + 1/Cs}$$

$$= \frac{1}{R_1 R_2 C}\left[\frac{1}{s + (R_1 + R_2)/R_1 R_2 C}\right],$$

which is the same as Eq. (8–49).

We now apply the convolution integral in Eq. (8–40):

$$i_2(t) = \int_0^t h(\tau)e(t - \tau)\, d\tau$$

$$= \int_0^t \left[\frac{1}{R_1 R_2 C}\, \epsilon^{-\alpha' \tau}\right] V_g \epsilon^{-\alpha(t-\tau)}\, d\tau$$

$$= \frac{V_g}{R_1 R_2 C}\, \epsilon^{-\alpha t} \int_0^t \epsilon^{-(\alpha' - \alpha)\tau}\, d\tau$$

$$= \frac{V_g}{R_1 R_2 C(\alpha' - \alpha)}\, [\epsilon^{-\alpha t} - \epsilon^{-\alpha' t}], \tag{8–52}$$

where α' is given in Eq. (8–51). Of course, we could have used the convolution integral in Eq. (8–39) instead of that in Eq. (8–40), to obtain the same result. In this problem, where $h(t)$ and $e(t)$ are of the same (exponential) form, it is immaterial whether Eq. (8–39) or Eq. (8–40) is used. In general, Eq. (8–39) would be simpler to use if $h(t)$ is simpler than $e(t)$, and Eq. (8–40) would be preferred if $e(t)$ is simpler than $h(t)$, the idea being to apply the shift to the simpler of the two functions under the convolution integral.

Before we leave the subject of convolution integrals, we introduce a theorem which is useful when we wish to find the convolution of a function with a unit impulse.

THEOREM. *The convolution of any function with the unit impulse function is the function itself.* We write

$$f * \delta = \int_0^t f(t - \tau)\, \delta(\tau)\, d\tau$$

$$= \int_0^t f(\tau)\, \delta(t - \tau)\, d\tau = f(t). \qquad (8\text{–}53)$$

That Eq. (8–53) is true can be seen by taking the Laplace transformation of both sides, which yields $F(s) \cdot 1 = F(s)$.

EXAMPLE 8–4. Find $\mathcal{L}^{-1}\left[\dfrac{s}{(s + 1)(s + 2)}\right]$ by using the convolution theorem.

Solution. Writing the given function of s as the product of two functions, we have

$$\frac{s}{(s + 1)(s + 2)} = \left(\frac{s}{s + 1}\right)\left(\frac{1}{s + 2}\right) = F_1(s)F_2(s), \qquad (8\text{–}54)$$

where

$$F_1(s) = \frac{s}{s + 1} = 1 - \frac{1}{s + 1}, \qquad (8\text{–}55)$$

$$F_2(s) = \frac{1}{s + 2}. \qquad (8\text{–}56)$$

Hence

$$f_1(t) = \mathcal{L}^{-1}[F_1(s)] = \delta(t) - \epsilon^{-t}, \qquad (8\text{–}57)$$

$$f_2(t) = \mathcal{L}^{-1}[F_2(s)] = \epsilon^{-2t}, \qquad (8\text{–}58)$$

and

$$\mathcal{L}^{-1}\left[\frac{s}{(s+1)(s+2)}\right] = f_1 * f_2 = \epsilon^{-2t} * \delta(t) - \epsilon^{-2t} * \epsilon^{-t}. \quad (8\text{-}59)$$

The first term on the right side of Eq. (8-59) is just ϵ^{-2t}, in view of the theorem expressed by Eq. (8-53). The second term is

$$\epsilon^{-2t} * \epsilon^{-t} = \int_0^t \epsilon^{-2\tau} \epsilon^{-(t-\tau)}\, d\tau$$

$$= \epsilon^{-t} \int_0^t \epsilon^{-\tau}\, d\tau = \epsilon^{-t} - \epsilon^{-2t}.$$

Finally, we obtain

$$\mathcal{L}^{-1}\left[\frac{s}{(s+1)(s+2)}\right] = \epsilon^{-2t} - (\epsilon^{-t} - \epsilon^{-2t}) = 2\epsilon^{-2t} - \epsilon^{-t}, \quad (8\text{-}60)$$

which can be readily checked by Heaviside's expansion theorem.

We note here that the function of s given in Example 8-4 is not entirely fictional. Consider the circuit in Fig. 8-9. If $CR = 1$, and the voltage $v_R(t)$ across the resistance R is desired for an applied voltage

$$v_g(t) = \epsilon^{-2t},$$

then

$$V_g(s) = \frac{1}{s+2} \quad (8\text{-}61)$$

and the transfer function

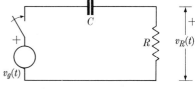

FIG. 8-9. Illustrating a situation in which $h(t)$ contains an impulse function.

$$H(s) = \frac{R}{R + 1/Cs} = \frac{s}{s + 1/CR} = \frac{s}{s+1}. \quad (8\text{-}62)$$

Therefore

$$V_R(s) = H(s)V_g(s) = \frac{s}{(s+1)(s+2)}$$

and

$$v_R(t) = \mathcal{L}^{-1}[V_R(s)] = \mathcal{L}^{-1}\left[\frac{s}{(s+1)(s+2)}\right],$$

which is exactly the same as the function given in Example 8-4. Inasmuch as $H(s)$ in Eq. (8-62) is identical with $F_1(s)$ in Eq. (8-55), $h(t)$ will be the same as $f_1(t)$ as given in Eq. (8-57) and will contain a unit impulse function. The impulse response of linear systems very frequently contains impulse functions. This fact underlines the usefulness of the theorem expressed by Eq. (8-53) when a system is analyzed by using the convolution theorem.

8–4 The superposition integral. The discussion of the preceding section has shown that the response of an initially relaxed linear system to any excitation can be completely determined once the impulse response of the system is known. In this section we shall show that a similar statement can be made when the response of the system to a unit step function is known.

Let us call the response of an initially relaxed linear system to a unit step input the *step response** and denote it by $w_u(t)$. Its transform will then be denoted by $W_u(s)$. With a unit step as the excitation function,

$$E(s) = \mathcal{L}[U(t)] = \frac{1}{s}.$$

From Eq. (8–38), we have

$$W_u(s) = \frac{1}{s} H(s) \tag{8–63}$$

or

$$H(s) = sW_u(s). \tag{8–64}$$

Now

$$\mathcal{L}\left[\frac{dw_u}{dt}\right] = sW_u(s) - w_u(0). \tag{8–65}$$

Cancelling $sW_u(s)$ between Eqs. (8–64) and (8–65), we have

$$H(s) = \mathcal{L}\left[\frac{dw_u}{dt}\right] + w_u(0). \tag{8–66}$$

Hence

$$h(t) = \mathcal{L}^{-1}[H(s)] = \frac{dw_u}{dt} + w_u(0)\,\delta(t). \tag{8–67}$$

Equation (8–67) relates the step response with the impulse response; $w_u(0)$ is understood to mean $w_u(0+)$ if $w_u(t)$ has a discontinuity at $t = 0$.

Let us now return to Eq. (8–38) for an initially relaxed system:

$$W(s) = H(s)E(s) = s\left[\frac{1}{s} H(s)E(s)\right] = s[W_u(s)E(s)]. \tag{8–68}$$

Equation (8–68) has been written in the form

$$W(s) = sF(s), \tag{8–69}$$

with

$$F(s) = W_u(s)E(s). \tag{8–70}$$

---

* The term *indicial response* is sometimes used, but the term *step response* is preferred.

We can find the inverse transform of $F(s)$ with the aid of the convolution theorem:

$$f(t) = \mathcal{L}^{-1}[F(s)] = w_u(t) * e(t) = \int_0^t w_u(t - \tau)e(\tau)\, d\tau. \quad (8\text{–}71)$$

Since it is apparent from Eq. (8–71) that $f(0) = 0$,† we conclude from Eq. (8–69), by virtue of the differentiation theorem, that $w(t)$ is the derivative of $f(t)$. Therefore

$$w(t) = \frac{d}{dt} f(t) = \frac{d}{dt} [w_u(t) * e(t)]$$

$$= \frac{d}{dt} \int_0^t w_u(t - \tau)e(\tau)\, d\tau \quad (8\text{–}72)$$

or, alternatively,

$$w(t) = \frac{d}{dt} \int_0^t w_u(\tau)e(t - \tau)\, d\tau. \quad (8\text{–}73)$$

Equations (8–72) and (8–73) express the important result that in order to find the response of an initially relaxed system to an arbitrary excitation (which is Laplace transformable), we need only know its response to a unit step input. *The step response then also characterizes the input-output relation of a system.*

The result obtained in Eqs. (8–72) and (8–73) can be put in several different equivalent forms that are sometimes easier to evaluate. This can be done quite simply by rearranging Eq. (8–68):

$$W(s) = W_u(s)[sE(s)]. \quad (8\text{–}74)$$

Since the right side of Eq. (8–74) is the product of two transforms, we can apply the convolution theorem directly. First let us write

$$\mathcal{L}^{-1}[W_u(s)] = w_u(t) \quad (8\text{–}75)$$

and

$$\mathcal{L}^{-1}[sE(s)] = \mathcal{L}^{-1}\{[sE(s) - e(0)] + e(0)\}$$

$$= \frac{de(t)}{dt} + e(0)\, \delta(t)$$

$$= e'(t) + e(0)\, \delta(t), \quad (8\text{–}76)$$

where we have used $e'(t)$ to represent the first derivative of $e(t)$ with respect to its argument t. Applying the convolution theorem to Eq. (8–74),

---

† This will not be true if either $w_u(t)$ or $e(t)$ contains an impulse function at the origin. However, these cases are unimportant and will be disregarded.

we now obtain

$$w(t) = \int_0^t w_u(t - \tau)[e'(\tau) + e(0)\,\delta(\tau)]\,d\tau$$

$$= e(0)w_u(t) + \int_0^t e'(\tau)w_u(t - \tau)\,d\tau, \qquad (8\text{–}77)$$

where the first term on the right side is the result of convolving the function $w_u(t)$ with the impulse function $e(0)\,\delta(t)$ in accordance with Eq. (8–53). Equation (8–77) is usually referred to as *Duhamel's integral*, although there has been some question about whether Duhamel should be credited for it. As we shall see presently, it is in reality a superposition integral, just as the convolution integral is a form of superposition integral. However, when the term *superposition integral* is used without qualification, it is customarily taken to mean Duhamel's integral. Other forms of the superposition integral are possible. For example, if we rearrange Eq. (8–74) as

$$W(s) = E(s)[sW_u(s)], \qquad (8\text{–}78)$$

we see at once that another form of the superposition integral can be obtained by simply interchanging the functions e and w_u in Eq. (8–77). Thus

$$w(t) = w_u(0)e(t) + \int_0^t w_u'(\tau)e(t - \tau)\,d\tau. \qquad (8\text{–}79)$$

Since a convolution integral is not changed by interchanging either the functions or the arguments of the two functions under the integral sign, two more forms of the superposition integral can obviously be derived from Eqs. (8–77) and (8–79). For all practical purposes, there is need only to recognize the existence of the two forms given in Eqs. (8–72) and (8–77); other forms are easily written from these two. We note that the form given in Eq. (8–77) cannot conveniently be used when the excitation function or its first derivative contains discontinuities.

We shall now attempt a physical interpretation of the superposition integral. In Fig. 8–10 is drawn the same excitation function $e(t)$ as shown in Fig. 8–7, but it is now being approximated by a series of step functions successively displaced in t by an amount $\Delta\tau$. The sum of all the step functions gives, in effect, a staircase approximation of the smooth curve, the approximation being better as $\Delta\tau$ becomes smaller. The typical step at $t = \tau$ (shaded in Fig. 8–10) is

$$\Delta e(\tau)U(t - \tau) = \left[\frac{\Delta e(\tau)}{\Delta\tau}\right]\Delta\tau U(t - \tau).$$

If the response to a unit step $U(t)$ which occurs at $t = 0$ is $w_u(t)$, the response to a step with height $\Delta e(\tau)$ at $t = \tau$ must be $\Delta e(\tau)w_u(t - \tau)$.

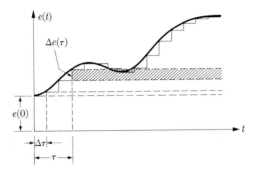

FIG. 8–10. Approximation of $e(t)$ by a series of step functions.

By the principle of superposition, the response to $e(t)$ up to time t is the sum of the responses to all the step functions up to that time. Hence

$$w(t) = e(0)w_u(t) + \lim_{\Delta\tau \to 0} \sum_{\tau=0}^{t} \left[\frac{\Delta e(\tau)}{\Delta\tau} \right] \Delta\tau \, w_u(t - \tau)$$

$$= e(0)w_u(t) + \int_0^t \frac{de(\tau)}{d\tau} \, w_u(t - \tau) \, d\tau, \qquad (8\text{–}80)$$

which is the same as the superposition integral, or Duhamel's integral, given in Eq. (8–77). Note that the first term on the right side of Eq. (8–80) must be added *separately* because the initial value $e(0)$ cannot be derived from the slope of the $e(t)$ curve. Similar care must be exercised when the given excitation possesses discontinuities.

Like the convolution integral, the superposition integral can be evaluated either graphically or numerically, and hence is useful even when the applied excitation is not expressible in a simple analytical form.

EXAMPLE 8–5. Solve the problem in Example 8–3 by using the superposition integral.

Solution. In order to use the superposition integral, we must first find the step response $w_u(t)$, which is the output for a unit step input. Equations (8–45) and (8–46) will apply if we change $\delta(t)$ and $h(t)$ to $U(t)$ and $w_u(t)$ respectively. However, since we already have $h(t)$ in Eq. (8–50), we can obtain $w_u(t)$ through the relation between $W_u(s)$ and $H(s)$ in Eq. (8–63),

$$W_u(s) = \frac{1}{s} H(s).$$

By the integration theorem, we have

$$w_u(t) = \int_0^t h(\tau) \, d\tau. \qquad (8\text{–}81)$$

For this problem,

$$h(t) = \frac{1}{R_1 R_2 C} \epsilon^{-\alpha' t}, \qquad \alpha' = \frac{R_1 + R_2}{R_1 R_2 C}.$$

Hence

$$w_u(t) = \frac{1}{R_1 R_2 C} \int_0^t \epsilon^{-\alpha' \tau} d\tau$$

$$= \frac{1}{R_1 + R_2} (1 - \epsilon^{-\alpha' t}). \qquad (8\text{-}82)$$

Any one of the different forms of the superposition integral can be used to find $i_2(t) = w(t)$. Let us use Eq. (8-77). In our notation,

$$i_2(t) = v_g(0) w_u(t) + \int_0^t v_g'(\tau) w_u(t - \tau) \, d\tau, \qquad (8\text{-}83)$$

$$v_g(t) = V_g \epsilon^{-\alpha t},$$

$$v_g(0) = V_g,$$

$$v_g'(t) = -\alpha V_g \epsilon^{-\alpha t}.$$

Thus we have

$$i_2(t) = \frac{V_g}{R_1 + R_2} (1 - \epsilon^{-\alpha' t}) - \int_0^t \frac{\alpha V_g}{R_1 + R_2} \epsilon^{-\alpha \tau} [1 - \epsilon^{-\alpha'(t-\tau)}] \, d\tau$$

$$= \frac{V_g}{R_1 R_2 C(\alpha' - \alpha)} [\epsilon^{-\alpha t} - \epsilon^{-\alpha' t}],$$

which is the same as the result in Eq. (8-52). More work is involved in reducing $i_2(t)$ to its final form than before. In general, this is true partly because $w_u(t)$ is usually more complicated than $h(t)$ and partly because it is necessary to combine the two terms in the result when the superposition integral is used.

8-5 Certain system concepts. In previous sections we have shown that the response of an initially relaxed linear system to any excitation can be completely determined when either the impulse response or the step response of the system is known. If we pause and think back a little, we realize that we encountered the same situation when Fourier methods were discussed in Chapter 5. It was pointed out there (in Example 5-6) that the response of a system to an arbitrary input could be found if the steady-state response to an applied sinusoid of unit amplitude and angular frequency ω at the input was known for all ω. The steady-state response to a unit sinusoid could be determined from the transfer function $H(j\omega)$, which could in turn be obtained by replacing s by $j\omega$ in the more general

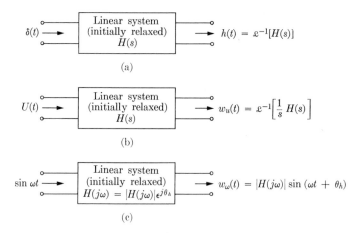

FIG. 8–11. Three ways of characterizing a linear system.

transfer function $H(s)$ introduced later in Chapter 7. The three different situations are illustrated in Fig. 8–11. For an arbitrary excitation $e(t)$, the response $w(t)$ of the system can be obtained in any one of the following ways:

(a) *From the impulse response $h(t)$:*

$$w(t) = h(t) * e(t) \qquad \text{(Convolution integral).} \qquad (8\text{–}84)$$

(b) *From the step response $w_u(t)$:*

$$w(t) = \frac{d}{dt}[w_u(t) * e(t)] \qquad \binom{\text{Derivative of}}{\text{convolution integral}}, \qquad (8\text{–}85)$$

or

$$w(t) = e(0)w_u(t) + e'(t) * w_u(t) \qquad \text{(Superposition integral),} \qquad (8\text{–}86)$$

or

$$w(t) = w_u(0)e(t) + w'_u(t) * e(t) \qquad \text{(Superposition integral).} \qquad (8\text{–}87)$$

(c) *From the steady-state transfer function $H(j\omega)$:*

$$w(t) = \mathfrak{F}^{-1}\{H(j\omega) \cdot \mathfrak{F}[e(t)]\} \qquad \text{(Fourier transforms).} \qquad (8\text{–}88)$$

Note that Eq. (8–88) does not apply when $e(t)$ is periodic, in which case $e(t)$ would be analyzed into its Fourier series components and $w(t)$ would be the *summation* of the responses of the system to these *discrete* components.

Another system concept which is quite useful in practice concerns the response of systems in cascade (in tandem). It will be assumed that the

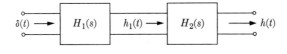

FIG. 8–12. Two systems in cascade.

performance of each individual system is not affected by its combination with other systems. In the terminology of electrical circuits, we say that the input impedance of each individual system is sufficiently high so that it will not "load" (change the operating characteristics of) another to which it is connected. When two systems (stages) having transfer functions $H_1(s)$ and $H_2(s)$ are connected in cascade,† the over-all transfer function $H(s)$ is simply the product of the individual transfer functions:

$$H(s) = H_1(s)H_2(s). \qquad (8-89)$$

Hence the impulse response of the two systems in cascade shown in Fig. 8–12 is

$$h(t) = \mathcal{L}^{-1}[H(s)] = \mathcal{L}^{-1}[H_1(s)H_2(s)]$$
$$= \{\mathcal{L}^{-1}[H_1(s)]\} * \{\mathcal{L}^{-1}[H_2(s)]\} = h_1(t) * h_2(t). \qquad (8-90)$$

For an excitation function $e(t)$, we have the response function

$$w(t) = h(t) * e(t) = h_1(t) * h_2(t) * e(t). \qquad (8-91)$$

We can also express the over-all system response in terms of the step responses $w_{u1}(t)$ and $w_{u2}(t)$ of the individual systems in the following way:

$$W(s) = sW_u(s)E(s) = s^2 W_{u1}(s)W_{u2}(s)E(s). \qquad (8-92)$$

It is important to note that $W_u(s) \neq W_{u1}(s)W_{u2}(s)$. Rather,

$$W_u(s) = sW_{u1}(s)W_{u2}(s). \qquad (8-93)$$

From Eq. (8–93) (assuming that the step responses of the individual stages contain no impulse functions at the origin; see footnote on p. 240), we find

$$w_u(t) = \frac{d}{dt}[w_{u1}(t) * w_{u2}(t)]. \qquad (8-94)$$

---

† Instead of assuming that the performance of each individual system is not affected by cascading, it would be exact to specify that $H_1(s)$ is the transfer function of the first system with the second system already connected in place.

In view of the superposition integral, we can also write Eq. (8–94) as

$$w_u(t) = w_{u1}(0)w_{u2}(t) + w'_{u1}(t) * w_{u2}(t). \qquad (8\text{–}95)$$

Of course the subscripts 1 and 2 in Eq. (8–95) could be interchanged. The over-all response of the two stages can be found from Eq. (8–92):

$$w(t) = \frac{d}{dt}[w_u(t) * e(t)]$$

$$= \frac{d^2}{dt^2}[w_{u1}(t) * w_{u2}(t) * e(t)]. \qquad (8\text{–}96)$$

In terms of steady-state transfer functions, $H_1(j\omega)$ and $H_2(j\omega)$, we have simply

$$H(j\omega) = H_1(j\omega)H_2(j\omega) \qquad (8\text{–}97)$$

and, from Eq. (8–88),

$$w(t) = \mathfrak{F}^{-1}\{H_1(j\omega) \cdot H_2(j\omega) \cdot \mathfrak{F}[e(t)]\}. \qquad (8\text{–}98)$$

EXAMPLE 8–6. It is found that when a d-c voltage of 10 volts is applied to the input terminals of an initially relaxed network at $t = 0$ the output voltage is $5\epsilon^{-t/2}$ volts. Determine the output voltage response of two such networks in cascade when an exponential voltage $v_i(t) = 20\epsilon^{-3t/2}U(t)$ volts is applied at the input. Assume that the input circuit of the network is such that the loading effect in cascading is negligible.

Solution. First we find the step response of one stage:

$$w_{u1}(t) = \tfrac{1}{10}(5\epsilon^{-t/2}) = \tfrac{1}{2}\epsilon^{-t/2}.$$

Now

$$W_{u1}(s) = \mathcal{L}[w_{u1}(t)] = \mathcal{L}[\tfrac{1}{2}\epsilon^{-t/2}] = \frac{1}{2(s + \tfrac{1}{2})},$$

$$H_1(s) = sW_{u1}(s) = \frac{s}{2s + 1}.$$

For two identical stages,

$$H(s) = H_1^2(s) = \left(\frac{s}{2s + 1}\right)^2 = \frac{1}{4}\left(1 - \frac{1}{2s + 1}\right)^2$$

$$= \frac{1}{4}\left[1 - \frac{2}{2s + 1} + \frac{1}{(2s + 1)^2}\right].$$

The over-all impulse response is

$$h(t) = \mathcal{L}^{-1}\frac{1}{4}\left[1 - \frac{1}{s + \tfrac{1}{2}} + \frac{1}{4(s + \tfrac{1}{2})^2}\right]$$

$$= \tfrac{1}{4}[\delta(t) - \epsilon^{-t/2} + \tfrac{1}{4}t\epsilon^{-t/2}]U(t).$$

For the given excitation

$$v_i(t) = 20\epsilon^{-3t/2}U(t),$$

the over-all output voltage response is found by taking the convolution of $h(t)$ and $v_i(t)$:

$$w(t) = \int_0^t v_i(t - \tau)h(\tau)\,d\tau$$

$$= \int_0^t [20\epsilon^{-3(t-\tau)/2}]U(t - \tau) \cdot \tfrac{1}{4}[\delta(\tau) - \epsilon^{-\tau/2} + \tfrac{1}{4}\tau\epsilon^{-\tau/2}]U(\tau)\,d\tau.$$

Note that $U(t - \tau) \cdot U(\tau) = G_0(t)$ is a gate function whose value is 1 in the interval $0 < \tau < t$ and 0 everywhere else; hence these two unit step functions can be dropped, since the interval of integration coincides with the gate. Multiplying out, we have

$$w(t) = 5\epsilon^{-3t/2}\int_0^t [\epsilon^{3\tau/2}\,\delta(\tau) - \epsilon^\tau + \tfrac{1}{4}\tau\epsilon^\tau]\,d\tau$$

$$= 5\epsilon^{-3t/2}[1 - (\epsilon^t - 1) + \tfrac{1}{4}(t\epsilon^t - \epsilon^t + 1)]U(t)$$

$$= \tfrac{5}{4}[9\epsilon^{-3t/2} + \epsilon^{-t/2}(t - 5)]U(t) \text{ volts,}$$

which is the desired result. We could have used Eq. (8–96) directly in this problem, but then we would have to perform a double convolution as well as a double differentiation.

8–6 Inverse Laplace transforms of some irrational functions. In Chapter 7 we explored the methods for finding the inverse Laplace transformation of rational algebraic functions of s. For irrational functions, that is, functions which contain nonintegral powers of s, Heaviside's expansion theorem does not apply because these functions cannot be expanded into simple partial fractions. In order to handle the inverse transformation of general irrational functions, a good, thorough knowledge of the theory of functions of a complex variable is necessary, which cannot be acquired from an abbreviated exposition. We shall not attempt to discuss the use of the theory of functions of a complex variable in this book. However, there are certain types of irrational functions whose inverse transforms can be found without the use of such function theory. We shall develop some of the techniques in this section by way of specific examples. But first let us introduce two special functions, the gamma function and the error function, which will appear in the following development. These functions appear quite frequently in engineering and applied mathematics; their values can be found from tables,* but we should know how they are defined.

The gamma function, $\Gamma(n)$. The *gamma function* of an argument n is defined by the following definite integral:

$$\Gamma(n) = \int_0^\infty x^{n-1}\epsilon^{-x}\,dx, \qquad n > 0. \tag{8-99}$$

Hence

$$\Gamma(n+1) = \int_0^\infty x^n \epsilon^{-x}\,dx. \tag{8-100}$$

Integrating by parts, we have

$$\Gamma(n+1) = (-x^n \epsilon^{-x})\Big|_0^\infty + n\int_0^\infty x^{n-1}\epsilon^{-x}\,dx$$

$$= 0 + n\int_0^\infty x^{n-1}\epsilon^{-x}\,dx,$$

or

$$\Gamma(n+1) = n\Gamma(n). \tag{8-101}$$

Equation (8-101) is the fundamental recursion relation satisfied by the gamma function. For integral values of n, the gamma function is definitely related to the factorial:

$$\Gamma(n+1) = n!, \qquad n = \text{an integer.} \tag{8-102}\dagger$$

Note that n does not have to be an integer. In this section, $\Gamma(\frac{1}{2})$ is of special interest to us. From Eq. (8-99),

$$\Gamma(\tfrac{1}{2}) = \int_0^\infty x^{-1/2}\epsilon^{-x}\,dx. \tag{8-103}$$

If we let

$$x = y^2,$$

then

$$dx = 2y\,dy$$

and Eq. (8-103) becomes

$$\Gamma(\tfrac{1}{2}) = 2\int_0^\infty \epsilon^{-y^2}\,dy = \sqrt{\pi}. \tag{8-104}$$

---

* E. Jahnke and F. Emde, *Table of Functions*, Dover Publications, New York; 1943.

† This can be seen by noting from Eq. (8-99) that

$$\Gamma(1) = \int_0^\infty \epsilon^{-x}\,dx = 1,$$

and from Eq. (8-101), $\Gamma(2) = 1\cdot\Gamma(1) = 1 = 1!,$
$\Gamma(3) = 2\cdot\Gamma(2) = 2\cdot1 = 2!,$
$\Gamma(4) = 3\cdot\Gamma(3) = 3\cdot2\cdot1 = 3!,$

and so on.

The definite integral in Eq. (8–104) is equal to $\sqrt{\pi}/2$; its proof can be found in almost any book on applied mathematics.

The error function, erf (x). The *error function,*[*] or the *probability integral,* of an argument x is defined by the following definite integral:

$$\operatorname{erf}(x) = \frac{2}{\sqrt{\pi}} \int_0^x \epsilon^{-u^2} \, du. \qquad (8\text{–}105)$$

The values of the error function at $x = 0$ and at $x = \infty$ are of some interest:

$$\operatorname{erf}(0) = 0 \qquad (8\text{–}106)$$

and

$$\operatorname{erf}(\infty) = \frac{2}{\sqrt{\pi}} \int_0^\infty \epsilon^{-u^2} \, du = \frac{2}{\sqrt{\pi}} \cdot \frac{\sqrt{\pi}}{2} = 1. \qquad (8\text{–}107)$$

It is clear from Eq. (8–107) that the factor $2/\sqrt{\pi}$ in front of the defining integral in Eq. (8–105) is a normalizing factor which makes erf $(\infty) = 1$. This remark has physical significance in the theory of probability. A graph of the error function is shown in Fig. 8–13.

We are now ready to consider the inverse transformation of some irrational functions of s, which will be important to our discussion in Chapter 11.

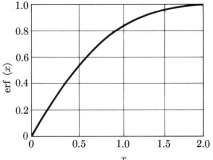

FIG. 8–13. Graph of error function, erf (x).

EXAMPLE 8–7. Find $\mathcal{L}^{-1}\left[\dfrac{1}{\sqrt{s}}\right]$.

Solution. Let us start with the Laplace transform of t^n. By definition,

$$\mathcal{L}[t^n] = \int_0^\infty t^n \epsilon^{-st} \, dt. \qquad (8\text{–}108)$$

The integral in Eq. (8–108) looks quite similar to the one that defines

---

[*] Sometimes also referred to as the *error integral.*

the gamma function. Comparison with Eq. (8–100) indicates that we should try setting

$$st = x, \quad t = \frac{x}{s}, \quad dt = \frac{1}{s}\, dx.$$

Substituting the above in Eq. (8–108), we get

$$\mathcal{L}[t^n] = \frac{1}{s^{n+1}} \int_0^\infty x^n \epsilon^{-x}\, dx = \frac{\Gamma(n+1)}{s^{n+1}}, \tag{8–109}$$

or

$$\mathcal{L}^{-1}\left[\frac{1}{s^{n+1}}\right] = \frac{t^n}{\Gamma(n+1)}. \tag{8–110}$$

When n is an integer, Eq. (8–109) reduces to the formula we already know (Eq. 6–34):

$$\mathcal{L}[t^n] = \frac{n!}{s^{n+1}}.$$

But n in Eq. (8–109) or Eq. (8–110) is not restricted to integral values. For the problem at hand, we set $n = -\frac{1}{2}$ in Eq. (8–110) and obtain

$$\mathcal{L}^{-1}\left[\frac{1}{\sqrt{s}}\right] = \frac{t^{-1/2}}{\Gamma(\frac{1}{2})} = \frac{1}{\sqrt{\pi t}}. \tag{8–111}$$

EXAMPLE 8–8. Find $\mathcal{L}^{-1}\left[\dfrac{1}{s\sqrt{s+\alpha}}\right]$.

Solution. Making use of the theorem expressed in Eq. (6–30), we write

$$\mathcal{L}^{-1}\left[\frac{1}{s\sqrt{s+\alpha}}\right] = \epsilon^{-\alpha t}\,\mathcal{L}^{-1}\left[\frac{1}{(s-\alpha)\sqrt{s}}\right]. \tag{8–112}$$

The convolution theorem can now be applied to the right side of Eq. (8–112), since we know that

$$\mathcal{L}^{-1}\left[\frac{1}{s-\alpha}\right] = \epsilon^{\alpha t}$$

and

$$\mathcal{L}^{-1}\left[\frac{1}{\sqrt{s}}\right] = \frac{1}{\sqrt{\pi t}}.$$

Thus

$$\mathcal{L}^{-1}\left[\frac{1}{(s-\alpha)\sqrt{s}}\right] = (\epsilon^{\alpha t}) * \left(\frac{1}{\sqrt{\pi t}}\right)$$

$$= \frac{\epsilon^{\alpha t}}{\sqrt{\pi}} \int_0^t \frac{\epsilon^{-\alpha \tau}}{\sqrt{\tau}}\, d\tau. \tag{8–113}$$

If we make the following substitutions:

$$\sqrt{\tau} = x, \qquad \tau = x^2, \qquad d\tau = 2x \, dx,$$

Eq. (8–113) becomes

$$\mathcal{L}^{-1}\left[\frac{1}{(s-a)\sqrt{s}}\right] = \frac{\epsilon^{at}}{\sqrt{\pi}}\int_0^{\sqrt{t}} 2\epsilon^{-ax^2}\, dx$$

$$= \frac{2}{\sqrt{a\pi}}\,\epsilon^{at}\int_0^{\sqrt{at}} \epsilon^{-(\sqrt{a}x)^2}\, d(\sqrt{a}x)$$

$$= \frac{\epsilon^{at}}{\sqrt{a}}\,\mathrm{erf}\,(\sqrt{at}). \tag{3–113a}$$

Substitution of Eq. (8–113a) in Eq. (8–112) gives us the desired answer:

$$\mathcal{L}^{-1}\left[\frac{1}{s\sqrt{s+a}}\right] = \frac{1}{\sqrt{a}}\,\mathrm{erf}\,(\sqrt{at}). \tag{8–114}$$

EXAMPLE 8–9. Find $\mathcal{L}^{-1}\left[\dfrac{1}{\sqrt{s^2+a^2}}\right].$

Solution. In this case we cannot associate the given function of s directly with any function we know so far. For a relatively simple irrational function like this one, the inverse transform can be found by first expanding it in a power series in $(1/s)$ by means of the binomial theorem:

$$\frac{1}{\sqrt{s^2+a^2}} = \frac{1}{s\sqrt{1+(a/s)^2}} = \frac{1}{s}\left[1+\left(\frac{a}{s}\right)^2\right]^{-1/2}$$

$$= \frac{1}{s} - \frac{a^2}{2s^3} + \frac{3a^4}{2!2^2s^5} - \frac{3\cdot5a^6}{3!2^3s^7} + \cdots \tag{8–115}$$

The inverse transform of each of the terms on the right side of Eq. (8–115) can be found by using Eq. (8–110). Hence

$$\mathcal{L}^{-1}\left[\frac{1}{\sqrt{s^2+a^2}}\right] = 1 - \frac{(at)^2}{2^2} + \frac{(at)^4}{2^24^2} - \frac{(at)^6}{2^24^26^2} + \cdots$$

$$= \sum_{k=0}^{\infty} \frac{(-1)^k}{(k!)^2}\left(\frac{at}{2}\right)^{2k} = J_0(at). \tag{8–116}$$

Here we have introduced a new function, $J_0(at)$.* It is called the *Bessel function* of the first kind of the zeroth order. The J_0 function is an important function in engineering whose values can also be found in tables.† The series in Eq. (8–116) is convergent for all values of the argument (at). Application of the initial- and final-value theorems yields the values of $J_0(0)$ and $J_0(\infty)$:

$$J_0(0) = \lim_{s \to \infty} \frac{s}{\sqrt{s^2 + a^2}} = 1, \tag{8–117}$$

$$J_0(\infty) = \lim_{s \to 0} \frac{s}{\sqrt{s^2 + a^2}} = 0. \tag{8–118}$$

A graph of $J_0(x)$ is shown in Fig. 8–14.

If a is imaginary, we can write

$$a = jb \tag{8–119}$$

and Eq. (8–116) becomes

$$\mathcal{L}^{-1}\left[\frac{1}{\sqrt{s^2 - b^2}}\right] = J_0(jbt) = I_0(bt), \tag{8–120}$$

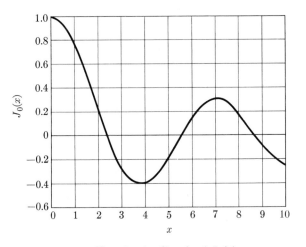

Fig. 8–14. Graph of $J_0(x)$.

---

* The J_0 function can also be expressed as a definite integral:

$$J_0(x) = \frac{1}{\pi} \int_0^\pi \cos (x \cos \theta) \, d\theta.$$

† E. Jahnke and F. Emde, *op. cit.*

where $I_0(bt)$ is called the *modified Bessel function* of the first kind of the zeroth order. It is also a tabulated function.* A graph of $I_0(x)$ is given in Fig. 8–15.

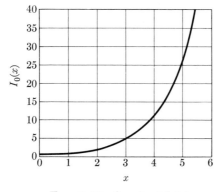

FIG. 8–15. Graph of $I_0(x)$.

---

* E. Jahnke and F. Emde, *op. cit.*

PROBLEMS

8–1. The switch in the circuit of Fig. 8–16 is initially closed, and there is a steady current V/R_1 in L_1 and R_1. (a) At time $t = 0$ the switch is opened. Find the current i_1 as a function of time. Explain why the value V/R_1 is not obtained by setting $t = 0$ in your expression of i_1. (b) Find the voltage across the inductance L_1 as a function of time.

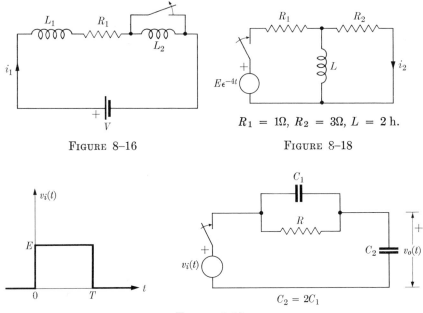

FIGURE 8–16

$R_1 = 1\Omega,\ R_2 = 3\Omega,\ L = 2\,\mathrm{h}.$

FIGURE 8–18

FIGURE 8–17

8–2. The switch in the circuit of Fig. 8–17 is closed at $t = 0$. Determine $v_o(t)$.

8–3. Find the inverse Laplace transforms of the following functions by using the convolution theorem:

(a) $\dfrac{1}{s^3}$
(b) $\left(\dfrac{s}{s+1}\right)^2$
(c) $\dfrac{s+1}{s(s^2+4)}$

(d) $\dfrac{s}{(s^2+\omega_1^2)(s^2+\omega_2^2)}$
(e) $\left(\dfrac{1-\epsilon^{-as}}{s}\right)^2$

8–4. The convolution theorem extended to three functions can be written as follows:

$$\mathcal{L}[f_1(t) * f_2(t) * f_3(t)] = F_1(s)F_2(s)F_3(s).$$

Find the inverse Laplace transform of

$$F(s) = \frac{s}{(s+1)(s+2)(s+3)} = \left(\frac{1}{s+1}\right)\left(\frac{1}{s+2}\right)\left(\frac{s}{s+3}\right)$$

by making use of the above relationship. Check your answer by Heaviside's expansion theorem.

8–5. Solve the following integral equation for $f(x)$ by the method of Laplace transforms:

$$2f(x) + \beta \int_0^x \sin \beta(x - u) \cdot f(u) \, du = \cos \beta x.$$

8–6. The switch in the circuit of Fig. 8–18 is closed at $t = 0$. The desired response is the current i_2 in R_2. Find (a) the impulse response, $h(t)$; (b) i_2 from $h(t)$ by using the convolution integral (Borel's theorem); (c) the step response, $w_u(t)$; and (d) i_2 from $w_u(t)$ by using the superposition (Duhamel's) integral.

8–7. The switch in the circuit of Fig. 8–19 is closed at $t = 0$. The desired response is the voltage v_L across the inductance. Find (a) the step response, $w_u(t)$; (b) $v_L(t)$ from $w_u(t)$ by using the superposition (Duhamel's) integral; (c) the impulse response, $h(t)$; and (d) $v_L(t)$ from $h(t)$ by using the convolution integral (Borel's theorem).

8–8. (a) A single half-sine force function is applied to a linear mechanical system (Fig. 8–20) at time t_1 and released at time $(t_1 + \pi/\omega)$. Find the force transmitted to mass M as a function of time, using step response and superposition integral. D_1 and D_2 are coefficients of linear damping. (b) If a spring is attached between the wall and the mass (between points a and b), what should be its compliance so that the force transmitted to the mass will be maximum when a periodic sinusoidal force of angular frequency ω (instead of the half-sine wave as shown) is applied?

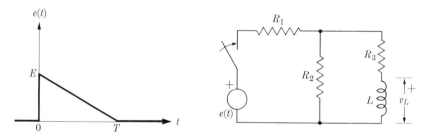

FIG. 8–19. $R_1 = R_2 = 2\Omega$, $R_3 = 5\Omega$, $L = 0.3$ h.

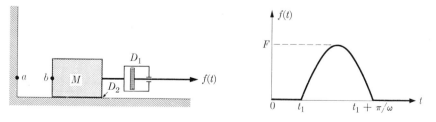

FIGURE 8–20

8–9. Starting from the relation $W(s) = sW_u(s)E(s)$, derive the following alternative form of the superposition integral:

$$w(t) = w_u(0)e(t) + \int_0^t e(\tau)\left[\frac{\partial}{\partial t}\, w_u(t - \tau)\right] d\tau.$$

8–10. The step response of a certain initially relaxed device is

$$w_u(t) = (1 - \tfrac{1}{2}\epsilon^{-t/3})U(t).$$

Determine the impulse response of a system consisting of two such devices in cascade, assuming that the loading effect due to cascading is negligible.

8–11. It is found that when a voltage $v_1(t) = tU(t)$ is applied at the input of an initially relaxed one-stage linear amplifier, the output voltage $v_2(t)$ is given by $v_2(t) = (\epsilon^{-2t} - 1)U(t)$. Suppose that two such identical stages are connected in cascade and a unit step voltage is applied at the input. Find the voltage at the output of the second stage.

8–12. Find the inverse Laplace transforms of the following functions:

(a) $s^{-5/2}$

(b) $\dfrac{\epsilon^{-2s}}{(s + 1)\sqrt{s + 5}}$

(c) $\dfrac{1}{\sqrt{s}\sqrt{s + 1}}$

(d) $\dfrac{1}{\sqrt{s}(\sqrt{s} + \sqrt{a})}$

(e) $\dfrac{1}{\sqrt{s + 1}}$

(f) $\dfrac{1}{\sqrt{s^2 + 4s + 13}}$

(g) $\dfrac{a}{(s + \sqrt{s^2 + a^2})\sqrt{s^2 + a^2}}$

(h) $\dfrac{\sqrt{s}}{\sqrt{s + a}} - 1$

CHAPTER 9

SYSTEMS WITH FEEDBACK

9-1 Introduction. A system with feedback is one in which the output is fed back to the input part of the system in such a way that it will affect its own value.* Thus a feedback path, as well as a forward path, exists within the system, resulting in what is referred to as a *closed-loop system.* Feedback is essential to the operation of an electric oscillator in which a part of the output voltage is brought back to the input circuit in a proper phase to further build up the output voltage. This process continues until oscillations are sustained, the amplitude of the oscillations being eventually limited by the nonlinearity of the characteristics of the active element (a vacuum tube or a transistor).

Feedback is also essential to the operation of automatic control systems. A familiar example is thermostatic control of room temperature. The desired temperature as set on the thermostat is the reference; the actual temperature is recorded on a thermometer. When the room temperature falls below the reference temperature, the thermostat is activated, a valve in the furnace is opened, and heat is sent into the room until the desired temperature is reached. In this closed-loop system the output (room temperature) is fed back into an error-detecting device (thermostat) and compared with a standard (desired temperature), and automatic control (of the valve in the furnace and hence the room temperature) is attained through the output of the error-detecting device. If an *open-loop system* (without the feedback) were used in this situation, the valve in the furnace (the forward path) would be set so that a constant amount of heat would be sent into the room, with the hope that it would just replenish the heat lost. But this open-loop system would be less satisfactory, because it cannot take account of the fluctuation of the weather, which determines the rate of heat loss, nor can it compensate for any undesired change in the valve setting.

Two conflicting factors, accuracy and stability, determine the performance of a feedback control system. To increase the accuracy of a feedback control system, the amplification (sensitivity) of the forward path must be increased, so that large corrective action of the controlled variable (output) can be actuated by a small error. However, because of inherent time delays in the system, the corrective action, once started,

---

* A *servo system* is a feedback system in which the output or some function of the output is fed back for comparison with the input and the difference between these quantities (the actuating error) is used to control a source of power.

may not stop in time to avoid overcorrections (overshoots of the controlled variable) even after the actuating error has subsided. When the extent of overcorrection exceeds the error which initiated the original control action, the system starts a larger corrective action in the opposite direction. This process of successive overcorrections may be built up to such an extent that violent oscillations are produced. The resulting unstable feedback system obviously could not perform any useful control function. A well-designed feedback control system must represent a good compromise between accuracy and stability. The Laplace transform method is particularly adaptable to the stability analysis of feedback systems.

In this chapter we shall first describe, in detail, the block-diagram method of representing circuits and systems. The signal flow graph method of representation will then be discussed, and finally the criteria for stability analysis of feedback systems will be examined. We shall not be concerned with feedback control system design.

9–2 Block-diagram representation. A system can be described in a number of different ways. For example, it can be described mathematically by a set of differential equations, or it can be represented by a detailed schematic diagram which shows all the components and their interconnections. However, when the system is fairly complicated, neither of these two methods of representation is particularly satisfactory. The purely mathematical representation tends to convert an engineering problem into a mathematical exercise; it is difficult to acquire from the equations and their manipulation a physical insight of the effect of the individual components and their interconnections on the over-all system response. On the other hand, detailed schematic diagrams are hard to draw and they do not give quantitative relationships unless equations are also written. The block-diagram method of describing a system is a combination of the two methods; i.e., it uses blocks ("black boxes") to indicate *mathematical* operations on the various *physical* quantities. A block may represent a single component or a group of components, but each block is completely characterized by a transfer function. Blocks in a complicated system may be combined and manipulated to yield a simplified equivalent block diagram by a set of transformation rules which we shall describe in Section 9–3. In fact, we came briefly in contact with the simplest type of block-diagram representation in Section 8–5 when we discussed systems in cascade.

Block diagrams of a system consist of *undirectional, operational* blocks; they show the direction of flow and the operations on the system variables in such a way that a relationship is established between the input and the output as the path is traced through the diagram. For simplicity, block diagrams are drawn as single-line diagrams; that is, a single

TABLE 9–1

VOLTAGE-CURRENT RELATIONSHIPS FOR PASSIVE ELECTRICAL
ELEMENTS WITHOUT INITIAL CONDITIONS

	Voltage-current relationship	Transformed relationship
	$v_R(t) = Ri_R(t)$	$V_R(s) = RI_R(s)$
	$v_L(t) = L\dfrac{di_L(t)}{dt}$	$V_L(s) = LsI_L(s)$
	$v_C(t) = \dfrac{1}{C}\displaystyle\int_0^t i_C(t)\,dt$	$V_C(s) = \dfrac{1}{Cs}\,I_C(s)$

line is used to show the *flow* of system variables from one block to another
even though some system variables (e.g., voltages) require two terminals
to specify.

The simplest block diagrams are those that can be drawn for single
electrical elements without initial conditions (by analogy we can, of
course, draw similar ones for mechanical elements). Table 9–1 lists the
voltage-current relationships and the corresponding transformed equations for resistive, inductive, and capacitive elements. We note that the
transformed equations in the last column are algebraic, and that they
can be inverted very simply to give the transformed currents in terms
of the respective transformed voltages. The block diagrams of Fig. 9–1
are drawn on the basis of the transformed relationships. In each case
*the transformed output is equal to the transformed input multiplied by the
transfer function given in the block.*

In order to represent more complicated systems by block diagrams, we
need to agree on two additional symbols: one for *summing points* and
one for *pickoff points*. A summing point is one at which several system
variables are added or subtracted. It is usually represented by a small
circle with a cross in it; two examples are given in Fig. 9–2. A pickoff
point is one from which the input variable may proceed unaltered along
several different paths to several different destinations. It is represented
by an ordinary junction without special symbolism, as shown in Fig. 9–3.

We are now able to draw the block diagrams for some relatively simple
circuits. We shall develop the techniques in the form of examples. After
we have mastered the basic techniques, we will find that the block diagrams for more complex circuits can be drawn in exactly the same manner,
the only difference being that more steps will be involved.

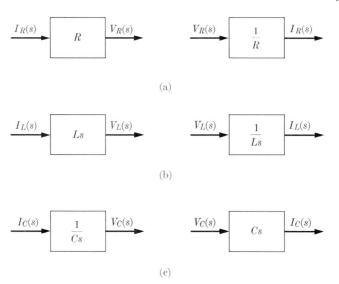

FIG. 9–1. Block-diagram representations of resistive, inductive, and capacitive elements. (a) For a resistive element. (b) For an inductive element. (c) For a capacitive element.

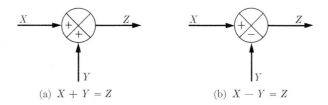

FIG. 9–2. Summing points.

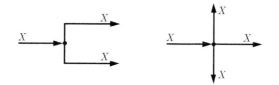

FIG. 9–3. Pickoff points.

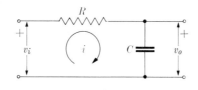

FIG. 9–4. Circuit for Example 9–1.

EXAMPLE 9–1. Draw a block diagram for the R-C circuit in Fig. 9–4, where v_i and v_o are the input and output variables respectively.

Solution. Let us start from the very beginning and use a block for each element. Since the block diagrams for the elements are drawn on the basis of their voltage-current relationships (Fig. 9–1), we need the current i in Fig. 9–4 as an intermediate variable. Two transformed equations can be written for the two elements in the circuit:

for R:

$$I(s) = \frac{1}{R} [V_i(s) - V_o(s)], \qquad (9\text{–}1)$$

for C:

$$V_o(s) = \frac{1}{Cs} I(s). \qquad (9\text{–}2)$$

The right side of Eq. (9–1) indicates that a summing point for $V_i(s)$ and $V_o(s)$ is necessary. The flow of the transformed variables proceeds from the input variable $V_i(s)$ to the intermediate variable $I(s)$ by Eq. (9–1), and then to the output variable $V_o(s)$ by Eq. (9–2). The result is the block diagram shown in Fig. 9–5. We mention in passing that the only change that is necessary if the output variable is the current i, instead of the voltage v_o, is to pick the output off at $I(s)$ instead of at $V_o(s)$.

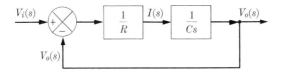

FIG. 9–5. Block diagram for circuit in Fig. 9–4.

EXAMPLE 9–2. Draw a block diagram for the two-section R-C circuit of Fig. 9–6.

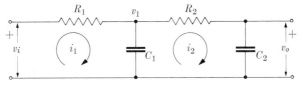

FIG. 9–6. Circuit for Example 9–2.

Solution. In this example we will have four blocks corresponding to the four elements. In order to write the transformed voltage-current relationship for each element, we introduce three intermediate variables v_1, i_1, and i_2, as shown.

The four transformed equations are:

for R_1:

$$I_1(s) = \frac{1}{R_1} [V_i(s) - V_1(s)], \tag{9-3}$$

for C_1:

$$V_1(s) = \frac{1}{C_1 s} [I_1(s) - I_2(s)], \tag{9-4}$$

for R_2:

$$I_2(s) = \frac{1}{R_2} [V_1(s) - V_o(s)], \tag{9-5}$$

for C_2:

$$V_o(s) = \frac{1}{C_2 s} I_2(s). \tag{9-6}$$

Equations (9–3), (9–4), and (9–5) indicate the need for summing points before the blocks representing R_1, C_1, and R_2. The flow of the transformed variables proceeds from the input variable $V_i(s)$ through the intermediate variables $I_1(s)$, $V_1(s)$, and $I_2(s)$ to the output variable $V_o(s)$. The required block diagram is shown in Fig. 9–7.

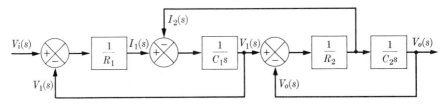

FIG. 9–7. Block diagram for circuit in Fig. 9–6.

It is important to note that the block diagram of a two-section R-C circuit *cannot* be obtained by putting the block diagrams of two single-section R-C circuits in cascade. This is evident from Fig. 9–7, which is not directly derivable from Fig. 9–5. The second section would "load" the first if this were attempted. In other words, v_1 and i_1 in the circuit of Fig. 9–6 are critically dependent upon what is connected across C_1.

We see that the block diagrams of both Fig. 9–5 and Fig. 9–7 contain feedback paths. In the next section we shall discuss the rules for simplifying block diagrams.

9–3 Block-diagram transformations. We shall now establish a few rules by which we can transform complex block diagrams into simpler equivalent ones. The simplest is a diagram consisting of a single block which has a transfer function that converts the transformed input variable into the same transformed output variable as the original block diagram. Simplification by transformation of the block diagram in Fig. 9–7, for example, performs effectively the same function as the solution of $V_o(s)$ in terms of $V_i(s)$ from the simultaneous equations (9–3) through (9–6).

Many rules for block-diagram transformation have been given in the literature,* but only a few are important. For easy reference, we have listed in Table 9–2 the few important rules, from which all others can be derived.

The validity of Rules 1 through 5 is obvious, as is shown in the Remarks column. The proof of Rule 6 may proceed as follows (refer to Fig. 9–8). First we write an equation for each of the two blocks:

for block $G(s)$:

$$O(s) = G(s)E(s)$$

$$= G(s)[I(s) \pm F(s)],$$

$$(9–7)$$

for block $H(s)$:

$$F(s) = H(s)O(s). \quad (9–8)$$

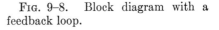

Fig. 9–8. Block diagram with a feedback loop.

Substituting the transformed feedback variable $F(s)$ given in Eq. (9–8) into Eq. (9–7), we have

$$O(s) = G(s)[I(s) \pm H(s)O(s)],$$

or

$$O(s) = \frac{G(s)}{1 \mp G(s)H(s)} I(s). \quad \text{Q.E.D.}$$

$$(9–9)$$

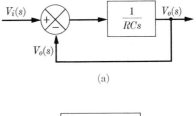

(a)

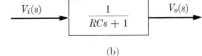

(b)

Now let us apply the rules in Table 9–2 to simplify the block diagrams of Figs. 9–5 and 9–7. The results, shown as the single-block, open-loop diagrams of Fig. 9–9(b) and Fig. 9–10(e), can be

Fig. 9–9. Reduction of block diagram in Fig. 9–5. (a) Combining the two cascaded blocks (Rule No. 1). (b) Eliminating the feedback loop (Rule No. 7).

---

* T. D. Graybeal, "Block Diagram Network Transformation," *Electrical Engineering*, **70**, pp. 985–990; November 1951. T. M. Stout, "A Block Diagram Approach to Network Analysis," *AIEE Transactions*, Part II, Applications and Industry, **71**, pp. 255–260; November 1952.

TABLE 9–2

SOME IMPORTANT BLOCK-DIAGRAM TRANSFORMATIONS

Original diagram	Equivalent diagram	Remarks
		1. Combining two blocks in cascade, $I \cdot G_1 \cdot G_2 = I(G_1 \cdot G_2)$
		2. Moving a summing point behind a block, $(I \pm F)G = (I \cdot G) \pm (F \cdot G)$
		3. Moving a summing point ahead of a block, $(I \cdot G) \pm F = \left(I \pm \dfrac{1}{G} \cdot F\right) G$
		4. Moving a pickoff point behind a block.
		5. Moving a pickoff point ahead of a block.
		6. Eliminating a feedback loop (see p. 263 for proof).
		7. Special case for Rule 6 above ($H = 1$).

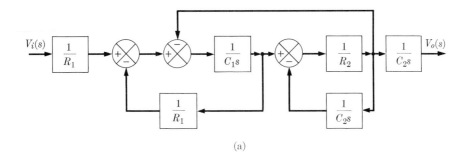

(a)

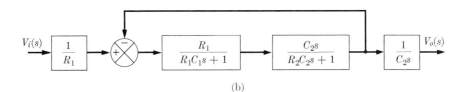

(b)

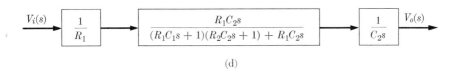

(c)

(d)

(e)

Fig. 9–10. Reduction of block diagram in Fig. 9–7. (a) Moving first summing point behind block $1/R_1$ (Rule No. 2) and last pickoff point ahead of block $1/C_2s$ (Rule No. 5). (b) Eliminating two feedback loops (Rule No. 6). The order of two consecutive summing points is interchangeable. (c) Combining two cascaded blocks (Rule No. 1). (d) Eliminating the last feedback loop (Rule No. 6). (e) Combining the three cascaded blocks in (d) (Rule No. 1).

checked for correctness by working with equations written directly from the circuits given in Figs. 9–4 and 9–6.

EXAMPLE 9–3. The switch across the constant-current source in the circuit of Fig. 9–11 is opened at $t = 0$. It is desired to determine the current in R_b. Find the transfer function by the block diagram method.

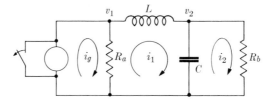

FIG. 9–11. Circuit for Example 9–3.

Solution. The transformed voltage-current relationships for the four elements R_a, L, C, and R_b are to be used for drawing the block diagram. i_2 is the desired output variable. Three intermediate variables, v_1, i_1, and v_2, have been introduced in the circuit. The four transformed equations are:

for R_a:
$$V_1(s) = R_a[I_g(s) - I_1(s)], \tag{9–10}$$

for L:
$$I_1(s) = \frac{1}{Ls}[V_1(s) - V_2(s)], \tag{9–11}$$

for C:
$$V_2(s) = \frac{1}{Cs}[I_1(s) - I_2(s)], \tag{9–12}$$

for R_b:
$$I_2(s) = \frac{1}{R_b}V_2(s). \tag{9–13}$$

Equations (9–10), (9–11), and (9–12) indicate the need for summing points before the blocks representing R_a, L, and C. The block diagram is drawn in Fig. 9–12.

Comparison of Figs. 9–7 and 9–12 reveals one thing thus far unexpected; that is, the circuits in Figs. 9–6 and 9–11 have similar block-diagram representations. This means that the flow of variables from the input to the output for these two circuits follows the same pattern. The reduction of the block diagram in Fig. 9–12 would proceed in exactly the same manner as in Fig. 9–10. However, the resulting over-all transfer function will be different because the circuits in Figs. 9–6 and 9–11 are

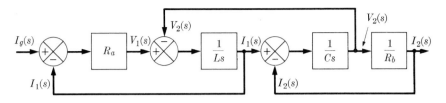

FIG. 9–12. Block diagram for circuit in Fig. 9–11.

not duals of each other. From this point of view, we can say that *two systems having the same pattern for the flow of variables will have similar block-diagram representations*, and that *analogous systems are systems with similar block-diagram representations in which the transfer functions of the corresponding individual blocks have the same forms.*

To complete this example, we obtain the required transfer function from that in Fig. 9–10(e),

$$\frac{1}{R_1 R_2 C_1 C_2 s^2 + (R_1 C_1 + R_2 C_2 + R_1 C_2)s + 1} ,$$

by making the following changes:

$$R_1 \to 1/R_a, \qquad C_1 s \to Ls, \qquad R_2 \to Cs, \qquad C_2 s \to R_b,$$

which yields

$$\frac{I_2(s)}{I_g(s)} = \frac{1}{(R_b/R_a)LCs^2 + [(Ls/R_a) + R_b Cs + (R_b/R_a)] + 1}$$

$$= \frac{R_a}{R_b LCs^2 + (L + R_a R_b C)s + (R_a + R_b)} . \qquad (9\text{--}14)$$

With a little experience one should be able to draw the block diagram from a given circuit directly, without first writing the equations relating the transformed variables. The rules given in Table 9–2 for reducing block diagrams are quite simple. System analysis by the method of block diagrams has the advantage of affording a better understanding of the part played by each component element than is possible to obtain by manipulation of equations.

9–4 Signal flow graph representation. The block-diagram representation of a system as described in the previous sections is an interconnection of unidirectional, operational blocks which shows the flow of system variables. A block may represent a single component or a group of several components, but each block is completely characterized by a transfer function. Construction of a block diagram is basically a matter of following a succession of causes and effects through a system, expressing each transformed variable in terms of itself and other transformed variables. It is a convenient way of describing a physical system and, with the aid of some simple rules of transformation as listed in Table 9–2, a complicated block diagram can be reduced and the over-all system response can be evaluated.

Although the block-diagram method of representing a system is convenient to use and is familiar to most engineers, a recently developed method of representation using signal flow graphs* is simpler and amenable to more elegant solutions. General signal flow graph algebra and formulas which permit writing functional relations between any two system variables have also been developed.† In this section we shall introduce the basic ideas about signal flow graphs and show how they can be used for the analysis of physical systems. Elementary rules of flow-graph transformation and a general formula for computing over-all transfer function will be presented and applied to typical problems.

A signal flow graph is a diagram consisting of *nodes* which are connected by *directed branches*. Each node j has associated with it a system variable (node signal) X_j, and each directed branch jk from node j to node k has an associated *branch transmittance* T_{jk}. *The nodes sum up all the signals that enter them, and transmit the sum signals (node signals) to all outgoing branches.* In this sense, the nodes are like repeater stations. If the node signals are *transformed* system variables, then the branch transmittance T_{jk} is essentially the transfer function which specifies the manner in which the signal at node k depends upon that at node j. The contribution of the signal at node j, X_j, to the signal at node k is equal to the product of X_j and the branch transmittance T_{jk}. Thus, at the kth node, we have

$$X_k = \sum_j T_{jk} X_j, \tag{9–15}$$

where the summation is taken over all branches *entering* the kth node. Because of the unidirectional property of the directed branches, T_{jk} carries no implication concerning the dependence of the signal at node j upon that at node k.

---

* S. J. Mason, "Feedback Theory—Some Properties of Signal Flow Graphs," *Proceedings of the Institute of Radio Engineers*, **41,** pp. 1144–1156; September 1953.

† S. J. Mason, "Feedback Theory—Further Properties of Signal Flow Graphs," *Proceedings of the Institute of Radio Engineers*, **44,** pp. 920–926; July 1956.

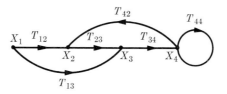

FIG. 9–13. A typical signal flow graph.

In Fig. 9–13 is shown a typical signal flow graph with four nodes and six directed branches. The equations connecting the (transformed) system variables at the four nodes form a set of linear algebraic equations:

$$X_1 = X_1, \tag{9–16}$$

$$X_2 = T_{12}X_1 + T_{42}X_4, \tag{9–17}$$

$$X_3 = T_{13}X_1 + T_{23}X_2, \tag{9–18}$$

$$X_4 = T_{34}X_3 + T_{44}X_4. \tag{9–19}$$

We can, of course, solve Eqs. (9–17), (9–18), and (9–19) simultaneously for X_2, X_3, and X_4 in terms of X_1. However, just as we simplify block diagrams, we can reduce the given flow graph by means of a few simple rules of transformation to one with a single directed branch going from the source (excitation) node X_1 to the dependent (response) node X_2, X_3, or X_4. A *source node* is a node which has only outgoing branches, and a *dependent node* is one which has one or more incoming branches. The overall transmittance between a source node and a specified dependent node is called a *graph transmittance*.

Before we establish the rules for flow-graph transformation, let us pause and compare the block-diagram and the flow-graph methods of representing a physical system. Apart from the fact that concise, general formulas for computing graph transmittances have been established for signal flow graphs, there are two obvious aspects which make the flow-graph method of representation superior to the block-diagram method: the omission of the "blocks," and the avoidance of the necessity for distinguishing summing and pickoff points. In a signal flow graph, simple directed branches instead of blocks are used to denote transmittances (transfer functions); nodes with one outgoing branch and more than one incoming branch are effectively summing points, while nodes with one incoming branch and more than one outgoing branch are effectively pickoff points. A general node which has several incoming as well as several outgoing branches therefore serves as a combined summing and pickoff point. The system represented by the signal flow graph of Fig. 9–13 in block-diagram representation will look like Fig. 9–14, which certainly has a more complicated appearance.

Several important rules for flow-graph transformation can be easily established and are tabulated in Table 9–3. Proofs for these rules are simple and are included in the Remarks column. With the aid of these rules we can reduce a complicated flow graph step by step and essentially compute the desired transmittance (transfer function) between an input variable (excitation) and an output variable (response) in the process.

EXAMPLE 9–4. Reduce the signal flow graph in Fig. 9–13, and find the graph transmittance from the source node X_1 to the dependent node $\dot{X}_4$.

Solution. Starting from the given graph, which is redrawn in Fig. 9–15(a), we reduce it in four steps, as shown in Figs. 9–15(b),(c),(d), and (e). Explanations accompany each step and the reader should have no difficulty in following the reduction process. The desired graph transmittance (transfer function between the transformed variables X_1 and X_4) is

$$T = \frac{X_4}{X_1} = \frac{T_{34}(T_{13} + T_{12}T_{23})}{1 - T_{44} - T_{42}T_{23}T_{34}}. \tag{9-20}$$

As a further illustration of the application of the signal flow graph technique, let us consider the following example.

EXAMPLE 9–5. Draw the flow graph for the two-section R-C circuit in Fig. 9–6 and determine the transfer function $V_o(s)/V_i(s)$.

Solution. As in Example 9–2, we can choose the five transformed circuit variables V_i, I_1, V_1, I_2, and V_o, and draw a five-node flow graph. Equations (9–3) through (9–6) are first rewritten as follows:

$$I_1(s) = \frac{1}{R_1} V_i(s) - \frac{1}{R_1} V_1(s), \tag{9-21}$$

$$V_1(s) = \frac{1}{C_1 s} I_1(s) - \frac{1}{C_1 s} I_2(s), \tag{9-22}$$

$$I_2(s) = \frac{1}{R_2} V_1(s) - \frac{1}{R_2} V_o(s), \tag{9-23}$$

$$V_o(s) = \frac{1}{C_2 s} I_2(s). \tag{9-24}$$

In constructing the flow graph, we simply follow the cause-and-effect procedure given by the above equations, from the input variable $V_i(s)$ to the output variable $V_o(s)$. The resulting flow graph is given in Fig. 9–16.

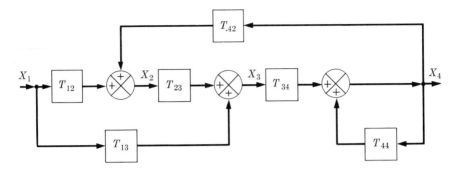

FIG. 9–14. Block-diagram representation of the system in Fig. 9–13.

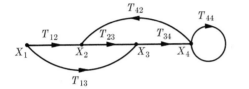

(a) The given flow graph.

(b) Absorbing node X_2 (Rule 3).

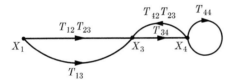

(c) Superposition of branches T_{13} and $T_{12}T_{23}$ (Rule 2) and reduction of feedback loop X_3-X_4-X_3 (Rule 5).

(d) Eliminating feedback loop at X_4 (Rule 6).

(e) Absorbing node X_3 (cascaded nodes, Rule 1).

FIG. 9–15. Reduction of a signal flow graph.

TABLE 9–3

SOME IMPORTANT FLOW-GRAPH TRANSFORMATIONS

Original graph	Equivalent graph	Remarks
		1. Cascaded nodes, $X_3 = T_{23}X_2$ $= T_{23}(T_{12}X_1)$ $= T_{12}T_{23}X_1$
		2. Superposition, $X_2 = T'_{12}X_1$ $+ T''_{12}X_1$ $= (T'_{12} + T''_{12})X_1$
		3. Absorbing a node, $X_5 = T_{15}X_1$ $+ T_{25}X_2$ $X_3 = T_{53}X_5,$ $X_4 = T_{54}X_5,$ or $X_3 = T_{15}T_{53}X_1$ $+ T_{25}T_{53}X_2$ $X_4 = T_{15}T_{54}X_1$ $+ T_{25}T_{54}X_2$
		4. Absorbing a node (special case of Rule 3 above), $T_{54} = 0,$ $X_2 = X_3$
		5. Reducing a feedback loop (special case of Rule 4 above), $T_{15} = 1$
		6. Eliminating a feedback loop, $X_2 = T_{12}X_1$ $+ T_{22}X_2,$ or $X_2 = \dfrac{T_{12}}{1 - T_{22}} X_1$

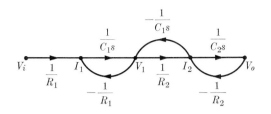

FIG. 9–16. Signal flow graph for R-C circuit in Fig. 9–6.

In order to determine the transfer function $V_o(s)/V_i(s)$, we reduce the flow graph (with arbitrary transmittances in a sequence of steps, as shown in Fig. 9–17. Two general rules for the reduction process are worthy of note: (1) In absorbing a node, all paths going through the node should be individually retained with proper path transmittances. Hence, in absorbing node I_1 in Fig. 9–17(b), we retain the two paths going through I_1 (V_i-I_1-V_1 with path transmittance T_1T_2, and V_1-I_1-V_1 with path transmittance T_2T_3). (2) In eliminating a feedback loop with loop transmittance T at a node, *either* the branch transmittances of all the branches entering the node *or* the branch transmittances of all the branches leaving the node should be divided by $(1 - T)$. Hence, in eliminating the feedback loop with loop transmittance T_2T_3 at node V_1 in Fig. 9–17(c), we divide the branch transmittance T_4 of the single outgoing branch V_1-I_2 by $(1 - T_2T_3)$. Alternatively, we could have divided the branch transmittances T_1T_2 and T_5 of both incoming branches V_i-V_1 and I_2-V_1 by $(1 - T_2T_3)$; there would be no change in the final result.

From Fig. 9–17(g) we have

$$\frac{V_o(s)}{V_i(s)} = \frac{T_1T_2T_4T_6}{(1 - T_2T_3 - T_4T_5)(1 - T_o)}$$

$$= \frac{T_1T_2T_4T_6}{1 - T_2T_3 - T_4T_5 - T_6T_7 + T_2T_3T_6T_7}. \qquad (9\text{–}25)$$

Substituting $T_1 = 1/R_1$, $T_2 = 1/C_1s$, $T_3 = -1/R_1$, $T_4 = 1/R_2$, $T_5 = -1/C_1s$, $T_6 = 1/C_2s$ and $T_7 = -1/R_2$ in Eq. (9–25), we obtain, finally,

$$\frac{V_o(s)}{V_i(s)} = \frac{1}{R_1R_2C_1C_2s^2 + (R_1C_1 + R_2C_2 + R_1C_2)s + 1}, \qquad (9\text{–}26)$$

which is the desired answer.

In the above example, the reader may wonder why the flow graph in Fig. 9–16 does not resemble the block-diagram representation shown in

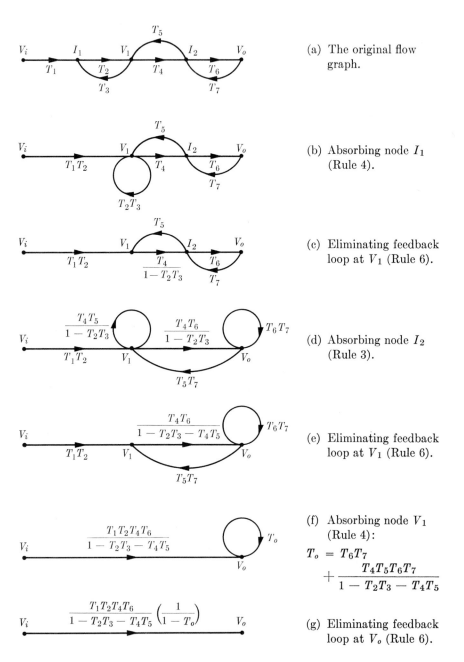

(a) The original flow graph.

(b) Absorbing node I_1 (Rule 4).

(c) Eliminating feedback loop at V_1 (Rule 6).

(d) Absorbing node I_2 (Rule 3).

(e) Eliminating feedback loop at V_1 (Rule 6).

(f) Absorbing node V_1 (Rule 4):

$$T_o = T_6 T_7 + \frac{T_4 T_5 T_6 T_7}{1 - T_2 T_3 - T_4 T_5}$$

(g) Eliminating feedback loop at V_o (Rule 6).

FIG. 9–17. Reduction of signal flow graph in Fig. 9–16.

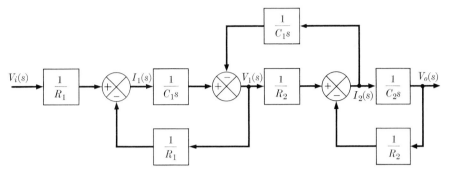

FIG. 9–18. Block diagram for R-C circuit in Fig. 9–6 based upon Eqs. (9–21) through (9–24).

Fig. 9–7 for the same R-C circuit. The answer lies in the fact that they were drawn from different sets of equations: the flow graph from Eqs. (9–21) through (9–24) and the block diagram from Eqs. (9–3) through (9–6). Although these two sets of equations are identical mathematically, they yield different representations because of the different ways in which the various terms are combined. If we were to draw the block-diagram representation of the given circuit based upon Eqs. (9–21) through (9–24), we would obtain Fig. 9–18.

The similarity between Figs. 9–18 and 9–16 is now obvious. Comparison of Fig. 9–7 with Fig. 9–18 reveals that the former can be transformed to the latter when all three summing points are moved behind a block in accordance with Rule 2 in Table 9–2. This points up the fact that neither the block-diagram nor the flow-graph representation of a physical system is unique, since there are different ways of writing the equations that describe a system.

We now present without proof a general formula for graph transmittance, that enables the transfer function between an input variable and an output variable to be written from a signal flow graph by inspection. An input variable is represented by a *source node* which has only outgoing branches; an output variable is represented by a *sink node* with only incoming branches. A path leading from a source node to a sink node without passing through any node more than once is called an *open path*.* V_i-I_1-V_1-I_2-V_o in Fig. 9–17(a) is then an open path. In a more general flow graph, there may be more than one open path between a source node and a sink node. For example, the flow graph in Fig. 9–13 has two open paths, X_1-X_3-X_4 and X_1-X_2-X_3-X_4, between the source node X_1 and the sink node X_4.

---

* Also called a *forward path*.

On the other hand, a *feedback loop* (or simply *loop*) is a closed path. There are three feedback loops in the flow graph shown in Fig. 9–17(a), namely, I_1-V_1-I_1, V_1-I_2-V_1, and I_2-V_o-I_2. The *loop transmittance* of a feedback loop is defined as the product of the transmittances of the branches forming the loop. Thus, the loop transmittances of the three feedback loops in Fig. 9–17(a) are T_2T_3, T_4T_5, and T_6T_7. The general formula for graph transmittance between a source node and a sink node in a signal flow graph may be written as follows:

$$T = \frac{1}{\Delta} \sum_k T_k \, \Delta_k, \qquad\qquad (9\text{--}27)^*$$

where $\Delta = 1 -$ (sum of all individual loop transmittances)

$\qquad\qquad + $ (sum of the products of loop transmittances of all possible nontouching feedback loops taken two at a time)

$\qquad\qquad - $ (sum of the products of loop transmittances of all possible nontouching feedback loops taken three at a time)

$\qquad + \dots,$

$\qquad \Delta_k = $ value of Δ for that part of the graph not touching the kth open path, and

$\qquad T_k = $ path transmittance of the kth open path.

The summation is taken over all open paths between the source node and the sink node under consideration.

The usefulness of Eq. (9–27) can be demonstrated very effectively by applying it to the flow graphs of Figs. 9–15(a) and 9–17(a).

For the flow graph in Fig. 9–15(a):

Two open paths: (a) X_1-X_3-X_4 $T_a = T_{13}T_{34}$

$\qquad\qquad\qquad\quad$ (b) X_1-X_2-X_3-X_4 $T_b = T_{12}T_{23}T_{34}$

Two feedback loops: (A) X_4-X_4 $T_A = T_{44}$

$\qquad\qquad\qquad\qquad$ (B) X_4-X_2-X_4 $T_B = T_{42}T_{23}T_{34}$

No nontouching feedback loops.

Both feedback loops touch open paths (a) and (b).

$$\Delta = 1 - (T_A + T_B) = 1 - T_{44} - T_{42}T_{23}T_{34},$$
$$\Delta_a = \Delta_b = 1.$$

---

* Δ is called the *graph determinant*, and Δ_k is called the *path factor* for the kth open path.

Hence

$$T = \frac{1}{\Delta} (T_a \Delta_a + T_b \Delta_b) = \frac{T_{34}(T_{13} + T_{12}T_{23})}{1 - T_{44} - T_{42}T_{23}T_{34}},$$

which is the same as Eq. (9–20).

For the flow graph in Fig. 9–17(a):

One open path: (a) V_i-I_1-V_1-I_2-V_o $T_a = T_1T_2T_4T_6$

Three feedback loops: (A) I_1-V_1-I_1 $T_A = T_2T_3$

                                         (B) V_1-I_2-V_1 $T_B = T_4T_5$

                                         (C) I_2-V_o-I_2 $T_C = T_6T_7$

Nontouching feedback loops: (A) and (C).

All three feedback loops touch the open path.

$$\Delta = 1 - (T_A + T_B + T_C) + T_A T_C$$
$$= 1 - T_2T_3 - T_4T_5 - T_6T_7 + T_2T_3T_6T_7,$$
$$\Delta_a = 1.$$

Hence

$$T = \frac{1}{\Delta} T_a \Delta_a = \frac{T_1T_2T_4T_6}{1 - T_2T_3 - T_4T_5 - T_6T_7 + T_2T_3T_6T_7},$$

which is the same as Eq. (9–25).

Note that the general formula for graph transmittance enables us to write the transfer function between an input variable and an output variable without the necessity of graphically reducing the signal flow graph, which can sometimes be a laborious task.

9–5 General stability requirements for feedback systems. In Section 9–1 we pointed out that an unstable feedback system cannot perform any useful control function. It is therefore of the utmost importance to be able to determine, from the characteristics of its forward and feedback paths, whether a system with feedback is stable. In this section we shall examine the fundamental requirements for system stability. Block-diagram representation will be used, since we can discuss stability in terms of the transfer functions represented by the blocks in a general way; we are not concerned with what types of components are contained in these blocks.

The block diagram of a servo system with a single feedback loop is shown in Fig. 9–19. The essential transformed relationships have been

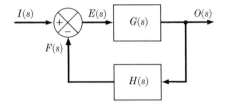

FIG. 9–19. A servo system with a single feedback loop.

derived in connection with Fig. 9–8, but they are repeated below because of their importance. The following notations are used:

$$I(s) = \mathcal{L}[i(t)] = \mathcal{L} \text{ [input function]}, \qquad (9\text{–}28)$$

$$O(s) = \mathcal{L}[o(t)] = \mathcal{L} \text{ [output function]}, \qquad (9\text{–}29)$$

$$F(s) = \mathcal{L}[f(t)] = \mathcal{L} \text{ [feedback function]}, \qquad (9\text{–}30)$$

$$E(s) = \mathcal{L}[e(t)] = \mathcal{L}[i(t) - f(t)] = \mathcal{L} \text{ [error signal]}. \qquad (9\text{–}31)$$

We have

$$O(s) = G(s)E(s), \qquad (9\text{–}32)$$

$$E(s) = I(s) - F(s), \qquad (9\text{–}33)$$

$$F(s) = H(s)O(s), \qquad (9\text{–}34)$$

and

$$O(s) = \frac{G(s)}{1 + G(s)H(s)} I(s). \qquad (9\text{–}35)$$

We can define the ratio of the transformed output to the transformed input, $O(s)/I(s)$, as the *over-all transfer function* or *the system transfer function:*

$$G_o(s) = \frac{O(s)}{I(s)} = \frac{G(s)}{1 + G(s)H(s)}. \qquad (9\text{–}36)$$

In general, both $G(s)$ and $H(s)$ may be ratios of polynomials in s:

$$G(s) = \frac{N_G(s)}{D_G(s)}, \qquad (9\text{–}37)$$

$$H(s) = \frac{N_H(s)}{D_H(s)}. \qquad (9\text{–}38)$$

Substituting Eqs. (9–37) and (9–38) in Eq. (9–36) and clearing fractions, we obtain

$$G_o(s) = \frac{N_G D_H}{N_G N_H + D_G D_H}$$

$$= \frac{a_m s^m + a_{m-1} s^{m-1} + \cdots + a_1 s + a_o}{b_n s^n + b_{n-1} s^{n-1} + \cdots + b_1 s + b_o}, \qquad (9\text{–}39)$$

where the polynomials in both the numerator and the denominator have been written out. In order to fix ideas, let us assume that the input function $i(t)$ is a unit step function $U(t)$, although the subsequent discussions will apply to a general input function. We have

$$I(s) = \mathcal{L}[U(t)] = \frac{1}{s}$$

and

$$O(s) = \frac{1}{s} G_o(s) = \frac{a_m s^m + a_{m-1} s^{m-1} + \cdots + a_1 s + a_0}{s(b_n s^n + b_{n-1} s^{n-1} + \cdots + b_1 s + b_0)}$$

$$= \frac{1}{b_n} \left[\frac{K_0}{s} + \frac{K_1}{s - s_1} + \frac{K_2}{s - s_2} + \cdots + \frac{K_n}{s - s_n} \right], \qquad (9\text{–}40)$$

where the constants $K_0, K_1, K_2, \ldots, K_n$ can be found by Heaviside's expansion theorem. The output or response function is then

$$o(t) = \mathcal{L}^{-1}[O(s)] = \frac{1}{b_n} [K_0 + K_1 \epsilon^{s_1 t} + K_2 \epsilon^{s_2 t} + \cdots + K_n \epsilon^{s_n t}] U(t).$$

$$(9\text{–}41)$$

In Eq. (9–41) the first term is a step function and represents the steady-state response. The exponential terms are transient responses for which the constants $K_1, K_2, \ldots, K_n$ depend upon the amplitude and the form of the input function, and the coefficients of t in the exponents, $s_1, s_2, \ldots, s_n$, are the roots of the polynomial in the denominator of the over-all transfer function $G_o(s)$.

Let us concentrate our discussion on a typical root, s_k. In general, it can be a complex number:

$$s_k = \sigma_k + j\omega_k. \qquad (9\text{–}42)$$

The term in the output function corresponding to the root s_k is

$$\frac{K_k}{b_n} \epsilon^{(\sigma_k + j\omega_k)t} = \left(\frac{K_k}{b_n} \epsilon^{\sigma_k t} \right) \epsilon^{j\omega_k t}.$$

It is clear from the above expression that the amplitude of this term will increase indefinitely with time when σ_k (the real part of s_k) is positive. On the other hand, when σ_k is negative the amplitude will decrease exponentially with time and the transient will subside. We therefore cannot tolerate a positive σ_k which will result in a runaway or an unstable situation. If $\sigma_k = 0$, we have a transitional case which will give rise to sustained oscillations of constant amplitude. This situation will also be considered unstable.

The roots of a polynomial are called the *zeros* of the polynomial for obvious reasons. $s_1, s_2, \ldots, s_k, \ldots, s_n$ are zeros of the denominator of $G_o(s)$; hence from Eq. (9–36) they are also the zeros of the polynomial $1 + G(s)H(s)$.* In view of the preceding discussions, we can state that *the necessary and sufficient condition that a feedback system be stable is that all the zeros of $1 + G(s)H(s)$ have negative real parts.*

The zeros of the polynomial in the denominator of a rational function make the function go to infinity; they are called the *poles* of the function; $s_1, s_2, \ldots, s_k, \ldots, s_n$ are poles of $G_o(s)$. Hence the stability requirement above can be restated as follows:

The necessary and sufficient condition that a feedback system be stable is that all the poles of its over-all transfer function have negative real parts.

The task of solving for all the roots of the equation

$$1 + G(s)H(s) = 0 \qquad (9\text{–}43)\dagger$$

is frequently a tedious one, especially when it is an equation of high degree in s. In many cases, it may not even be possible to write the analytical equation (9–43) accurately for more complicated systems. The remainder of this chapter will be devoted to methods of determining whether the stability requirement is satisfied without actually finding all the roots of Eq. (9–43).

9–6 The Routh-Hurwitz criterion.

In 1877 E. J. Routh developed a method which can be used to determine the number of roots of an algebraic equation having positive real parts without actually finding the roots. This method can also be used to detect the number of roots having zero real parts. In 1895 A. Hurwitz independently arrived at the same method in a different form. We shall use this method on the characteristic equation of a feedback system to establish the Routh-Hurwitz criterion for stability.

---

* We assume here that we do not have to worry about the zeros of $D_G(s)$ [the poles of the forward transfer function $G(s)$]. In systems with multiple feedback loops, this assumption may not be true. We shall discuss this possibility in Section 9–8.

† Note that Eq. (9–43) is the characteristic equation of the over-all system. From Eqs. (9–37) and (9–38) we also see that the roots of Eq. (9–43) are the same as the roots of $N_G N_H + D_G D_H = 0$.

Before discussing the Routh-Hurwitz criterion, let us review some of the properties of the roots of an algebraic equation. Suppose that $s_1, s_2, s_3, \ldots, s_n$ are the roots (which may be real or complex, and may have positive, negative, or zero real parts) of an equation of the nth degree. Then

$$(s - s_1)(s - s_2)(s - s_3) \cdots (s - s_n) = 0. \tag{9-44}$$

Multiplying the factors out, we have, in general, an nth-degree equation in s as follows:

$$s^n - \text{(sum of all the roots) } s^{n-1}$$
$$+ \text{(sum of the products of roots taken two at a time) } s^{n-2}$$
$$- \text{(sum of the products of roots taken three at a time) } s^{n-3}$$
$$+ \cdots$$
$$+ (-1)^n \text{(product of all } n \text{ roots) } s^0 = 0, \tag{9-45}$$

or

$$s^n - (s_1 + s_2 + s_3 + \cdots + s_n) s^{n-1}$$
$$+ (s_1 s_2 + s_2 s_3 + s_1 s_3 + \cdots) s^{n-2}$$
$$- (s_1 s_2 s_3 + s_1 s_2 s_4 + \cdots) s^{n-3}$$
$$+ \cdots$$
$$+ (-1)^n s_1 s_2 s_3 \cdots s_n = 0. \tag{9-46}$$

From Eq. (9–45) or (9–46), we can make the following important observations about the roots:

1. For an equation with real coefficients complex or purely imaginary roots must occur in conjugate pairs. Hence if one of the roots is $s_k = \sigma_k + j\omega_k$, there must be another root $s_k^* = \sigma_k - j\omega_k$.

2. A necessary (but not sufficient) condition for the real parts of all the roots to be negative is that *all* the coefficients of the equation have the same sign.

3. A second necessary (but not sufficient) condition for the real parts of all the roots to be negative is that *none* of the coefficients of the equation is zero; in other words, for an nth-degree equation, there must be $n + 1$ terms (powers of s from 0 to n with no missing terms).

It is then obvious from observations 2 and 3 above that *some* roots will have positive real parts and consequently *a system is unstable if not all the coefficients of its characteristic equation have the same sign or if not all the terms are present;* we need not look further. On the other hand, even if all the coefficients are nonzero and have the same sign, we have no guarantee that the system will be stable. Here, then, the Routh-Hurwitz criterion will be useful. The proof of the Routh-Hurwitz criterion is rather involved.* We shall present the procedure below without proof.

Consider the following equation:

$$a_n s^n + a_{n-1} s^{n-1} + \cdots + a_1 s + a_0 = 0, \qquad (9\text{-}47)$$

where the coefficients $a_n, a_{n-1}, \ldots, a_1, a_0$ are all of the same sign and none is zero.

Step 1. Arrange the coefficients of the given equation in two rows in the following fashion:

$$\begin{array}{llll} \text{Row 1:} & a_n & a_{n-2} & a_{n-4} \ldots \\ \text{Row 2:} & a_{n-1} & a_{n-3} & a_{n-5} \ldots \end{array}$$

where the arrow directions indicate the order of arrangement.

Step 2. Form a third row from the first and second rows as follows:

$$\begin{array}{llll} \text{Row 1:} & a_n & a_{n-2} & a_{n-4} \ldots \\ \text{Row 2:} & a_{n-1} & a_{n-3} & a_{n-5} \ldots \\ \text{Row 3:} & b_{n-1} & b_{n-3} & b_{n-5} \ldots \end{array}$$

where

$$b_{n-1} = -\frac{1}{a_{n-1}} \begin{vmatrix} a_n & a_{n-2} \\ a_{n-1} & a_{n-3} \end{vmatrix}, \qquad (9\text{-}48)$$

$$b_{n-3} = -\frac{1}{a_{n-1}} \begin{vmatrix} a_n & a_{n-4} \\ a_{n-1} & a_{n-5} \end{vmatrix}, \qquad (9\text{-}49)$$

$$\vdots$$

Step 3. Form a fourth row from the second and third rows:

$$\text{Row 4:} \quad c_{n-1} \quad c_{n-3} \quad c_{n-5} \quad \ldots$$

where

$$c_{n-1} = -\frac{1}{b_{n-1}} \begin{vmatrix} a_{n-1} & a_{n-3} \\ b_{n-1} & b_{n-3} \end{vmatrix}, \qquad (9\text{-}50)$$

$$c_{n-3} = -\frac{1}{b_{n-1}} \begin{vmatrix} a_{n-1} & a_{n-5} \\ b_{n-1} & b_{n-5} \end{vmatrix}, \qquad (9\text{-}51)$$

$$\vdots$$

---

* Interested readers are referred to E. A. Guillemin, *The Mathematics of Circuit Analysis*, Chap. VI, Art. 26, John Wiley and Sons, Inc., New York; 1949.

Step 4. Continue this procedure of forming a new row from the two preceding rows until only zeros are obtained. In general, an array of $n + 1$ rows will result, with the last two rows each containing a single element.

The Routh-Hurwitz criterion states that the system whose characteristic equation is Eq. (9–47) *is stable if and only if all the elements in the first column of the array* formed by the coefficients in the above manner *have the same algebraic sign.* If the elements in the first column are not all of the same sign, then *the number of changes of sign in that column equals the number of roots with positive real parts.*

We shall now illustrate the application of the Routh-Hurwitz criterion by examples. In the process we will also show that there are certain special situations for which suitable modifications of the standard procedure are necessary in order to use the criterion.

EXAMPLE 9–6. Apply the Routh-Hurwitz criterion to the following equation:

$$(s + 1)^2(s + 2)(s + 3) = s^4 + 7s^3 + 17s^2 + 17s + 6 = 0.$$

Solution. Following the procedure outlined above, we obtain an array of coefficients as follows:

1st row:	1	17	6
2nd row:	7	17	
3rd row:	14.58	6	
4th row:	14.12		
5th row:	6		

Since all the coefficients in the first column are of the same sign (positive), the given equation has no roots with positive real parts (as we knew from the outset).

EXAMPLE 9–7. Apply the Routh-Hurwitz criterion to the following equation:

$$(s + 1)^2(s - 2)(s - 1 + j2)(s - 1 - j2)$$
$$= s^5 - 2s^4 + 2s^3 + 4s^2 - 11s - 10 = 0.$$

Solution. Even if the equation were given without the roots, we would detect immediately that the system is unstable because the coefficients of the terms in the characteristic equation are not all of the same sign. Application of the Routh-Hurwitz procedure would reveal the *number* of roots with positive real parts.

1st row:	1	2	−11
2nd row:	−2	4	−10
3rd row:	4	−16	
4th row:	−4	−10	
5th row:	−26		
6th row:	−10		

There are three changes of sign in the first column, which correctly correspond to the three roots with positive real parts, namely, 2, $(1 - j2)$, and $(1 + j2)$.

It is obvious that the coefficients in any *row* may be multiplied or divided by a *positive* number without disturbing the changes of sign. This is sometimes done to simplify the numerical work in finding the coefficients of the succeeding rows. This step is illustrated below, although there is probably not much to be gained in this particular example because the problem itself is relatively simple. The new array may be as follows:

1	2	−11	
−1	2	−5	(after dividing by 2)
1	−4		(after dividing by 4)
−2	−5		
−6.5			
−5			

We note that the first column of coefficients still has three changes of sign.

EXAMPLE 9–8.   Apply the Routh-Hurwitz criterion to the following equation:

$$s^5 + 2s^4 + 2s^3 + 4s^2 + 11s + 10 = 0.$$

Solution.   Here all the coefficients are positive and none is zero. We proceed to write the Routh-Hurwitz array as follows:

1st row:	1	2	11
2nd row:	2	4	10
3rd row:	0	6	

We apparently cannot proceed further with this procedure because the first element in the third row is zero. However, this situation can be remedied by multiplying the original equation by a factor $(s + a)$, where $a > 0$, and then applying the Routh-Hurwitz procedure. The reasoning behind this modification is that multiplication by $(s + a)$ will change the vanishing element without changing the number of roots with positive real parts. Any positive real number can be used for a; the simplest choice is $a = 1$.

Multiplying the given equation by $(s + 1)$, we have the new equation

$$s^6 + 3s^5 + 4s^4 + 6s^3 + 15s^2 + 21s + 10 = 0.$$

The Routh-Hurwitz array now becomes:

$$
\begin{array}{llll}
1 & 4 & 15 & 10 \\
1 & 2 & 7 & \text{(after dividing by 3)} \\
1 & 4 & 5 & \text{(after dividing by 2)} \\
-1 & 1 & & \text{(after dividing by 2)} \\
1 & 1 & & \text{(after dividing by 5)} \\
2 \\
1
\end{array}
$$

There are two changes of sign in the first column of this array, which indicates that there are two roots having positive real parts in the given equation.

EXAMPLE 9–9. We shall now illustrate another situation which requires special handling.* Consider the following equation:

$$(s - 1 + j\sqrt{6})(s - 1 - j\sqrt{6})(s + j\sqrt{3})(s - j\sqrt{3})(s + 3)$$
$$= s^5 + s^4 + 4s^3 + 24s^2 + 3s + 63 = 0.$$

Solution. All coefficients in the given equation are nonzero and positive. The Routh-Hurwitz array goes as follows:

$$
\begin{array}{lll}
1 & 4 & 3 \\
1 & 24 & 63 \\
-1 & -3 & \text{(after dividing by 20)} \\
1 & 3 & \text{(after dividing by 21)} \\
0 \\
?
\end{array}
$$

We cannot complete this array because of the zero in the 5th row. We are then tempted to use the technique introduced in the last example,

---

* M. F. Gardner and J. L. Barnes, *Transients in Linear Systems*, Chap. VII, Sec. 9, John Wiley and Sons, Inc., New York; 1942.

i.e., to multiply the original equation by a factor $(s + 1)$. However, this technique is not helpful in this instance because the *whole* 5th row (which happens to have only one element) vanishes. Nevertheless, let us proceed and see what happens. Multiplying the given equation by $(s + 1)$, we have the new equation

$$s^6 + 2s^5 + 5s^4 + 28s^3 + 27s^2 + 66s + 63 = 0.$$

The Routh-Hurwitz array of coefficients for the new equation becomes

1	5	27	63
2	28	66	
−3	−2	21	(after dividing by 3)
1	3		(after dividing by 80/3)
1	3		(after dividing by 7)
0			
?			

We are faced with the same dilemma as before! Examination reveals that this situation will occur whenever the array has two consecutive rows in which the ratio of the corresponding elements is a constant (either positive or negative). When this happens, *all* the elements in the following row will vanish. This is an indication that the given equation has at least one pair of roots which lie radially opposite each other and equidistant from the origin. (That is, if one of the roots is $\sigma_1 + j\omega_1$, there must be another root which is $-\sigma_1 - j\omega_1$, where either σ_1 or ω_1 may be zero.) Therefore *a system is unstable if the Routh-Hurwitz array of its characteristic equation contains two consecutive rows of equal numbers of elements such that the corresponding elements are proportional to each other.*

The array can be completed by first forming an auxiliary polynomial in s^2, using the elements in the last nonvanishing row as its coefficients; the coefficients of the derivative of this auxiliary polynomial are to be taken for the elements in the following row. For our example, the elements in the last nonvanishing row are 1 and 3; hence the auxiliary polynomial in s^2 is

$$s^2 + 3.$$

The derivative of this polynomial is $2s$, whose coefficient 2 is to be taken as the element in the row following the last nonvanishing row. The Routh-Hurwitz array of the given equation now becomes

1	4	3
1	24	63

$$-1 \quad -3 \qquad \text{(after dividing by 20)}$$
$$1 \quad 3 \qquad \text{(after dividing by 21)}$$
$$2$$
$$3$$

There are two changes of sign in the first column, which correctly correspond to the two roots with positive real parts, namely, $(1 - j\sqrt{6})$ and $(1 + j\sqrt{6})$.

The roots of the equation formed by the auxiliary polynomial

$$s^2 + 3 = 0$$

are also roots of the original equation. In this problem they are

$$s = \pm j\sqrt{3},$$

which have zero real parts. Hence *a necessary* (but not sufficient) *condition that a given equation have conjugate roots with zero real parts is that its Routh-Hurwitz array contains two consecutive rows of equal numbers of elements such that the corresponding elements are proportional to each other.*

9–7 Mapping of contours. Although the Routh-Hurwitz criterion enables us to determine the number of roots of the characteristic equation of a feedback system having positive or zero real parts without actually finding the roots, it is impractical to use in most experimental situations since the coefficients of the characteristic equation of a closed-loop system cannot be conveniently determined. Moreover, the Routh-Hurwitz criterion does not provide an indication of the degree of stability when a system is stable, nor does it suggest ways of stabilizing a system that is unstable. For these reasons the stability criterion developed by H. Nyquist[*] in 1932, which is based upon a plot of the open-loop steady-state response of the system for sinusoidal inputs and which overcomes the shortcomings of the Routh-Hurwitz method, has received wide acceptance. A rigorous development of the Nyquist criterion must rely on the theory of the functions of a complex variable. In this section we shall discuss the mapping of contours in complex planes; our discussion will be based upon a simplified explanation of complex-plane integration.

Consider the denominator of the over-all transfer function

$$D_o(s) = 1 + G(s)H(s) = K \frac{(s - s_1)(s - s_2) \cdots (s - s_m)}{(s - s_a)(s - s_b) \cdots (s - s_n)}, \quad (9\text{–}52)$$

---

[*] H. Nyquist, Regeneration Theory, *Bell System Technical Journal*, **11**, p. 126; 1932.

where $s_1, s_2, \ldots, s_m$ are zeros of $D_o(s)$ [poles of the over-all transfer function $G_o(s)$], $s_a, s_b, \ldots, s_n$ are poles of $D_o(s)$ [zeros of $G_o(s)$], and s is a complex variable:

$$s = \sigma + j\omega. \tag{9-53}$$

The function $D_o(s)$ apparently will also be complex. Let us write

$$D_o(s) = u(s) + jv(s). \tag{9-54}$$

When s takes on different values, both $u(s)$ and $v(s)$ will vary. It is manifestly impossible to represent the function $D_o(s)$ graphically in a single diagram for all complex values of s. However, since the zeros, $s_1, s_2, \ldots, s_m$, and the poles, $s_a, s_b, \ldots, s_n$, completely characterize the function $D_o(s)$ (K is merely a scale factor), a simple plot in the s-plane ($\sigma\omega$-plane) showing the location of the zeros and the poles [the dots and crosses in Fig. 9-20(a)] provides a neat way of specifying the properties of $D_o(s)$ graphically. For each value, s_0, of s, there corresponds a value $D_o(s_0) = u(s_0) + jv(s_0)$ which can be represented by a point in the $D_o(s)$-plane (uv-plane). The contribution of a typical factor $(s - s_1)$ in the numerator of Eq. (9-52) at $s = s_0$ is

$$(s_0 - s_1) = \rho_1 \epsilon^{j\phi_1}. \tag{9-55}$$

The contribution of a typical factor $(s - s_a)$ in the denominator of Eq. (9-52) at $s = s_0$ is

$$\frac{1}{(s_0 - s_a)} = \frac{1}{\rho_a \epsilon^{j\phi_a}} = \frac{1}{\rho_a} \epsilon^{-j\phi_a}. \tag{9-56}$$

Equations (9-55) and (9-56) are illustrated graphically in Fig. 9-20(a).

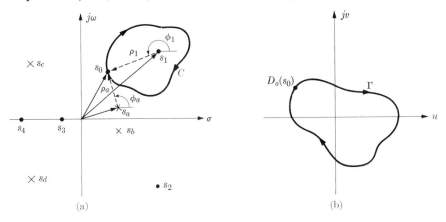

FIG. 9-20. Mapping of a contour C in the s-plane onto a contour Γ in the $D_o(s)$-plane. (a) s-plane. (b) $D_o(s)$-plane.

In computing $D_o(s_0)$, lines should be constructed from all the zeros and all the poles to the point s_0, and in the notation of Eqs. (9–55) and (9–56), we write

$$D_o(s_0) = K \frac{\rho_1 \rho_2 \ldots \rho_m}{\rho_a \rho_b \ldots \rho_n} \epsilon^{j[(\phi_1 + \phi_2 + \cdots + \phi_m) - (\phi_a + \phi_b + \cdots + \phi_n)]}. \quad (9\text{–}57)$$

If we now allow s to take on successive values along a closed contour C in the s-plane, a corresponding closed contour Γ, as shown in Fig. 9–20(b), will be traced. We say that the contour C in the s-plane is *mapped* onto the $D_o(s)$-plane as the contour Γ by the functional relationship given in Eq. (9–52). Two things about this mapping are of particular interest, namely, the direction of tracing and the general location of the contour Γ relative to a given contour C. We know that the exact shape of Γ depends critically on that of C; but why have we drawn contour Γ clockwise for a clockwise contour C as shown? Is there any particular reason that contour Γ should encircle the origin of the $D_o(s)$-plane? In order to answer these questions we have to examine the dependence of the direction of tracing and the general location of contour Γ to those of contour C more closely. Let us consider four different cases: (1) contour C encircles neither zeros nor poles of $D_o(s)$, (2) contour C encircles zeros but not poles, (3) contour C encircles poles but not zeros, and (4) contour C encircles both zeros and poles.

(1) *Contour C encircles neither zeros nor poles.* It is clear that a closed contour in the s-plane maps onto a closed contour in the $D_o(s)$-plane. Contours C and Γ essentially divide the s- and $D_o(s)$-planes into two regions, one inside and one outside the contours. If the contour C in the s-plane does not encircle any zeros or poles, then the contour Γ in the $D_o(s)$-plane cannot encircle either the origin or the point at infinity. As we trace the contour C in Fig. 9–21(a) in the direction of the arrow (clockwise), the interior of the contour which contains neither zeros nor poles is always on the right of the contour. Correspondingly, in the $D_o(s)$-plane

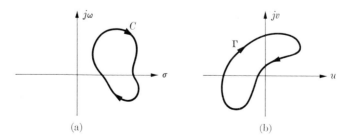

(a)   (b)

Fig. 9–21. Mapping of a contour C which encircles neither zeros nor poles of $D_o(s)$. (a) s-plane. (b) $D_o(s)$-plane.

of Fig. 9–21(b) the region which contains neither the origin nor the point at infinity should also be on the right of the contour as Γ is traced.* In other words, the interior of the contour C in the s-plane is mapped onto the $D_o(s)$-plane as the interior of the contour Γ, and if C is described in a clockwise direction, so is Γ.

(2) *Contour C encircles zeros but not poles.* As we trace the contour C in Fig. 9–22(a) in the direction of the arrow, the interior of the contour, which is always on the right of the contour and which contains a zero, s_1, maps onto the $D_o(s)$-plane as the contour Γ in Fig. 9–22(b). Γ is correspondingly traced with its interior, which contains the origin, always on its right. Hence if C is described in a clockwise direction, so is Γ.

In Fig. 9–22(b), the contour Γ encircles the origin of the $D_o(s)$-plane *once* in the clockwise direction because the contour C encircles *one* zero in the s-plane in the clockwise direction. This can also be seen from Eq. (9–55) and Fig. 9–20(a) since each encirclement of the zero s_1 (each circuit of the contour C made by the tip of the phasor $s - s_1$) in the clockwise direction corresponds to a change of -2π in the angle ϕ_1. It is easy to extend this to the situation where the contour C encircles two zeros, s_1 and s_2:

$$(s - s_1)(s - s_2) = \rho_1\rho_2\epsilon^{j(\phi_1+\phi_2)}. \tag{9–58}$$

When the contour C is described once in the clockwise direction the total change in the phase angle $(\phi_1 + \phi_2)$ is -4π, which means that the contour Γ in the $D_o(s)$-plane encircles the origin twice in the clockwise direction. This is illustrated in Fig. 9–23. In general, *if the contour C encircles*

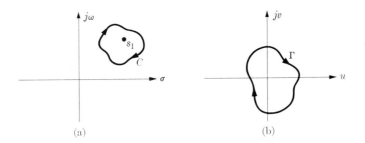

(a)   (b)

FIG. 9–22. Mapping of a contour C which encircles a zero of $D_o(s)$. (a) s-plane. (b) $D_o(s)$-plane.

---

* A mapping or transformation that preserves this property is said to be *conformal* in the language of the theory of the functions of a complex variable. It only requires that the function $D_o(s)$ is analytic and $D_o'(s) \neq 0$ along the contour.

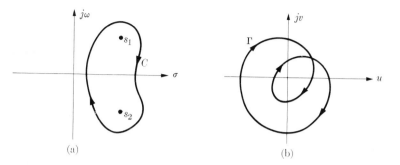

FIG. 9–23. Mapping of a contour C which encircles two zeros of $D_o(s)$. (a) s-plane. (b) $D_o(s)$-plane.

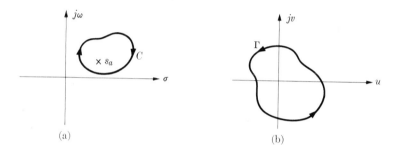

FIG. 9–24. Mapping of a contour C which encircles a pole of $D_o(s)$. (a) s-plane. (b) $D_o(s)$-plane.

Z zeros in the clockwise direction, the corresponding contour Γ encircles the origin of the $D_o(s)$-plane Z times, also in the clockwise direction. A zero of multiplicity k counts as k zeros.

(3) *Contour C encircles poles but not zeros.* In this case, as we trace the contour C in Fig. 9–24(a) in the direction of the arrow, the region on the right which contains a pole, s_a, maps onto the $D_o(s)$-plane as the region also on the right of the contour Γ, which contains the point at infinity as Γ is described. Hence the interior of the contour C maps onto the exterior of the contour Γ. Since the interior of C does not contain zeros, the exterior of Γ cannot contain the origin. Consequently, if the contour C encircles a pole in the clockwise direction, the contour Γ encircles the origin once in the *counterclockwise* direction.

The above conclusion can also be drawn from Eq. (9–56) and Fig. 9–20(a), since each encirclement of the pole s_a in the clockwise direction corresponds to a change of $+2\pi$ in the angle $-\phi_a$. In general, *if the contour C encircles P poles in the clockwise direction, the corresponding contour Γ*

encircles the origin of the $D_o(s)$-*plane* P *times in the counterclockwise direction.* A pole of multiplicity k counts as k poles.

(4) *Contour C encircles both zeros and poles.* Combining the results in cases (2) and (3) above, we can make the following general statement: *If the contour C encircles Z zeros and P poles in the clockwise direction, the corresponding contour* Γ *encircles the origin of the* $D_o(s)$-*plane*

$$N = Z - P \tag{9-59}$$

times in the clockwise direction.

In the following section we shall see that this mapping theorem is the heart of the Nyquist stability criterion.

9–8 The Nyquist criterion. In Section 9–5 we made the statement that the necessary and sufficient condition for a feedback system to be stable is that all the zeros of the function $D_o(s) = 1 + G(s)H(s)$ have negative real parts. Since the entire complex s-plane is divided in half by the $j\omega$-axis, and all points in the left half-plane have negative real parts, those in the right half-plane have positive real parts, and those on the imaginary axis have zero real parts, this condition is the same as requiring that no zeros of $D_o(s)$ should lie either in the right half-plane or on the imaginary axis. The Nyquist stability criterion is a statement of this requirement in terms of the mapping theorem developed in the preceding section; there is no need to determine the exact location of the zeros.

If we construct in the s-plane a *clockwise* contour which extends along the imaginary axis from $\omega = -\infty$ to $\omega = +\infty$ and then folds back around a semicircular path of infinite radius enclosing the entire right half-plane, this contour will certainly either encircle or pass through *all* the zeros and poles in the right half-plane or on the imaginary axis. Such a contour is shown in Fig. 9–25. If this contour is mapped onto the $D_o(s)$-plane as a contour Γ, then, in view of Eq. (9–59), we can state at once that *a system is unstable if* Γ *either encircles or passes through the origin in a clockwise direction.* This is so because a *clockwise* encirclement of (or trace through) the origin is possible only when some *zeros* of $D_o(s)$ lie in the right half-plane (or on the imaginary axis). Since $D_o(s) = 1 + G(s)H(s)$, or $G(s)H(s) = D_o(s) - 1$, a contour Γ in the $D_o(s)$-plane can be further mapped into a contour Γ' in the $G(s)H(s)$-plane by shifting itself horizontally to the left by one unit. Consequently, we can modify the above statement to say that *a system is unstable if the contour* Γ' *in the* $G(s)H(s)$-*plane either encircles or passes through the* $(-1, 0)$ *point in a clockwise direction.*

Let us now examine the contour C in the s-plane more closely. We shall discuss the two portions of C separately: (1) the semicircular path with infinite radius, and (2) the path along the $j\omega$-axis.

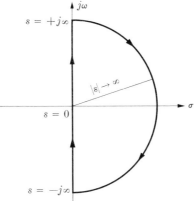

FIG. 9–25. Contour C which encloses the entire right half of the complex s-plane.

(1) *The semicircular path with infinite radius.* Preparatory to the formal introduction of the Nyquist criterion, we will assume that *the product $G(s)H(s)$ approaches zero or a constant as s approaches ∞.* As a result, $D_o(s)$ will also be either zero or a constant as $s \to \infty$. If $D_o(s)$ is expressed in the form of Eq. (9–52), this means that $m \leq n$. This condition must be satisfied when we apply the Nyquist criterion.

Along the semicircular path, $|s| \to \infty$. Hence, because of the above restriction, both $D_o(s)$ and $G(s)H(s)$ remain constant, and there is no rotation of the contour Γ in the $D_o(s)$-plane or of the contour Γ' in the $G(s)H(s)$-plane as s varies along the infinite semicircular path. In other words, *all the rotations or encirclements of contour Γ and contour Γ' occur while the $j\omega$-axis is being traversed;* the infinite semicircular path contributes nothing in this respect. In mapping the contour C in the s-plane onto the $G(s)H(s)$-plane as the contour Γ', all we need do is to plot the graph of the function $G(j\omega)H(j\omega)$ as ω varies from $-\infty$ to $+\infty$. But $G(j\omega)H(j\omega)$ is the open-loop (the forward and the feedback blocks in cascade) steady-state response of the system. We will call its graph, as ω varies from $-\infty$ to $+\infty$, the *open-loop transfer locus.* In view of Eq. (9–59) we state the requirement that no zeros of $D_o(s)$ lie in the right half of the s-plane or on the $j\omega$-axis as the *Nyquist criterion for stability:*

> *A feedback system is stable if and only if its open-loop transfer locus does not pass through the $(-1, 0)$ point and the number of counterclockwise encirclements about the $(-1, 0)$ point equals the number of poles of $G(s)H(s)$ with positive real parts.*

This follows from the relation $N = -P$ by putting $Z = 0$ in Eq. (9–59); $-P$ encirclements in the clockwise direction is the same as P encirclements in the *counterclockwise* direction. The above assumes that $G(s)H(s)$ has no poles on the $j\omega$-axis. (We shall presently discuss the case where this is not true.) In many single-loop feedback systems, $G(s)H(s)$ has no poles in the right half-plane ($P = 0$), and the Nyquist criterion then simplifies to the following:

> *A feedback system with $P = 0$ is stable if and only if its open-loop transfer locus does not pass through or encircle the $(-1, 0)$ point.*

The open-loop transfer locus in the $G(s)H(s)$-plane is often referred to as the *Nyquist diagram*.

(2) *The path along the $j\omega$-axis.* The fact that the Nyquist diagram is a plot in the complex plane of the open-loop steady-state response of a system to sinusoidal excitation is an important advantage of the Nyquist criterion because the diagram can be determined by experimental methods and because it lends physical meaning to the procedure. Quite frequently the open-loop transfer function $G(s)H(s)$ of a feedback system will have a pole at the origin; $G(s)H(s)$ will then become infinite there. As the $j\omega$-axis is traversed, there will be a discontinuity at $\omega = 0$ and the open-loop transfer locus is broken into two segments. The connection between these segments can be properly defined by modifying the path along the $j\omega$-axis to include a semicircular detour of very small radius, ρ, at the pole *in the positive half-plane.* On the semicircular detour around the origin

$$s = \rho\epsilon^{j\phi}. \tag{9-60}$$

When ρ is very small, the predominant term in the partial-fraction expansion of $G(s)H(s)$ will be that having s in the denominator:

$$G(s)H(s) \cong \frac{K_0}{s} = \frac{K_0}{\rho}\,\epsilon^{-j\phi}, \qquad -\frac{\pi}{2} \le \phi \le \frac{\pi}{2}. \tag{9-61}$$

As $\rho \to 0$, $|G(s)H(s)| \to \infty$ and the angle of $G(s)H(s)$ goes from $+\pi/2$ to $-\pi/2$ as ϕ changes from $-\pi/2$ to $\pi/2$ along the detour. Hence the small semicircular detour around the simple pole at $s = 0$ is mapped onto the $G(s)H(s)$-plane as a clockwise semicircular path of infinite radius covering the whole right half-plane. If the pole at $s = 0$ is a double pole, then on the small semicircular detour

$$G(s)H(s) \cong \frac{K_0}{s^2} = \frac{K_0}{\rho^2}\,\epsilon^{-j2\phi}, \qquad -\frac{\pi}{2} \le \phi \le \frac{\pi}{2}. \tag{9-62}$$

The detour around $s = 0$ is now mapped onto the $G(s)H(s)$-plane as a clockwise circular path of infinite radius starting from an angle $+\pi$ and

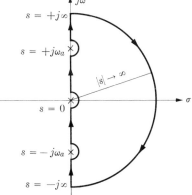

FIG. 9–26. Modified contour C in s-plane to by-pass poles on $j\omega$-axis.

ending at $-\pi$. This technique of making small semicircular detours as we traverse the $j\omega$-axis can be extended to other poles on the imaginary axis. A modified contour C in the s-plane is shown in Fig. 9–26. The semicircular detours connect the segments of the open-loop transfer locus, and since $\rho \to 0$, contour C still encircles *all* the zeros and poles, if any, of $D_o(s)$ with positive real parts.

There is one more point which is worthy of note with regard to the open-loop transfer locus. Since the systems with which we are concerned can be described by differential equations with real coefficients, we have

$$[G(j\omega)H(j\omega)]^* = G(-j\omega)H(-j\omega). \tag{9-63}$$

This means that the plot of $G(j\omega)H(j\omega)$ for the range $-\infty < \omega < 0$ will be the complex conjugate of the plot for the range $0 < \omega < +\infty$; in other words, the open-loop transfer locus will be symmetric about the real axis (u-axis). Hence *there is need to construct the open-loop transfer locus for the range $0 < \omega < +\infty$ only; the part for the range $-\infty < \omega < 0$ can be obtained by reflection about the real axis.*

EXAMPLE 9–10. Determine by the Nyquist criterion whether a feedback system which has the following open-loop transfer function is stable:

$$G(s)H(s) = \frac{K}{s(Ts+1)}. \tag{9-64}$$

Solution. We see from Eq. (9–64) that the given function has a pole at $s = 0$. Hence we would make a small semicircular detour at the origin toward the right as we traverse the $j\omega$-axis:

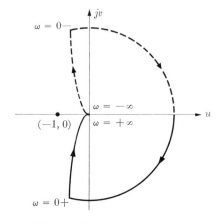

FIG. 9–27. Nyquist diagram based upon Eq. (9–65).

$$G(j\omega)(Hj\omega) = \frac{K}{j\omega(j\omega T + 1)} = -\frac{K}{\omega^2 T^2 + 1}\left[T + j\frac{1}{\omega}\right]. \quad (9\text{–}65)$$

A rough sketch of the open-loop transfer locus (Nyquist diagram) based on Eq. (9–65) is shown in Fig. 9–27. The part for $0 < \omega < +\infty$ is drawn as a solid line and that for $-\infty < \omega < 0$ is shown as a broken line which is the mirror image of the solid line about the u-axis. The arrows indicate the direction in which the locus is traced as ω increases from $-\infty$ through 0 to $+\infty$; as $|\omega| \to 0$,

$$\text{Re}\,[G(j\omega)H(j\omega)] \to -KT$$

and

$$|\text{Im}\,[G(j\omega)H(j\omega)]| \to \infty.$$

The slope of the open-loop transfer locus is $1/\omega T$, which approaches zero as $|\omega|$ tends to ∞. The semicircular detour around the origin in the s-plane is mapped into a semicircular path of infinite radius representing a change of phase from $+\pi/2$ to $-\pi/2$. The system is obviously stable because the point $(-1, 0)$ is not encircled by the locus (irrespective of the values of K and T, so long as they are both positive). This can be checked readily by the Routh-Hurwitz criterion. The characteristic equation to be considered is

$$D_o(s) = 1 + G(s)H(s) = 1 + \frac{K}{s(Ts + 1)} = 0, \quad (9\text{–}66)$$

which reduces to

$$s^2 + \frac{1}{T}s + \frac{K}{T} = 0. \quad (9\text{–}67)$$

The Routh-Hurwitz array is

$$1 \qquad K/T$$
$$1/T$$
$$K/T$$

There are no rows with vanishing elements and no changes of sign in the first column. Thus the roots of $D_o(s) = 0$ have no zero or positive real parts and the system is stable.

EXAMPLE 9–11. Determine by the Nyquist criterion whether a feedback system which has the following open-loop transfer function is stable:

$$G(s)H(s) = \frac{K}{s(T_1 s + 1)(T_2 s + 1)}. \qquad (9\text{–}68)$$

Solution. From Eq. (9–68), we have

$$G(j\omega)H(j\omega) = \frac{K}{j\omega(j\omega T_1 + 1)(j\omega T_2 + 1)}$$

$$= -j\left(\frac{K}{\omega}\right) \frac{(1 - \omega^2 T_1 T_2) - j\omega(T_1 + T_2)}{(1 - \omega^2 T_1 T_2)^2 + \omega^2(T_1 + T_2)^2}$$

$$= -K \frac{(T_1 + T_2) + (j/\omega)(1 - \omega^2 T_1 T_2)}{1 + \omega^2(T_1^2 + T_2^2) + \omega^4 T_1^2 T_2^2}. \qquad (9\text{–}69)$$

As $|\omega| \to 0$,

$$\text{Re}\,[G(j\omega)H(j\omega)] \to -K(T_1 + T_2)$$

and

$$|\text{Im}\,[G(j\omega)H(j\omega)]| \to \infty.$$

The Nyquist diagrams for two values of K are plotted in Fig. 9–28. The system represented by the diagram in Fig. 9–28(a) is stable because the locus does not encircle the point $(-1, 0)$. However, the locus in Fig. 9–28(b) for a larger K encircles the $(-1, 0)$ point twice in the clockwise direction, making the system unstable. The value of ω at which the locus crosses the negative real axis can be found by setting the imaginary part of $G(j\omega)H(j\omega)$ to zero:

$$\text{Im}\,[G(j\omega_0)H(j\omega_0)] = \text{Im}\left[\frac{K}{j\omega_0(j\omega_0 T_1 + 1)(j\omega_0 T_2 + 1)}\right] = 0,$$

which yields

$$\omega_0^2 T_1 T_2 = 1 \qquad (9\text{–}70)$$

or

$$\omega_0 = \frac{1}{\sqrt{T_1 T_2}}. \qquad (9\text{–}71)$$

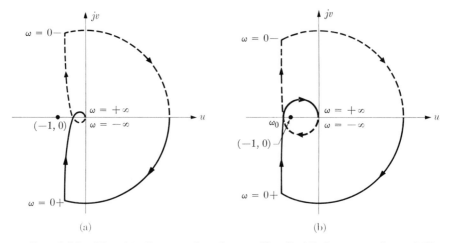

FIG. 9–28. Nyquist diagrams based upon Eq. (9–69) for two values of K. (a) A stable system. (b) An unstable system.

The value of the open-loop transfer function at ω_0 is

$$G(j\omega_0)H(j\omega_0) = -K\,\frac{T_1 T_2}{T_1 + T_2}. \tag{9-72}$$

For stability, we must have

$$|G(j\omega_0)H(j\omega_0)| < 1. \tag{9-73}$$

Equations (9–72) and (9–73) combine to give the condition for stability as

$$K < \left(\frac{1}{T_1} + \frac{1}{T_2}\right). \tag{9-74}$$

This example demonstrates that the Nyquist criterion provides an indication of how a system which has been found unstable can be made stable.

EXAMPLE 9–12. Determine by the Nyquist criterion whether a feedback system which has the following open-loop transfer function is stable:

$$G(s)H(s) = \frac{K}{s^2(Ts + 1)}. \tag{9-75}$$

Solution. From Eq. (9–75), we have

$$G(j\omega)H(j\omega) = \frac{K}{(j\omega)^2(j\omega T + 1)} = -\,\frac{K(1 - j\omega T)}{\omega^2(1 + \omega^2 T^2)}. \tag{9-76}$$

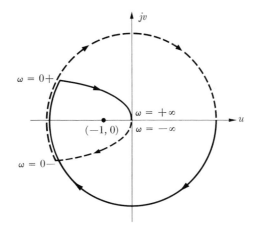

FIG. 9–29. Nyquist diagram for Eq. (9–76).

Because of the double pole at $s = 0$, a small semicircular detour at the origin should be made when the $j\omega$-axis is traversed. Both the real part and the imaginary part of $G(j\omega)H(j\omega)$ tend to infinity as $|\omega| \to 0$. The slope of the open-loop transfer locus is $-\omega T$, which tends to zero as $|\omega| \to 0$. The complete Nyquist diagram is shown in Fig. 9–29. The system is obviously unstable, since the open-loop transfer locus encircles the $(-1, 0)$ point twice in the clockwise direction. In this case a stable system cannot be obtained by reducing K.

PROBLEMS

9–1. Draw a block diagram for the circuit in Fig. 9–30, where v_g and i_L are the input and output variables respectively.

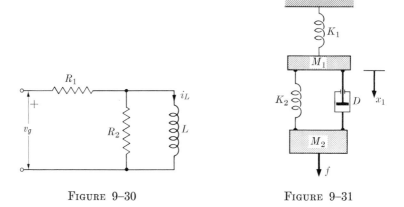

FIGURE 9–30   FIGURE 9–31

9–2. Draw a block diagram for the mechanical system in Fig. 9–31, where f and x_1 are the input and output variables respectively.

9–3. Draw a block diagram for the circuit in Fig. 9–32, where v_i and v_o are the input and output variables respectively.

9–4. Draw a block diagram for the bridge circuit in Fig. 9–33, where v_i and i_o are the input and output variables respectively.

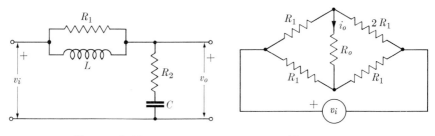

FIGURE 9–32   FIGURE 9–33

9–5. Derive the rules for the following transformation in block-diagram simplification: (a) moving a pickoff point ahead of a summing point; (b) moving a pickoff point behind a summing point.

9–6. Simplify the block diagram obtained in problem 9–1 to a single-block open-loop diagram.

9–7. Simplify the block diagram obtained in problem 9–2 to a single-block open-loop diagram.

9–8. Simplify the block diagram obtained in problem 9–3 to a single-block open-loop diagram.

9–9. Simplify the block diagram obtained in problem 9–4 to a single-block open-loop diagram.

9–10. Draw a signal flow graph for the circuit in Fig. 9–30 and find the transfer function between the input variable v_g and the output variable i_L by (a) simplifying the flow graph, and (b) applying the general formula for graph transmittance.

9–11. Draw a signal flow graph for the mechanical system in Fig. 9–31 and find the transfer function between the input variable f and the output variable x_1 by (a) simplifying the flow graph, and (b) applying the general formula for graph transmittance.

9–12. Draw a signal flow graph for the circuit in Fig. 9–32 and find the transfer function between the input variable v_i and the output variable v_o by (a) simplifying the flow graph, and (b) applying the general formula for graph transmittance.

9–13. Draw a signal flow graph for the bridge circuit in Fig. 9–33 and find the transfer function between the input variable v_i and the output variable i_o by (a) simplifying the flow graph, and (b) applying the general formula for graph transmittance.

Find the over-all transfer function $O(s)/I(s)$, for each of the systems in Figs. 9–34, 9–35, and 9–36.

9–14.

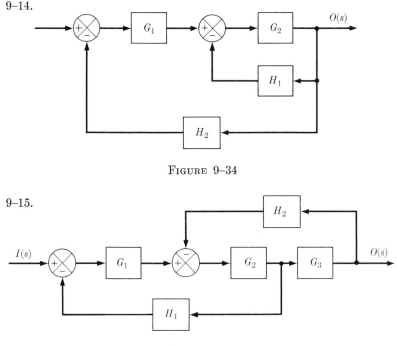

FIGURE 9–34

9–15.

FIGURE 9–35

9–16.

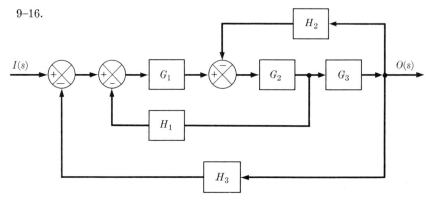

FIGURE 9–36

9–17. Apply the Routh-Hurwitz criterion to determine (i) the number of roots with positive real parts, (ii) the number of roots with zero real parts, and (iii) the number of roots with negative real parts for the following equations:

(a) $s^4 + 5s^3 + 2s + 10 = 0$
(b) $s^5 + 5.5s^4 + 14.5s^3 + 8s^2 - 19s - 10 = 0$
(c) $2s^5 + s^4 + 6s^3 + 3s^2 + s + 1 = 0$
(d) $s^4 + 2s^3 + 7s^2 + 10s + 10 = 0$
(e) $s^5 - s^4 - 2s^3 + 2s^2 - 8s + 8 = 0$

For each of the following open-loop transfer functions (a) sketch the Nyquist diagram, (b) determine whether the system is stable, and (c) check your answer to part (b) by applying the Routh-Hurwitz criterion.

9–18. $G(s)H(s) = \dfrac{K}{(T_1s + 1)(T_2s + 1)}$

9–19. $G(s)H(s) = \dfrac{K(s + 1)}{s^2(s + 4)(s + 5)}$

9–20. $G(s)H(s) = \dfrac{K(s + 2)}{(s + 1)(s - 3)}$

9–21. $G(s)H(s) = \dfrac{K}{(Ts + 1)^3}$

9–22. $G(s)H(s) = \dfrac{K}{s(s + 2)^2}$

CHAPTER 10

SAMPLED-DATA SYSTEMS; THE z TRANSFORMATION

10–1 Introduction. There is an important class of linear systems for which the input signal (excitation function) is in the form of discrete samples (pulses) of short duration. Such systems are referred to as linear *sampled-data systems*. Examples of sampled-data systems include tracking radar, where data regarding targets are collected in the form of discrete pulses at the pulse repetition frequency, and multichannel, time-division, pulse-modulated telemetering or communication systems. There are also sampled-data feedback systems in which the actuating error is supplied intermittently at discrete instants of time. Control systems incorporating a digital computer as a component belong to this category.

Although the conventional Laplace transformation can be used, the analysis of sampled-data systems is greatly facilitated by the introduction of the z transformation, especially when responses only at the sampling instants are desired. In this chapter we shall first define the z transformation and discuss its properties. This will be followed by the development of a table listing the z transforms of some important functions and operations. The transformation is then applied to the solution of difference equations and to the analysis of linear open-loop and closed-loop systems. Finally, the stability of sampled-data feedback systems and the method of determining the responses between sampling instants will be discussed.

10–2 The z transformation. An essential component in the block diagram of any sampled-data system is the *sampler*, which converts a continuous signal to a train of regularly spaced, amplitude-modulated, narrow pulses. The function of a sampler is illustrated in Fig. 10–1, where the sampler is represented by a simple switch. If the duration of the sampling pulses is small in comparison with the time constants of the

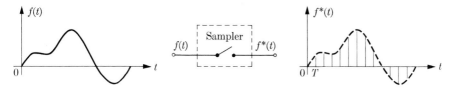

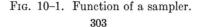

Fig. 10–1. Function of a sampler.

303

system of which the sampler is a part, the sampled output $f^*(t)$ can be considered as a sequence of impulses occurring at the sampling instants $0, T, 2T, 3T, \ldots$, the strengths of (areas under) the individual impulses being equal to the values of the input function $f(t)$ at the respective instants. Thus, we can write†

$$f^*(t) = f(t)\, \delta_T(t), \tag{10-1}$$

where the superscript $*$ is used to indicate that the function $f^*(t)$ is a discrete, sampled function, and $\delta_T(t)$ represents a *periodic* train of unit impulses spaced T seconds apart:

$$\delta_T(t) = \sum_{n=-\infty}^{\infty} \delta(t - nT). \tag{10-2}$$

If $f(t) = 0$ for $t < 0$, Eq. (10-1) becomes

$$f^*(t) = \sum_{n=0}^{\infty} f(nT)\, \delta(t - nT). \tag{10-3}$$

Let us now take the Laplace transform of the sampled output function $f^*(t)$, and denote the result by $F^*(s)$:

$$F^*(s) = \mathcal{L}[f^*(t)] = \mathcal{L}\left[\sum_{n=0}^{\infty} f(nT)\, \delta(t - nT)\right] = \sum_{n=0}^{\infty} f(nT)\epsilon^{-nTs}. \tag{10-4}$$

Since s appears in Eq. (10-4) only in the exponential factor, it is convenient to introduce a new symbol:

$$z = \epsilon^{Ts}. \tag{10-5}$$

We have, from Eqs. (10-4) and (10-5),

$$F(z) = \sum_{n=0}^{\infty} f(nT)z^{-n} = Z[f^*(t)]. \tag{10-6}$$

Although the notation in Eq. (10-6) has been widely accepted for the sake of convenience, $F(z)$ is not $F(s)$ with s replaced by z. Strictly speaking,

$$Z[f^*(t)] = F^*(s)\Big|_{s=(\ln z)/T} = F^*\left(\frac{1}{T}\ln z\right).$$

---

† J. R. Ragazzini and L. A. Zadeh, "The Analysis of Sampled-Data Systems," *Transactions of AIEE*, **71**, Part II (Applications and Industry), pp. 225–232; November 1952.

Some authors use an asterisk on $F(z)$ and write

$$Z[f^*(t)] = F^*(z),$$

but this is not necessary because the appearance of z as the argument already implies that the transform is for a sampled function. As a matter of fact, we can write

$$F(z) = Z[f(t)],$$

since $Z[f(t)] = Z[f^*(t)]$, but the asterisk on $f(t)$ must be retained on the expression for the inverse transformation. Thus,

$$Z^{-1}[F(z)] = f^*(t) \neq f(t).$$

Equation (10–6) defines the Z *transform*, $F(z)$, of the sampled function $f^*(t)$. $F(z)$ is seen to be an infinite series in powers of z^{-1}. If $F(z)$ is given in the form of a ratio of two polynomials in z or z^{-1}, all we have to do to find its inverse transform is to expand it in a series in powers of z^{-1} by long division; the coefficients in the series will then correspond to the values of $f^*(t)$ at the sampling instants. An integral formula for the inversion of Z transforms corresponding to the inversion integral, Eq. (6–16), for Laplace transformation exists.† However, we shall not use it in this book.

EXAMPLE 10-1. Find $Z^{-1} \left[\dfrac{Tz}{(z-1)^2} \right]$.

Solution. Before we perform the long division, it is advantageous to arrange both the numerator and the denominator in ascending powers of z^{-1}:

$$\frac{Tz}{(z-1)^2} = T\frac{z^{-1}}{(1-z^{-1})^2} = T\frac{z^{-1}}{1-2z^{-1}+z^{-2}}$$

$$= T(z^{-1} + 2z^{-2} + 3z^{-3} + \cdots + kz^{-k} + \cdots)$$

$$= \sum_{n=0}^{\infty} (nT)z^{-n}.$$

Hence

$$f^*(t) = Z^{-1}\left[\frac{Tz}{(z-1)^2}\right] = \sum_{n=0}^{\infty} (nT)\,\delta(t-nT). \qquad (10\text{–}7)$$

A graph of the $f^*(t)$ given by Eq. (10–7) is shown in Fig. 10–2. It is a train of equally spaced impulses whose strength increases directly with the time of sampling. It is to be noted that although the graph suggests

---

† See, for instance, E. I. Jury, "Analysis and Synthesis of Sampled-Data Control Systems," *Communication and Electronics*, pp. 1–15; November 1954.

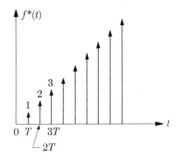

that $f(t) = tU(t)$, we *cannot* determine $f(t)$ from a known $f^*(t)$ because *the transition from a sampled function $f^*(t)$ to a continuous function $f(t)$ is not unique;* there are an infinite number of $f(t)$'s which will have the same $f^*(t)$. In the present example all $f(t)$'s which are equal to nT at $t = nT$ will have the same $f^*(t)$, and therefore the same $F(z)$, irrespective of what values the functions may assume in between the sampling instants.

FIG. 10–2. Graph for $f^*(t) = \sum_{n=0}^{\infty} (nT)\, \delta(t - nT)$.

We shall now develop the Z transforms for some important functions and operations, with a view to constructing a table of transforms which will be useful later on. All functions are assumed to be 0 for $t < 0$.

(1) *A shifted impulse function,* $\delta(t - nT)$. For this particular case, $f^*(t) = f(t)$. We have

$$\mathcal{L}[\delta(t - nT)] = \epsilon^{-nTs},$$

$$\mathsf{Z}[\delta(t - nT)] = z^{-n}. \tag{10–8}$$

If $n = 0$, we have a unit impulse at $t = 0$ and

$$\mathsf{Z}[\delta(t)] = z^{-0} = 1. \tag{10–9}$$

(2) *An exponential function,* ϵ^{at}. Using the fundamental relation in Eq. (10–6),

$$F(z) = \sum_{n=0}^{\infty} f(nT)z^{-n},$$

we can write directly

$$\mathsf{Z}[\epsilon^{at}] = \sum_{n=0}^{\infty} (\epsilon^{aT}z^{-1})^n$$

$$= 1 + \epsilon^{aT}z^{-1} + \epsilon^{2aT}z^{-2} + \epsilon^{3aT}z^{-3} + \cdots$$

$$= \frac{1}{1 - \epsilon^{aT}z^{-1}} = \frac{z}{z - \epsilon^{aT}}. \tag{10–10}$$

(3) *The unit step function,* $U(t)$. The Z transform of $U(t)$ can be obtained from Eq. (10–10) by setting $a = 0$. Thus

$$\mathsf{Z}[U(t)] = \frac{1}{1 - z^{-1}} = \frac{z}{z - 1}. \tag{10–11}$$

(4) *Sine and cosine functions.* Putting $a = j\omega$ in Eq. (10–10), we have

$$Z[\epsilon^{j\omega t}] = \frac{z}{z - \epsilon^{j\omega T}} = \frac{z}{(z - \cos \omega T) - j \sin \omega T}$$

$$= \frac{z(z - \cos \omega T) + jz \sin \omega T}{z^2 - 2z \cos \omega T + 1}$$

$$= Z[\cos \omega t + j \sin \omega t].$$

The Z transforms of $\sin \omega t$ and $\cos \omega t$ are obtained from the imaginary and real parts of the above equation, respectively:

$$Z[\sin \omega t] = \frac{z \sin \omega T}{z^2 - 2z \cos \omega T + 1}, \qquad (10\text{–}12)$$

$$Z[\cos \omega t] = \frac{z(z - \cos \omega T)}{z^2 - 2z \cos \omega T + 1}. \qquad (10\text{–}13)$$

(5) *Hyperbolic sine and cosine functions.* Since

$$\sinh bt = -j \sin (jb)t \qquad (10\text{–}14)$$

and

$$\cosh bt = \cos (jb)t, \qquad (10\text{–}15)$$

the Z transform of $\sinh bt$ can be obtained from Eq. (10–12) as follows:

$$Z[\sinh bt] = -jZ[\sin (jb)t]$$

$$= \frac{-jz \sin (jb) T}{z^2 - 2z \cos (jb) T + 1}$$

$$= \frac{z \sinh bT}{z^2 - 2z \cosh bT + 1}. \qquad (10\text{–}16)$$

Similarly, from Eq. (10–13), we have

$$Z[\cosh bt] = \frac{z(z - \cosh bT)}{z^2 - 2z \cosh bT + 1}. \qquad (10\text{–}17)$$

(6) *A ramp function,* $f(t) = t.$

$$F(z) = \sum_{n=0}^{\infty} f(nT)z^{-n}$$

$$= Tz^{-1} + 2Tz^{-2} + 3Tz^{-3} + \cdots$$

$$= Tz^{-1}(1 + 2z^{-1} + 3z^{-2} + \cdots)$$

$$= \frac{Tz^{-1}}{(1 - z^{-1})^2} = \frac{Tz}{(z - 1)^2}.$$

Hence,

$$Z[t] = \frac{Tz}{(z-1)^2}. \tag{10-18}$$

We have already encountered this in Example 10–1.

(7) *Function with a damping factor,* $\epsilon^{-\alpha t} f(t)$. Letting

$$f_1(t) = \epsilon^{-\alpha t} f(t),$$

we have

$$F_1(z) = \sum_{n=0}^{\infty} f_1(nT)z^{-n} = \sum_{n=0}^{\infty} f(nT)(\epsilon^{\alpha T}z)^{-n}. \tag{10-19}$$

Comparison of Eq. (10–19) with Eq. (10–6) reveals that $F_1(z)$ can be obtained from $F(z)$ if z is replaced by $(\epsilon^{\alpha T}z)$. Thus

$$Z[\epsilon^{-\alpha t} f(t)] = F_1(z) = F(\epsilon^{\alpha T}z). \tag{10-20}$$

Table 10–1 tabulates some of the important Z transform pairs.

EXAMPLE 10–2. Find $Z^{-1}\left[\dfrac{z}{3z^2 - 4z + 1}\right]$.

Solution. This problem could be solved by the method of long division, by arranging the result in ascending powers of z^{-1} as was done in Example 10–1. However, in doing it that way, we would have no assurance that we would be able to arrive at a closed form for the answer. Since the denominator of the given fraction can be factored, we will use instead the method of partial fractions. Anticipating that we will need a z in the numerator in order to find the inverse transforms from Table 10–1, we expand $F(z)/z$:

$$\frac{F(z)}{z} = \frac{1}{3z^2 - 4z + 1} = \frac{1}{3(z-1)(z-\frac{1}{3})} = \frac{1}{2}\left(\frac{1}{z-1} - \frac{1}{z-\frac{1}{3}}\right),$$

or

$$F(z) = \frac{1}{2}\left(\frac{z}{z-1} - \frac{z}{z-\frac{1}{3}}\right). \tag{10-21}$$

From Table 10–1, or Eq. (10–11), we have

$$Z^{-1}\left[\frac{z}{z-1}\right] = U^*(t) = \sum_{n=0}^{\infty} \delta(t-nT). \tag{10-22}$$

From Table 10–1, or Eq. (10–10),

$$Z^{-1}\left[\frac{z}{z-\frac{1}{3}}\right] = \left(\frac{1}{3^{t/T}}\right)^* = \sum_{n=0}^{\infty} 3^{-n} \delta(t-nT). \tag{10-23}$$

TABLE 10–1

SOME IMPORTANT Z TRANSFORM PAIRS†

$f(t)$	$F(z) = \mathrm{Z}[f(t)] = \displaystyle\sum_{n=0}^{\infty} f(nT)z^{-n}$
$\delta(t - nT)$	$\dfrac{1}{z^n}$
$U(t)$	$\dfrac{z}{z - 1}$
ϵ^{at}	$\dfrac{z}{z - \epsilon^{aT}}$
$\sin \omega t$	$\dfrac{z \sin \omega T}{z^2 - 2z \cos \omega T + 1}$
$\cos \omega t$	$\dfrac{z(z - \cos \omega T)}{z^2 - 2z \cos \omega T + 1}$
$\sinh bt$	$\dfrac{z \sinh bT}{z^2 - 2z \cosh bT + 1}$
$\cosh bt$	$\dfrac{z(z - \cosh bT)}{z^2 - 2z \cosh bT + 1}$
t	$\dfrac{Tz}{(z - 1)^2}$
t^2	$\dfrac{T^2 z(z + 1)}{(z - 1)^3}$
$\epsilon^{-\alpha t} f(t)$	$F(\epsilon^{\alpha T} z)$
$f(t + T)$	$z[F(z) - f(0)]$
$f(t + 2T)$	$z^2[F(z) - f(0)] - zf(T)$
$f(t - nT)U(t - nT)$	$z^{-n}F(z)$

† Note that $\mathrm{Z}^{-1}[F(z)] \neq f(t)$, but $\mathrm{Z}^{-1}[F(z)] = f^*(t) = f(t)\,\delta_T(t)$.

Substituting Eqs. (10–22) and (10–23) in Eq. (10–21), we obtain

$$Z^{-1}\left[\frac{z}{3z^2 - 4z + 1}\right] = \frac{1}{2}\left[(1 - 3^{-t/T})U(t)\right]^*$$

$$= \frac{1}{2}\sum_{n=0}^{\infty}(1 - 3^{-n})\,\delta(t - nT). \quad (10\text{–}24)$$

Three useful theorems will now be proved: the shifting theorem, the initial-value theorem, and the final-value theorem.

THE SHIFTING THEOREM. *If* $Z[f(t)] = F(z)$, *then*

$$Z[f(t + T)] = z[F(z) - f(0)]. \quad (10\text{–}25)$$

Proof. By definition,

$$Z[f(t + T)] = \sum_{n=0}^{\infty}f[(n + 1)T]z^{-n}$$

$$= z\sum_{n=0}^{\infty}f[(n + 1)T]z^{-(n+1)}$$

$$= z\sum_{k=1}^{\infty}f(kT)z^{-k}, \quad (10\text{–}26)$$

where $k = n + 1$. By adding and subtracting an $f(0)$ term under the summation sign in Eq. (10–26), we can write the summation over the range from $k = 0$ to $k = \infty$. Thus

$$Z[f(t + T)] = z\left[\sum_{k=0}^{\infty}f(kT)z^{-k} - f(0)\right]$$

$$= z[F(z) - f(0)]. \qquad \text{Q.E.D.}$$

This theorem is very useful in solving difference equations (see Section 10–3).

COROLLARY I. *If* $Z[f(t)] = F(z)$, *then*

$$Z[f(t + 2T)] = z^2[F(z) - f(0)] - zf(T). \quad (10\text{–}27)$$

Proof. Using Eq. (10–25), we have

$$Z[f(t + 2T)] = Z[f(\overline{t + T} + T)]$$

$$= z\{Z[f(t + T)] - f(t + T)|_{t=0}\}$$

$$= z^2[F(z) - f(0)] - zf(T). \qquad \text{Q.E.D.}$$

The extension to $Z[f(t + mT)]$, m a positive integer, is obvious.

COROLLARY II. *If* $Z[f(t)] = F(z)$, *then*

$$Z[f(t - nT)U(t - nT)] = z^{-n}F(z). \tag{10–28}$$

Proof. By definition,

$$Z[f(t - nT)U(t - nT)] = \sum_{m=0}^{\infty} f[(m - n)T]U[(m - n)T]z^{-m}$$

$$= z^{-n} \sum_{m=0}^{\infty} f[(m - n)T]U[(m - n)T]z^{-(m-n)}.$$

$$\tag{10–29}$$

On letting $m - n = k$, Eq. (10–29) becomes

$$Z[f(t - nT)U(t - nT)] = z^{-n} \sum_{k=-n}^{\infty} f(kT)U(kT)z^{-k}$$

$$= z^{-n} \sum_{k=0}^{\infty} f(kT)z^{-k} = z^{-n}F(z). \quad \text{Q.E.D.}$$

EXAMPLE 10–3. Find $Z[t^2]$.

Solution. If we used the fundamental relation in Eq. (10–6), we would have

$$Z[t^2] = \sum_{n=0}^{\infty} (nT)^2 z^{-n} = T^2[z^{-1} + 2^2 z^{-2} + 3^2 z^{-3} + \cdots]. \tag{10–30}$$

However, it is not obvious that the series in Eq. (10–30) can be expressed in a closed form. Let us proceed in the following way:

$$f(t) = t^2,$$

$$f(t + T) = (t + T)^2.$$

Subtracting, we get

$$f(t + T) - f(t) = (t + T)^2 - t^2 = T(2t + T). \tag{10–31}$$

From Eq. (10–25), the Z transform of the left side of Eq. (10–31) is

$$Z[f(t + T) - f(t)] = (z - 1)F(z), \tag{10–32}$$

since $f(0) = 0$. The Z transform of the right side of Eq. (10–31), from Eqs. (10–18) and (10–11), is

$$Z[T(2t + T)] = T\left[\frac{2Tz}{(z - 1)^2} + \frac{Tz}{z - 1}\right] = T^2 z \frac{z + 1}{(z - 1)^2}. \tag{10–33}$$

Equating Eqs. (10–32) and (10–33), we obtain the desired answer:

$$Z[t^2] = F(z) = T^2 \frac{z(z+1)}{(z-1)^3}. \qquad (10\text{–}34)$$

THE INITIAL-VALUE THEOREM.

$$f(0) = \lim_{z \to \infty} F(z). \qquad (10\text{–}35)$$

Proof. Equation (10–35) can be readily proved by noting that

$$F(z) = \sum_{n=0}^{\infty} f(nT)z^{-n} = f(0) + f(T)z^{-1} + f(2T)z^{-2} + \cdots \qquad (10\text{–}36)$$

The initial-value theorem is obtained by letting $z \to \infty$ in the above expression. From Eq. (10–36) it is also seen that if the initial value is zero, then

$$f(T) = \lim_{z \to \infty} zF(z), \quad \text{if} \quad f(0) = 0. \qquad (10\text{–}37)$$

THE FINAL-VALUE THEOREM.

$$\lim_{t \to \infty} f(t) = \lim_{z \to 1} (z-1)F(z). \qquad (10\text{–}38)$$

Proof. From Eq. (10–25), we have

$$Z[f(t+T) - f(t)] = (z-1)F(z) - zf(0). \qquad (10\text{–}39)$$

But, by definition,

$$Z[f(t+T) - f(t)] = \lim_{k \to \infty} \sum_{n=0}^{k} \{f[(n+1)T] - f(nT)\}z^{-n}.$$

On letting $z \to 1$, we have

$$\lim_{z \to 1} Z[f(t+T) - f(t)] = \lim_{k \to \infty} \{[f(T) - f(0)] + [f(2T) - f(T)]$$

$$+ [f(3T) - f(2T)] + \cdots + [f(\overline{k+1}T) - f(kT)]\}$$

$$= f(\infty) - f(0). \qquad (10\text{–}40a)$$

From Eq. (10–39),

$$\lim_{z \to 1} Z[f(t+T) - f(t)] = \lim_{z \to 1} (z-1)F(z) - f(0). \qquad (10\text{–}40b)$$

The final-value theorem in Eq. (10–38) is obtained by equating the right sides of Eqs. (10–40a) and (10–40b).

10–3 Solution of difference equations. Difference equations arise in electrical and mechanical systems where there is a recurrence of structure. For example, an electrical network consisting of a chain of identical sections (wave filters, multistage amplifiers, string insulators, etc.) is best analyzed by the use of difference equations because, instead of solving as many differential equations as there are sections, the solution of one difference equation for a typical section would suffice. Similarly, mechanical systems such as periodically loaded strings and multicylinder engine shafts are solvable most simply by the use of difference equations. In this section we shall show how the techniques of Z transformation can be used to solve linear difference equations.

Whereas differential equations contain derivatives of the dependent variable, difference equations contain differences of the values of the dependent variable at equally spaced, discrete values of the independent variable. The following is a typical second-order linear difference equation:

$$w^*(t + 2T) + b_1 w^*(t + T) + b_0 w^*(t) = e^*(t), \qquad (10\text{–}41)$$

where the coefficients (some of which may be negative) are assumed to be constants for simplicity, and the functions with an asterisk are defined *only* at $t = nT$ for $n = 0, 1, 2, \ldots$ It is an equation of the second order because it contains a second-order ordinate (value of the dependent variable when the independent variable is two regular periods away), $w^*(t + 2T)$; and it is *linear* because it does not contain powers (except first powers) or products of the dependent ordinates. In problems involving recurrent structures, t would represent the position index of the component structure under consideration. It is often convenient to set $T = 1$ and write $w^*(t)$ as $w(n)$. Thus Eq. (10–41) becomes

$$w(n + 2) + b_1 w(n + 1) + b_0 w(n) = e(n). \qquad (10\text{–}42)$$

Equation (10–42) is an important special case of Eq. (10–41).

The classical method† for solving linear difference equations is quite similar to that for solving linear differential equations. Briefly, the solution consists of two parts:

(1) *The complementary function.* The complementary function, $w_c(n)$, is the general solution of the *homogeneous* equation with the excitation function $e(n)$ set to zero:

$$w_c(n + 2) + b_1 w_c(n + 1) + b_0 w_c(n) = 0. \qquad (10\text{–}43)$$

---

† T. von Kármán and M. A. Biot, *Mathematical Methods in Engineering*, McGraw-Hill Book Company, New York, Chapter XI; 1940. L. A. Pipes, *Applied Mathematics for Engineers*, McGraw-Hill Book Company, New York, Chapter X; 1946.

To solve Eq. (10–43) for $w_c(n)$, we assume a solution of the exponential form ϵ^{sn}. Thus

$$w_c(n) = \epsilon^{sn}, \tag{10–44}$$

$$w_c(n+1) = \epsilon^{s(n+1)} = \epsilon^s \epsilon^{sn} = z\epsilon^{sn}, \tag{10–45}$$

$$w_c(n+2) = \epsilon^{s(n+2)} = \epsilon^{2s}\epsilon^{sn} = z^2\epsilon^{sn}. \tag{10–46}$$

Substituting Eqs. (10–44) through (10–46) in Eq. (10–43), we have

$$(z^2 + b_1 z + b_0)\epsilon^{sn} = 0. \tag{10–47}$$

Since ϵ^{sn} does not vanish, Eq. (10–47) can be satisfied only if

$$z^2 + b_1 z + b_0 = 0. \tag{10–48}$$

This second-degree algebraic equation in z has two roots. Let us call the roots z_1 and z_2. Corresponding to z_1 and z_2, there are two values of s such that

$$z_1 = \epsilon^{s_1} \tag{10–49}$$

and

$$z_2 = \epsilon^{s_2}. \tag{10–50}$$

Hence the two possible solutions for $w_c(n)$ are

$$\epsilon^{s_1 n} = z_1^n \tag{10–51}$$

and

$$\epsilon^{s_2 n} = z_2^n. \tag{10–52}$$

The general solution for the homogeneous equation (10–43) is then

$$w_c(n) = c_1 z_1^n + c_2 z_2^n. \tag{10–53}$$

If the characteristic equation (10–48) has a double root, $z_1 = z_2$, the complementary function should be modified to the following, in exactly the same manner as for a differential equation whose characteristic equation possesses multiple roots (see Section 2–5):

$$w_c(n) = (c_1 + c_2 n)z_1^n. \tag{10–54}$$

(2) *The particular integral.* The particular integral is a particular solution of the given nonhomogeneous equation (10–42). In general, the particular integral can be obtained most simply by the method of undetermined coefficients, and the techniques developed in Section 2–4 are directly applicable; we shall not engage in a long discussion of these techniques again here. Moreover, a majority of the difference equations encountered in engineering problems are of the homogeneous variety.

If $w_p(n)$ is the particular integral, then the complete solution for the second-order linear difference equation (10–42) with constant coefficients is

$$w(n) = w_c(n) + w_p(n). \tag{10–55}$$

The two arbitrary constants, c_1 and c_2, in $w_c(n)$ are to be determined from known initial conditions.

To solve the difference equation (10–42) by the method of z transforms, we take the z transformation of the entire equation. Let

$$z[w(n)] = W(z)$$

and

$$z[e(n)] = E(z).$$

We then have

$$z^2[W(z) - w(0)] - zw(1) + b_1 z[W(z) - w(0)] + b_0 W(z) = E(z)$$

or

$$(z^2 + b_1 z + b_0)W(z) = E(z) + (z^2 + b_1 z)w(0) + zw(1). \tag{10–56}$$

Thus, by the use of z transformation, we have transformed the given difference equation into an algebraic equation. From Eq. (10–56), we obtain

$$W(z) = \frac{1}{z^2 + b_1 z + b_0} [E(z) + (z^2 + b_1 z)w(0) + zw(1)]. \tag{10–57}$$

Comparison of Eq. (10–57) with Eq. (7–9) representing the Laplace-transformed solution of a second-order differential equation reveals the close resemblance of the two expressions. We see also that Eq. (10–57) contains the two given values $w(0)$ and $w(1)$; no separate operation is necessary to determine arbitrary constants. The desired solution is

$$w(n) = z^{-1}[W(z)], \tag{10–58}$$

which has meaning only for $n = 0, 1, 2, \ldots$

EXAMPLE 10–4. Solve the following first-order linear difference equation, given $w(0) = -1$:

$$w^*(t + T) + 2w^*(t) = 5t^*. \tag{10–59}$$

Solution. Taking the z transformation of Eq. (10–59), we have

$$z[W(z) + 1] + 2W(z) = 5 \frac{Tz}{(z - 1)^2}$$

or

$$(z + 2)W(z) = \frac{5Tz}{(z - 1)^2} - z,$$

from which $W(z)$ can be solved:

$$W(z) = \frac{5Tz}{(z+2)(z-1)^2} - \frac{z}{z+2}. \qquad (10\text{--}60)$$

To find the inverse transform of the first term, we expand it (without the constants and the z in the numerator) in partial fractions:

$$\frac{1}{(z+2)(z-1)^2} = \frac{K_1}{z+2} + \frac{K_{22}}{(z-1)^2} + \frac{K_{21}}{z-1}. \qquad (10\text{--}61)$$

Using the method in Section 7–3, we find

$$K_1 = \frac{1}{(z-1)^2}\bigg|_{z=-2} = \frac{1}{9},$$

$$K_{22} = \frac{1}{z+2}\bigg|_{z=1} = \frac{1}{3},$$

$$K_{21} = \left[\frac{d}{dz}\left(\frac{1}{z+2}\right)\right]_{z=1} = -\frac{1}{9}.$$

Substituting these constants in Eq. (10–61), we can now write Eq. (10–60) in its partial-fraction form:

$$W(z) = \frac{5T}{9}\left[\frac{z}{z+2} + \frac{3z}{(z-1)^2} - \frac{z}{z-1}\right] - \frac{z}{z+2}. \qquad (10\text{--}62)$$

Finally, we obtain

$$w^*(t) = \frac{5T}{9}\left[\left(1 - \frac{9}{5T}\right)(-2)^{t/T} + 3\,\frac{t}{T} - 1\right]^*$$

$$= \frac{5T}{9}\sum_{n=0}^{\infty}\left[\left(1 - \frac{9}{5T}\right)(-2)^n + 3n - 1\right]\delta(t - nT). \qquad (10\text{--}63)$$

The solution in Eq. (10–63) is given in the form of a train of impulses of varying strength. At a particular sampling instant $t = kT$, the value of the dependent variable is

$$w(kT) = \frac{5T}{9}\left[\left(1 - \frac{9}{5T}\right)(-2)^k + 3k - 1\right]. \qquad (10\text{--}64)$$

EXAMPLE 10–5.   Twenty-one 1-ohm resistors are connected to a d-c voltage source of 10 volts in a ladder arrangement, as shown in Fig. 10–3. Determine the current in the last resistor, R_{21}.

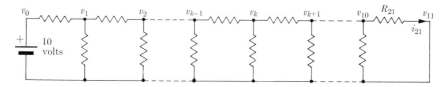

FIG. 10–3. Ladder network consisting of twenty-one 1-ohm resistors.

Solution. Let us write the node equation for the typical section consisting of the three nodes at which the node-pair voltages referring to the common datum node are v_{k-1}, v_k, and v_{k+1}:

$$-v_{k-1} + 3v_k - v_{k+1} = 0. \qquad (10\text{–}65)$$

Equation (10–65) applies for k equal to $1, 2, \ldots, 10$. It is in a form slightly different from that of Eq. (10–42) but the difference is only superficial, for if we write $k = n + 1$, we have

$$-v(n) + 3v(n + 1) - v(n + 2) = 0, \qquad (10\text{–}66)$$

where n ranges from 0 to 9. This is a second-order linear difference equation.

$\mathbb{Z}$ transformation of Eq. (10–66) yields

$$-V(z) + 3z[V(z) - v(0)] - z^2[V(z) - v(0)] + zv(1) = 0. \quad (10\text{–}67)$$

From Fig. 10–3, we see that

$$v(0) = 10 \text{ volts} \qquad (10\text{–}68)$$

but, for the moment, we do not know the numerical value of $v(1)$. Substituting Eq. (10–68) in Eq. (10–67) and rearranging, we obtain

$$(z^2 - 3z + 1)V(z) = 10z(z - 3) + zv(1)$$

or

$$V(z) = \frac{10z(z - 3) + zv(1)}{z^2 - 3z + 1}$$

$$= \frac{10z(z - \frac{3}{2})}{z^2 - 3z + 1} - \frac{[15 - v(1)]z}{z^2 - 3z + 1}. \qquad (10\text{–}69)$$

Comparison of the terms on the right side of Eq. (10–69) with Eqs. (10–17) and (10–16) indicates that they are exact $\mathbb{Z}$ transforms of hyperbolic cosine and hyperbolic sine functions, respectively, for $T = 1$, and that

$$\cosh b = \tfrac{3}{2} \tag{10-70}$$

$$\sinh b = \sqrt{\cosh^2 b - 1} = \frac{\sqrt{5}}{2}, \tag{10-71}$$

$$b = 0.9624. \tag{10-72}$$

Therefore

$$v(n) = \mathrm{Z}^{-1}[V(z)] = 10 \cosh 0.9624n - \frac{2}{\sqrt{5}} [15 - v(1)] \sinh 0.9624n. \tag{10-73}$$

To determine $v(1)$, we note that $v_{11} = 0$. Thus, on substituting $n = 11$ in Eq. (10-73), we have

$$\frac{2}{\sqrt{5}} [15 - v(1)] \sinh 10.5864 = 10 \cosh 10.5864,$$

$$v(1) = 15 - 5\sqrt{5} \coth 10.5864. \tag{10-74}$$

The general solution for the difference equation (10-66) is then

$$v(n) = 10[\cosh 0.9624n - \coth 10.5864 \sinh 0.9624n]. \tag{10-75}$$

The desired current in the last resistor, R_{21}, is numerically equal to $v(10)$:

$$i_{21} = 10[\cosh 9.624 - \coth 10.5864 \sinh 9.624]. \tag{10-76}$$

Because of the nature of the problem, it is necessary to carry many decimal places† in order to get a meaningful answer for i_{21}, which is extremely small. Rough calculations by means of a slide rule and a short table for hyperbolic functions would yield a vanishing i_{21}.

The types of difference equations we have discussed so far will describe simple systems in which the variables are sampled at equally spaced instants of time, or systems in which time is not an independent variable but in which there are recurrent identical component structures. In a more general case, we may encounter situations that involve not only recurrent structures but also time-dependent variables. An example for such a situation is obtained if the shunt resistors in Fig. 10-3 are replaced by capacitors. A typical section of the structure will then look like Fig. 10-4. The node equation for the kth node is

---

† It may be helpful to point out here that hyperbolic functions can be expressed in terms of exponential functions, which in turn can be evaluated by using a comprehensive table of logarithms.

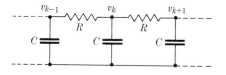

FIG. 10-4. A typical section of a recurrent network which yields differential-difference equations.

$$-\frac{1}{R}\,v_{k-1} + \frac{2}{R}\,v_k + C\,\frac{dv_k}{dt} - \frac{1}{R}\,v_{k+1} = 0. \qquad (10\text{-}77)$$

We now have a *differential-difference equation* in which there are two independent variables, time t and position index k. If we make the same change in the positional index as we did in Eq. (10-66) by writing k as $n + 1$, we obtain

$$C\,\frac{d}{dt}\,v(n+1,\,t) - \frac{1}{R}\,[v(n,\,t) - 2v(n+1,\,t) + v(n+2,\,t)] = 0. \qquad (10\text{-}78)$$

Equation (10-78) is a first-order differential equation with respect to the continuous variable t, and at the same time it is a second-order difference equation with respect to the discrete variable n. We already have all the essential techniques at our command for the solution of differential-difference equations. However, since we shall discuss similar techniques in detail when we take up the solution of partial differential equations in Chapter 11, we shall not go further than outlining the procedure for handling equations like (10-78):

1. Take the Laplace transformation of Eq. (10-78). The differential-difference equation in n and t is transformed into a difference equation in n and s. For simplicity, we assume that the capacitor C is initially uncharged:

$$CsV(n+1,\,s) - \frac{1}{R}\,[V(n,\,s) - 2V(n+1,\,s) + V(n+2,\,s)] = 0$$

or

$$-\frac{1}{R}\,V(n,\,s) + \left(Cs + \frac{2}{R}\right) V(n+1,\,s) - \frac{1}{R}\,V(n+2,\,s) = 0, \qquad (10\text{-}79)$$

where

$$V(n,\,s) = \mathcal{L}[v(n,\,t)].$$

2. Take the Z transformation of Eq. (10-79). The difference equation in n and s is transformed into an algebraic equation. Let

$$Z[V(n,\,s)] = V^*(z,\,s).$$

We then have

$$-\frac{1}{R}V^*(z, s) + z\left(Cs + \frac{2}{R}\right)[V^*(z, s) - V(0, s)]$$

$$-\frac{1}{R}z^2[V^*(z, s) - V(0, s)] + \frac{1}{R}zV(1, s) = 0,$$

or

$$[z^2 - (2 + RCs)z + 1]V^*(z, s) = [z^2 - (2 + RCs)z]V(0, s) + zV(1, s),$$

$$(10\text{--}80)$$

from which $V^*(z, s)$ can be solved. $V(0, s)$ and $V(1, s)$ can be determined by the Laplace transforms of the fixed (boundary) conditions at the first and last sections of the given structure.

3. Find the inverse Z transform of $V^*(z, s)$:

$$V(n, s) = Z^{-1}[V^*(z, s)]. \qquad (10\text{--}81)$$

4. Find the inverse Laplace transform of $V(n, s)$:

$$v(n, t) = \mathcal{L}^{-1}[V(n, s)], \qquad (10\text{--}82)$$

which is the desired solution for the differential-difference equation (10–78).

10–4 Application of Z transformation to open-loop systems. The most important application of Z transformation lies in the analysis of linear sampled-data systems. It is a natural tool in such applications, and gives the desired answer in a simple and convenient manner. In this section we shall show how Z transformation can be used to analyze open-loop sampled-data systems.

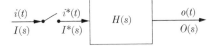

FIG. 10–5. A system with sampled input function.

Figure 10–5 is a block-diagram representation of a simple sampled-data system in which the input function $i(t)$ is periodically sampled by the sampler (switch). The transfer function, $H(s)$, is the ratio of the Laplace transform of the output function to the Laplace transform of the input function, irrespective of whether the input function is continuous or sampled. Thus, for the situation in Fig. 10–5,

$$O(s) = H(s)I^*(s). \qquad (10\text{--}83)$$

The output function $o(t)$, which is the inverse Laplace transform of $O(s)$, is a continuous function in spite of the fact that $i^*(t)$ is sampled.

Suppose now that the output function is sampled in synchronism with the input function. We wish to inquire: Can a transformed relationship similar to Eq. (10–83) be established between $O^*(s)$ and $I^*(s)$? Before we answer this important question, let us examine in general the relationship between the Laplace transform of a continuous function $f(t)$ and that of the sampled $f^*(t)$. From Eqs. (10–1) and (10–2), we have

$$f^*(t) = f(t)\, \delta_T(t) \tag{10–84}$$

and

$$\delta_T(t) = \sum_{n=-\infty}^{\infty} \delta(t - nT). \tag{10–85}$$

We note that the Fourier series expansion of the periodic train of unit impulse represented by $\delta_T(t)$ can be derived directly from Example 5–3 by letting

$$T_p \rightarrow 0$$

and

$$A T_p \rightarrow 1.$$

The complex coefficient α_n is then, from Eq. (5–66),

$$\alpha_n = \lim_{\substack{T_p \rightarrow 0 \\ A T_p \rightarrow 1}} \frac{A T_p}{T} \left[\frac{\sin\,(n\pi T_p/T)}{n\pi T_p/T} \right] \epsilon^{-jn\pi T_p/T} = \frac{1}{T}. \tag{10–86}$$

Consequently,

$$\delta_T(t) = \frac{1}{T} \sum_{n=-\infty}^{\infty} \epsilon^{jn(2\pi/T)t}. \tag{10–87}$$

Substituting Eq. (10–87) in Eq. (10–84) and letting $2\pi/T = \omega_0$, we have

$$f^*(t) = \frac{1}{T} f(t) \sum_{n=-\infty}^{\infty} \epsilon^{jn\omega_0 t}. \tag{10–88}$$

Laplace transformation of Eq. (10–88) yields directly

$$F^*(s) = \frac{1}{T} \sum_{n=-\infty}^{\infty} F(s + jn\omega_0), \tag{10–89}$$

since $\mathcal{L}[\epsilon^{-\alpha t}f(t)] = F(s + \alpha)$. Equation (10–89) expresses $F^*(s)$ in terms of a double-infinite summation of shifted $F(s)$ functions. We also see from Eq. (10–89) that $F^*(s)$ is a *periodic function with a period* $j\omega_0 = j2\pi/T$, since

$$F^*(s) = F^*(s + jk\omega_0), \tag{10–90}$$

where k is any integer.

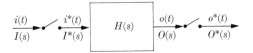

FIG. 10–6.   A system with both input and output functions sampled in synchronism.

Now let us go back to the original query: What is the relation between $O^*(s)$ and $I^*(s)$? The situation is depicted in Fig. 10–6.

From the general relationship given in Eq. (10–89), we can write for the sampled output function

$$O^*(s) = \frac{1}{T} \sum_{n=-\infty}^{\infty} O(s + jn\omega_0). \qquad (10\text{–}91)$$

Substituting Eq. (10–83) in Eq. (10–91), we have

$$O^*(s) = \frac{1}{T} \sum_{n=-\infty}^{\infty} H(s + jn\omega_0) I^*(s + jn\omega_0)$$

$$= I^*(s) \left[\frac{1}{T} \sum_{n=-\infty}^{\infty} H(s + jn\omega_0) \right], \qquad (10\text{–}92)$$

in view of the periodicity of the starred transformed functions as expressed by Eq. (10–90). From Eq. (10–89), we note that the expression in brackets in Eq. (10–92) is simply $H^*(s)$. Hence

$$O^*(s) = H^*(s) I^*(s), \qquad (10\text{–}93)$$

which is the relation we have been seeking. From Eq. (10–4), we see that the variable s appears in starred transformed functions only in the form of an exponential factor ϵ^{Ts}; it is convenient to use the notations of Z transformation as given in Eqs. (10–5) and (10–6), and write Eq. (10–93) in the following alternative form:

$$O(z) = H(z) I(z). \qquad (10\text{–}94)$$

Equation (10–94) is an important result. It tells us that *the Z-transformed relationship for a system in which both the input and the output are sampled in synchronism is exactly like the $\mathcal{L}$-transformed relationship for the system for continuous functions.* The output or response function found by the inverse Z transformation of $O(z)$ from Eq. (10–94), of course, has meaning *only* at the sampling instants.

Only one thing bothers us about Eqs. (10–93) and (10–94). While we know that $I^*(s)$, $O^*(s)$ and $I(z)$, $O(z)$ are respectively the Laplace and Z

transforms of the sampled input and output functions, what significance can we attach to the functions $H^*(s)$ and $H(z)$? It is semantically wrong to say that they are respectively the Laplace and Z transforms of the sampled network, because the network itself is neither sampled nor transformed. A little reflection of our adopted notation will reveal that, by definition,

$$H^*(s) = \mathcal{L}[h^*(t)] \tag{10–95}$$

and, correspondingly,

$$H(z) = Z[h^*(t)], \tag{10–96}$$

where $h^*(t)$ is the *sampled impulse response*. Consequently, it is customary to call $H^*(s)$ and $H(z)$ the *sampled transfer function* and the *z-transfer function* of the network, respectively.†

To examine the analogy between the Laplace-transformed relationship for a continuous system and the Z-transformed relationship for a sampled-data system further, let us consider the corresponding relations in the time domain.

(a) *For the continuous system:*

$$O(s) = H(s)I(s). \tag{10–97}$$

By virtue of the convolution integral or Borel's theorem, we have

$$o(t) = h(t) * i(t) = \int_0^t h(t - \tau)i(\tau)\, d\tau. \tag{10–98}$$

(b) *For the sampled-data system:*

$$O(z) = H(z)I(z). \tag{10–99}$$

The sampled output function at any sampling instant $t = nT$ can be expressed as a *convolution summation:*

$$o(nT) = \sum_{k=0}^{n} h(nT - kT)i(kT). \tag{10–100}$$

This can also be seen from Eq. (10–99) by recalling that the coefficients of the Z transform of a time function expressed as an ascending power series in z^{-1} are the values of the time function at the sampling instants. Thus, when both $H(z)$ and $I(z)$ in Eq. (10–99) are expressed in ascending powers of z^{-1}, the coefficient of the z^{-n} term in $O(z)$, $O_{(n)}$, is a summation of the products of the coefficients in $H(z)$ and $I(z)$ such that

$$O_{(n)} = \sum_{k=0}^{n} H_{(n-k)}I_{(k)}, \tag{10–101}$$

———————

† They are also referred to as the *pulsed transfer functions*.

which tells us essentially the same thing as Eq. (10–100). Note that in Eq. (10–101) the sum of the subscripts on H and I is always equal to the subscript on O.

EXAMPLE 10–6. A sampled input function $i^*(t)$ is applied to a network which has an impulse response $h(t)$. The Z transforms of $i^*(t)$ and $h^*(t)$ are as follows:

$$I(z) = 4z^{-1} + 6z^{-2} + 7z^{-3} + 5z^{-4} + \cdots, \qquad (10\text{–}102)$$

$$H(z) = 2 + z^{-1} + \tfrac{2}{3}z^{-2} + \tfrac{1}{2}z^{-3} + \tfrac{1}{3}z^{-4} + \cdots \qquad (10\text{–}103)$$

Determine the values of the output function at the first five sampling instants.

Solution. The required answers can be obtained by multiplying the given series of $I(z)$ and $H(z)$ together as a direct application of Eq. (10–99). Thus,

	z^0	z^{-1}	z^{-2}	z^{-3}	z^{-4} ...
$I(z) = $	0,	4,	6,	7,	5, ...
$H(z) = $	2,	1,	$\tfrac{2}{3}$,	$\tfrac{1}{2}$,	$\tfrac{1}{3}$, ...
	0,	8,	12,	14,	10, ...
		0,	4,	6,	7, ...
			0,	$\tfrac{8}{3}$,	4, ...
				0,	2, ...
					0, ...
$O(z) = $	0,	8,	16,	$22\tfrac{2}{3}$,	23, ...

The above result can be checked by using Eq. (10–101):

$$O_{(0)} = H_{(0)}I_{(0)} = 2 \times 0 = 0,$$

$$O_{(1)} = \sum_{k=0}^{1} H_{(1-k)}I_k = 1 \times 0 + 2 \times 4 = 8,$$

$$O_{(2)} = \sum_{k=0}^{2} H_{(2-k)}I_k = \tfrac{2}{3} \times 0 + 1 \times 4 + 2 \times 6 = 16,$$

$$O_{(3)} = \sum_{k=0}^{3} H_{(3-k)}I_k = \tfrac{1}{2} \times 0 + \tfrac{2}{3} \times 4 + 1 \times 6 + 2 \times 7 = 22\tfrac{2}{3},$$

$$O_{(4)} = \sum_{k=0}^{4} H_{(4-k)}I_k = \tfrac{1}{3} \times 0 + \tfrac{1}{2} \times 4 + \tfrac{2}{3} \times 6 + 1 \times 7 + 2 \times 5 = 23.$$

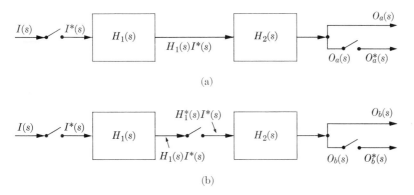

FIG. 10–7. Two different cascade connections. (a) Component blocks in direct cascade. (b) Component blocks in cascade through a sampler.

We now turn our attention to sampled-data systems in cascade (tandem). There are two different ways in which two component blocks in a sampled-data system can be connected in cascade: they may be directly connected, one after the other, as in Fig. 10–7(a), or they may be connected through a sampler which operates in synchronism with that in the input circuit, as in Fig. 10–7(b). The transformed output functions are obviously different in these two cases.

For Fig. 10–7(a):

Transformed continuous output, $O_a(s) = H_1(s)H_2(s)I^*(s)$.   (10–104)

Transformed sampled output, $O_a^*(s) = \overline{H_1H_2}^*(s)I^*(s)$,   (10–105)

or $O_a(z) = \overline{H_1H_2}(z)I(z)$.   (10–106)

For Fig. 10–7(b):

Transformed continuous output, $O_b(s) = H_1^*(s)H_2(s)I^*(s)$.   (10–107)

Transformed sampled output, $O_b^*(s) = H_1^*(s)H_2^*(s)I^*(s)$,   (10–108)

or $O_b(z) = H_1(z)H_2(z)I(z)$.   (10–109)

It is important to note that $\overline{H_1H_2}(z)$ in Eq. (10–106) is *not* equal to $H_1(z)H_2(z)$. In fact,

$$\overline{H_1H_2}(z) = \mathcal{Z}[\mathcal{L}^{-1}H_1(s)H_2(s)] = \mathcal{Z}\left[\int_0^t h_1(t-\tau)h_2(\tau)\,d\tau\right]$$ (10–110)

and

$$H_1(z)H_2(z) = \mathcal{Z}[h_1(t)] \cdot \mathcal{Z}[h_2(t)].$$ (10–111)

10–5 Application of Z transformation to closed-loop systems. Closed-loop systems which contain samplers are called *sampled-data feedback systems.* Servos in which the actuating error data are supplied intermittently at discrete, equally spaced instants and control and simulation systems which employ digital computers are such examples. Z transformation is particularly useful for the analysis of sampled-data feedback systems when only the response at the sampling instants is desired.

There are many different ways† in which the sampler or samplers may be located in a sampled-data feedback system. We shall analyze a few important cases. For each case, both the Laplace transform of the continuous output and the Z transform of the sampled output will be given.

1. *Sampler in forward path* (Fig. 10–8).

(a) Transformed continuous output, $O(s)$.

$$\text{Summing point: } E(s) = I(s) - F(s) \qquad (10\text{–}112)$$

$$\text{Forward path: }\quad O(s) = G(s)E^*(s) \qquad (10\text{–}113)$$

$$\text{Feedback path: } F(s) = H(s)O(s) \qquad (10\text{–}114)$$

Substitution of Eq. (10–113) in Eq. (10–114) and then in Eq. (10–112) yields

$$E(s) = I(s) - G(s)H(s)E^*(s). \qquad (10\text{–}115)$$

We wish to find $E^*(s)$ from Eq. (10–115) so that it can be substituted into Eq. (10–113) in order to obtain $O(s)$. If $E^*(s)$ is written in place of $E(s)$ on the left side of Eq. (10–115), then, on the right side, $I^*(s)$ must be written in place of $I(s)$ and $\overline{GH}^*(s)E^*(s)$ in place of $G(s)H(s)E^*(s)$, following the same reasoning by which Eq. (10–93) was derived from

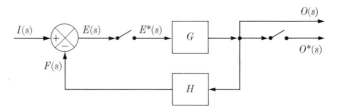

Fɪɢ. 10–8. Feedback system with sampler in forward path.

---

† E. I. Jury, "Additions to the Modified z-Transform Methods," 1957 *IRE WESCON Convention Record*, part 4, pp. 136–156.

Eq. (10–83). Thus

$$E^*(s) = I^*(s) - \overline{GH}^*(s)E^*(s), \tag{10–116}$$

from which

$$E^*(s) = \frac{I^*(s)}{1 + \overline{GH}^*(s)}. \tag{10–117}$$

Finally, from Eq. (10–113), we have

$$O(s) = \frac{G(s)I^*(s)}{1 + \overline{GH}^*(s)}. \tag{10–118}$$

(b) Transformed sampled output, $O^*(s)$ and $O(z)$.

From Eq. (10–113), we can write

$$O^*(s) = G^*(s)E^*(s). \tag{10–119}$$

Substituting Eq. (10–117) in Eq. (10–119), we obtain

$$O^*(s) = \frac{G^*(s)I^*(s)}{1 + \overline{GH}^*(s)}. \tag{10–120}$$

In terms of Z transforms:

$$O(z) = \frac{G(z)I(z)}{1 + \overline{GH}(z)}. \tag{10–121}$$

We emphasize here again that $\overline{GH}^*(s) \neq G^*(s)H^*(s)$, and consequently $\overline{GH}(z) \neq G(z)H(z)$.

2. *Sampler in feedback path* (Fig. 10–9).

(a) Transformed continuous output, $O(s)$.

$$\text{Summing point: } E(s) = I(s) - F(s). \tag{10–122}$$

$$\text{Forward path: } \quad O(s) = G(s)E(s) \tag{10–123}$$

$$\text{Feedback path: } F(s) = H(s)O^*(s) \tag{10–124}$$

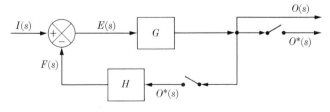

FIG. 10–9. Feedback system with sampler in feedback path.

Substitution of Eq. (10–124) in Eq. (10–122) and then in Eq. (10–123) yields

$$O(s) = G(s)I(s) - G(s)H(s)O^*(s).   \tag{10–125}$$

Converting Eq. (10–125) into an equation of transformed sampled functions, we have

$$O^*(s) = \overline{GI}^*(s) - \overline{GH}^*(s)O^*(s),   \tag{10–126}$$

from which

$$O^*(s) = \frac{\overline{GI}^*(s)}{1 + \overline{GH}^*(s)}.   \tag{10–127}$$

Substituting the above $O^*(s)$ into Eq. (10–125), we obtain

$$O(s) = G(s)\left[I(s) - \frac{H(s)\overline{GI}^*(s)}{1 + \overline{GH}^*(s)} \right].   \tag{10–128}$$

(b) Transformed sampled output, $O^*(s)$ and $O(z)$.

The expression for $O^*(s)$ has already been given in Eq. (10–127). In terms of Z transforms, we have directly

$$O(z) = \frac{\overline{GI}(z)}{1 + \overline{GH}(z)}.   \tag{10–129}$$

3. *Samplers in both forward and feedback paths* (Fig. 10–10).

(a) Transformed continuous output, $O(s)$.

$$\text{Summing point: } E(s) = I(s) - F(s)   \tag{10–130}$$

$$\text{Forward path: } \quad O(s) = G(s)E^*(s)   \tag{10–131}$$

$$\text{Feedback path: } F(s) = H(s)O^*(s)   \tag{10–132}$$

From Eq. (10–132), we can write

$$F^*(s) = H^*(s)O^*(s).   \tag{10–133}$$

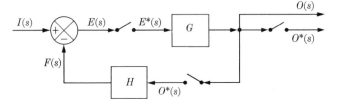

FIG. 10–10. Feedback system with samplers in both forward and feedback paths.

Substitution of this $F^*(s)$ into the Z transformation of Eq. (10–130) yields

$$E^*(s) = I^*(s) - H^*(s)O^*(s). \qquad (10\text{–}134)$$

From Eqs. (10–131) and (10–134), we have

$$O(s) = G(s)[I^*(s) - H^*(s)O^*(s)]. \qquad (10\text{–}135)$$

Converting Eq. (10–135) into an equation of transformed sampled functions, we get

$$O^*(s) = G^*(s)[I^*(s) - H^*(s)O^*(s)], \qquad (10\text{–}136)$$

from which

$$O^*(s) = \frac{G^*(s)I^*(s)}{1 + G^*(s)H^*(s)}. \qquad (10\text{–}137)$$

Substituting the above $O^*(s)$ into Eq. (10–135), we finally obtain

$$O(s) = G(s)\left[I^*(s) - \frac{G^*(s)H^*(s)I^*(s)}{1 + G^*(s)H^*(s)} \right]. \qquad (10\text{–}138)$$

(b) Transformed sampled output, $O^*(s)$ and $O(z)$.

The expression for $O^*(s)$ has already been given in Eq. (10–137). In terms of Z transforms, we have directly

$$O(z) = \frac{G(z)I(z)}{1 + G(z)H(z)}. \qquad (10\text{–}139)$$

10–6 Stability of sampled-data feedback systems. Sampled-data feedback systems, like systems in which continuous data flow, should be examined for stability, since no useful control function can be performed by an unstable feedback system. In Section 9–5 it was pointed out that a feedback system will be stable if and only if all the poles of the over-all transfer function $G_o(s) = O(s)/I(s)$ [i.e., all the zeros of its denominator $D_o(s) = 1 + G(s)H(s)$] lie in the left half of the s-plane. Two methods, the Routh-Hurwitz criterion and the Nyquist criterion, for determining whether this stability requirement is satisfied without actually finding all the roots of the equation $1 + G(s)H(s) = 0$ were discussed in Chapter 9.

Examination of the transformed output functions for the sampled-data feedback systems we considered in Section 10–5 reveals that in general they tend to be more complicated than those for the corresponding systems without the samplers. The poles of the transformed output function, and therefore of the over-all transfer function, are seen to be the roots of an equation which takes one of the following two forms:

$$1 + \overline{GH}^*(s) = 0, \tag{10-140}$$

$$1 + G^*(s)H^*(s) = 0. \tag{10-141}$$

Since the complex variable s appears in the starred functions in the exponential form ϵ^{sT}, (10-140) and (10-141) are *transcendental* equations in s. One immediate consequence is that the Routh-Hurwitz stability criterion, which detects the existence of the roots of an *algebraic* equation with positive or zero real parts, cannot be directly applied to the characteristic equation of a sampled-data feedback system.

The natural thing to do now is to see whether any useful information can be obtained from the characteristic equation of the system in z. In $\mathcal{Z}$-transform notation Eqs. (10-140) and (10-141) become

$$1 + \overline{GH}(z) = 0, \tag{10-142}$$

$$1 + G(z)H(z) = 0. \tag{10-143}$$

Both (10-142) and (10-143) are algebraic equations in z. In order to interpret the stability requirement in terms of z, it is necessary to know what the transformation

$$z = \epsilon^{sT} \tag{10-144}$$

does to the right half of the s-plane. Substituting

$$s = \sigma + j\omega \tag{10-145}$$

in Eq. (10-144), we have

$$z = \epsilon^{\sigma T}\epsilon^{j\omega T}, \tag{10-146}$$

where T is the sampling period. On the imaginary axis of the s-plane, $\sigma = 0$. Hence

$$z|_{\sigma=0} = \epsilon^{j\omega T}. \tag{10-147}$$

Equation (10-147) shows very clearly that *the $j\omega$-axis of the s-plane maps onto the z-plane as a unit circle centered at the origin*. As the angular frequency ω is increased from $-\infty$ through 0 to $+\infty$, the unit circle is traced over and over again every time ωT goes through an angle of 2π radians. In other words, *the graph in the z-plane is periodic in ω with a period $\omega_0 = 2\pi/T$*. This is depicted in Fig. 10-11. It will be recalled that transformed sampled functions are periodic in s with a period $j\omega_0 = j2\pi/T$. Hence *the graphs of transformed sampled functions as s varies along the $j\omega$-axis are completely described by their values in any interval $\Delta\omega = 2\pi/T$.*†

---

† As noted previously in connection with Eq. (9-63), because of symmetry about the real axis, the graphs of transformed sampled functions as s varies along the $j\omega$-axis are actually fixed by their values in the half-interval $0 \leq \omega \leq \pi/T$.

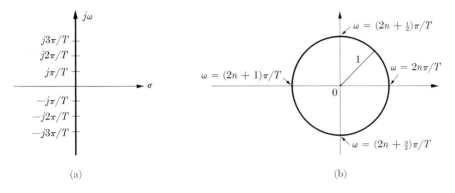

Fig. 10–11. Mapping of $j\omega$-axis in s-plane onto z-plane. (a) $j\omega$-axis in s-plane. (b) Unit circle in z-plane (n = any integer from $-\infty$ to $+\infty$).

For points in the right half of the s-plane, $\sigma > 0$ and, from Eq. (10–146), $|z| > 1$. Thus, *the right half of the s-plane maps onto the z-plane as the region exterior to the unit circle;* correspondingly, the left half of the s-plane maps over as the interior of the unit circle. We can now state the general stability requirement for a sampled-data feedback system as follows:

The necessary and sufficient condition for a sampled-data feedback system to be stable is that all the poles of its over-all transfer function lie inside the unit circle in the z-plane.

The above statement is quite similar to the one for continuous-data feedback systems except that for sampled-data systems it is more convenient to restrict the location of the poles in the z-plane rather than in the s-plane. An alternative statement for the stability requirement above is: *the necessary and sufficient condition for a sampled-data feedback system to be stable is that all the roots of its characteristic equation in z,* (10–142) or (10–143) as the case may be, *have an absolute value less than one.* It is clear that the absolute value of a root has to be less than one in order for it to lie inside the unit circle in the z-plane.

The Routh-Hurwitz criterion described in Section 9–6, which can be used to determine whether or not the roots of an algebraic equation have positive or zero real parts, is paralleled by a criterion (the Schur-Cohn criterion†) which can be used to detect the existence of the roots of an algebraic equation of an arbitrary degree whose absolute value is greater than or equal to one. However, the development of the latter criterion is

† M. Marden, The Geometry of the Zeros of a Polynomial in a Complex Variable, *American Mathematical Society Mathematical Survey No. III*, Chapter X; 1949.

quite involved and will be omitted here. In any case, a characteristic equation in z can be converted into one in another variable, say w, by the following transformation:

$$z = \frac{w + 1}{w - 1}.$$ (10–148)

Equation (10–148) maps the exterior of the unit circle $|z| > 1$ in the z-plane into the right half of the w-plane. The Routh-Hurwitz criterion can then be applied to the transformed characteristic equation in w for the detection of the existence and the number of roots with positive or zero real parts.

When the characteristic equation is of the second degree,

$$D_o(z) = z^2 + Az + B = 0,$$ (10–149)

where A and B are real, we can establish the following: Both roots of Eq. (10–149) have an absolute value less than one (i.e., the system is stable), if and only if the following three conditions are satisfied:

(a) $|D_o(0)| = |B| < 1,$ (10–150a)

(b) $D_o(1) = 1 + A + B > 0,$ (10–150b)

(c) $D_o(-1) = 1 - A + B > 0.$ (10–150c)

Condition (a) is obviously necessary because $|B|$ is the absolute value of the product of the two roots of Eq. (10–149). If the roots are a complex conjugate pair, condition (a) is all that is necessary. However, when the roots are real, condition (a) alone does not guarantee the exclusion of the possibility that one root lies inside and the other outside the unit circle, with the absolute value of their product less than unity. Conditions (b) and (c) are necessary in order to exclude real roots whose absolute values are greater than or equal to one. The proof is left as a problem at the end of this chapter. In terms of the coefficients of the characteristic equation, Eqs. (10–150) may be rewritten as:

(a) $|B| < 1,$ (10–151a)

(b) $1 + B > -A,$ (10–151b)

(c) $1 + B > A.$ (10–151c)

In general, coefficients A and B will contain terms in the exponential form.

The stability of a sampled-data feedback system can also be investigated graphically by means of a Nyquist diagram. The function $D_o(j\omega)$ can be plotted as ω is varied from $-\pi/T$ to $+\pi/T$; values of ω outside this range need not be considered, since $D_o(j\omega)$ is periodic. Inasmuch as each period of change in ω as ω is *increased* along the $j\omega$-axis corresponds to a *counterclockwise* trace of the unit circle in the z-plane, the graph of $D_o(j\omega)$ just described is the same as that of $D_o(z)$ by varying z along the unit circle in a *counterclockwise* direction. If $D_o(z)$ has no poles outside the unit circle, then we can state the Nyquist stability criterion as follows:

A sampled-data feedback system is stable if and only if the graph of $D_o(z)$ does not pass through or encircle the origin as z is varied along the unit circle.

Depending upon the location of the sampler, $D_o(z) = 1 + \overline{GH}(z)$ or $D_o(z) = 1 + G(z)H(z)$, we can alternatively state that *a sampled-data feedback system is stable if and only if the graph of $\overline{GH}(z)$ or $G(z)H(z)$, as the case may be, does not pass through or encircle the $(-1, 0)$ point as z is varied along the unit circle.*

As pointed out in Chapter 9 in connection with continuous feedback systems, the graphical approach of the Nyquist diagram not only tells us whether a system is stable but also gives us an indication of the degree of stability or instability. The amount by which the graph of $D_o(z)$ misses encircling the origin is indicative of the degree of stability. If a system is found unstable, the Nyquist diagram furnishes the information as to how system parameters should be altered in order to make the system stable.

EXAMPLE 10–7.   Examine the sampled-data feedback system in Fig. 10–12 for stability, given

$$G(s) = \frac{K}{s(T_1s + 1)}, \qquad (10\text{–}152)$$

where T_1 is a time constant of the system and should not be confused with the sampling period T.

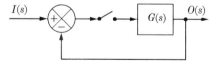

FIG. 10–12. Sampled-data feedback system for Example 10–7.

Solution. Here we have a direct-feedback path, $H(s) = 1$. According to Eq. (10–121), the denominator of the over-all transfer function is

$$D_o(z) = 1 + G(z). \qquad (10\text{–}153)$$

Hence $G(z)$ must first be found from the given $G(s)$ in Eq. (10–152):

$$G(s) = \frac{K}{s(T_1s + 1)} = K\left[\frac{1}{s} - \frac{1}{s + (1/T_1)}\right]. \qquad (10\text{–}154)$$

The inverse Laplace transforms of the two terms on the right side of Eq. (10–154) are $KU(t)$ and $-K\epsilon^{-t/T_1}$ respectively. From Table 10–1, we have

$$G(z) = K\left(\frac{z}{z-1} - \frac{z}{z - \epsilon^{-T/T_1}}\right) = \frac{Kz(1 - \epsilon^{-T/T_1})}{(z-1)(z - \epsilon^{-T/T_1})} \cdot \quad (10\text{–}155)$$

Substituting this $G(z)$ into Eq. (10–153), we obtain the characteristic equation of the given system:

$$D_o(z) = 1 + \frac{Kz(1 - \epsilon^{-T/T_1})}{(z-1)(z - \epsilon^{-T/T_1})} = 0,$$

or

$$z^2 + [K(1 - \epsilon^{-T/T_1}) - (1 + \epsilon^{-T/T_1})]z + \epsilon^{-T/T_1} = 0. \quad (10\text{–}156)$$

This is a quadratic equation in z. Let us examine the stability of this system both analytically and graphically.

(1) *Analytical approach.* The three conditions in Eqs. (10–151) that have to be satisfied for stability are

$$\epsilon^{-T/T_1} < 1, \quad\quad\quad\quad\quad (10\text{–}157a)$$

$$K(1 - \epsilon^{-T/T_1}) > 0, \quad\quad\quad\quad (10\text{–}157b)$$

$$2(1 + \epsilon^{-T/T_1}) > K(1 - \epsilon^{-T/T_1}). \quad\quad (10\text{–}157c)$$

Equation (10–157a) is always satisfied. Equation (10–157b) is satisfied so long as K is positive. Equation (10–157c) sets a limit on the maximum allowable value of K:

$$K < 2\left(\frac{1 + \epsilon^{-T/T_1}}{1 - \epsilon^{-T/T_1}}\right)$$

or

$$K < 2 \coth\left(\frac{T}{2T_1}\right). \quad\quad\quad\quad (10\text{–}158)$$

Hence, for stability, we require

$$0 < K < 2 \coth\left(\frac{T}{2T_1}\right). \quad\quad\quad\quad (10\text{–}159)$$

(2) *Graphical approach.* Here we plot $G(z)$ in Eq. (10–155) for values of z around the unit circle. Figure 10–13 shows such a graph in the $G(z)$-plane. The magnitude of z is maintained constant at unity when the angle, θ_z, of z is varied from $-180°$ through $0°$ to $+180°$. An exact

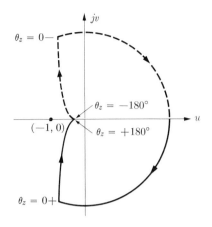

FIG. 10–13. Graph of $G(z)$ for Example 10–7.

graph of $G(z)$ requires the knowledge of K and T/T_1. At $\theta_z = 180°$, the intercept on the negative real axis is

$$\text{Re}\,[G(z)]_{|z|=1} = -\frac{K}{2}\left(\frac{1 - \epsilon^{-T/T_1}}{1 + \epsilon^{-T/T_1}}\right).$$

As $\theta_z \to 0$,

$$\text{Re}\,[G(z)]_{|z|=1} \to -\frac{K}{2}\left(\frac{1 + \epsilon^{-T/T_1}}{1 - \epsilon^{-T/T_1}}\right)$$

and

$$\left|\text{Im}\,[G(z)]\right|_{|z|=1} \to \infty.$$

The system whose graph of $G(z)$ is as shown in Fig. 10–13 is stable because the locus of $G(z)$ does not encircle the $(-1, 0)$ point. When K is increased the locus will expand to the left and eventually encircle the $(-1, 0)$ point, making the system unstable.

For the particular problem in Example 10–7, where the characteristic equation is quadratic in z, the analytical approach gives us the bounding values for K for stable operation quite simply. However, with a more complex system, the characteristic equation will be of a degree higher than the second. The three conditions in Eqs. (10–150) or (10–151) will no longer be applicable, and we must resort to graphical analysis if we do not wish to determine all the roots of the characteristic equation.

The result of Example 10–7 should be compared with that of Example 9–10 in Section 9–8. In Example 9–10, the Nyquist diagram for a continuous feedback system with the same open-loop transfer function as given in Eq. (10–152) was plotted in Fig. 9–27. We note that a continuous feedback system which is stable for all values of K (>0) has been made conditionally stable for $0 < K < 2 \coth (T/2T_1)$ by the introduction of a sampler in the forward path. The sampled-data feedback system will be unstable if K is larger than $2 \coth (T/2T_1)$, which decreases as the ratio T/T_1 is increased.

The problem of inserting a corrective network in the feedback loop in order to improve the stability of a sampled-data system is inherently more difficult than the corresponding one for a continuous system because, as we have emphasized in Section 10–4, the sampled transfer function of two networks in cascade is not equal to the product of the sampled transfer functions of the individual networks. A new set of calculations including the determination of new sampled transfer functions will be required with the insertion of any corrective network.

10–7 The modified Z transformation. Calculations using the Z-transform method that we have developed so far yield information about the output function only at the sampling instants. The inverse Z transformation is not unique and therefore cannot be used to determine the response between sampling instants without modifications. Since most sampled-data control systems have a continuous output, we naturally wish to inquire whether the above restriction could be removed. The answer lies in an extended technique called the *modified Z transformation.*†

We recall from Section 10–4 that the sampled output function at any sampling instant can be calculated from a convolution summation involving the sampled transfer function of the network and the sampled input function. With the same input function and the same network, if we wish to determine the response at a new set of points that occur regularly at ηT ($0 < \eta < 1$) seconds after the sampling instants, we would expect to convolve the sampled input function with a "delayed" sampled transfer function. Our problem is then, in effect, one of finding this new sampled transfer function which, when multiplied by the Z-transformed input function, gives the Z transform of the output function sampled at $(n + \eta)T$. By varying η from zero to unity, we can explore the entire output function.

---

† R. H. Barker, The Pulse Transfer Function and Its Applications to Sampling Servo Systems, *Proceedings of the Institution of Electrical Engineers,* **99,** Part IV, pp. 302–317; 1952.

G. V. Lago, Additions to Z-Transformation Theory for Sampled-Data Systems, *Trans. AIEE,* **73,** Part II — Applications and Industry, pp. 403–408; January 1955.

E. I. Jury, Additions to the Modified Z-Transform Methods, 1957 *IRE WESCON Convention Record,* Part 4, pp. 136–156.

Instead of Eq. (10–100), the output function at time $t = (n + \eta)T$ (ηT seconds behind the sampling instant nT) is

$$o(nT + \eta T) = \sum_{k=0}^{n} h(nT + \eta T - kT)i(kT). \qquad (10\text{–}160)$$

If we denote the delayed, sampled impulse response by $h^*(t, \eta)$, then

$$h^*(t, \eta) = \sum_{n=0}^{\infty} h(nT + \eta T)\, \delta(t - nT). \qquad (10\text{–}161)$$

The Laplace transform of Eq. (10–161) is

$$H^*(s, \eta) = \sum_{n=0}^{\infty} h(nT + \eta T)\epsilon^{-nTs}. \qquad (10\text{–}162)$$

Writing z for ϵ^{Ts}, we have

$$H(z, \eta) = \sum_{n=0}^{\infty} h(nT + \eta T)z^{-n} = Z_\eta[h^*(t)]. \qquad (10\text{–}163)$$

Equation (10–163) defines the *modified* Z *transform*,† $H(z, \eta)$, of the function $h(t)$, or $h^*(t)$, with respect to a delay ηT. We see that, with $\eta = 0$, $H(z, \eta)$ correctly reduces to the conventional Z transform of $h(t)$; that is,

$$H(z) = \lim_{\eta \to 0} H(z, \eta). \qquad (10\text{–}164)$$

For an open-loop system such as the one shown in Fig. 10–5, for which the transformed continuous output is

$$O(s) = H(s)I^*(s), \qquad (10\text{–}165)$$

we have for the transformed delayed and sampled output

$$O(z, \eta) = H(z, \eta)I(z), \qquad (10\text{–}166)$$

where $O(z, \eta)$ is the modified Z transform of the output function with respect to the delay ηT. Following the technique used in Section 10–5,

---

† The modified Z transform as defined here is different from that used by Barker and Jury (references in the footnote on page 336). Those authors used a *negative* artificial delay ΔT and then introduced an $m = 1 - \Delta$. The result was that the first sample $h(\eta T)$ at $t = \eta T$ would be the coefficient of z^{-1} in the polynomial expansion of the modified Z transform, instead of being the constant term (the z^0 term) as in the case for the conventional Z transform. The tabulated $W(z, m)$ in Barker's paper and $G^*(z, m)$ in Jury's paper can be converted to our $H(z, \eta)$ by multiplication by z.

we can derive corresponding formulas for the modified Z transforms of the output or response functions for closed-loop systems. The output at $t = (n + \eta)T$, $n = 0, 1, 2, \ldots$, can be found by taking the inverse Z transformation of $O(z, \eta)$:

$$o^*(t, \eta) = Z^{-1}[O(z, \eta)]. \tag{10–167}$$

Note that $o^*(t, \eta)$ is *not* the inverse *modified* Z transform of $\overset{*}{O}(z, \eta)$ because, by the definition in Eq. (10–163),

$$Z_\eta^{-1}[O(z, \eta)] = o^*(t) \neq o^*(t, \eta). \tag{10–168}$$

The operation in Eq. (10–167) can be evaluated either by partial-fraction expansion or by arranging $O(z, \eta)$ as a polynomial in ascending powers of z^{-1} through long division if necessary.

EXAMPLE 10–8. A unit step function is applied to the sampled-data feedback system shown in Fig. 10–14. The sampling period T is 1 second. Find the response (a) at $t = nT$, and (b) at $t = nT + 0.2$ seconds, where $n = 0, 1, 2, \ldots$

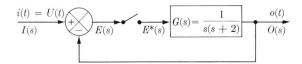

FIG. 10–14. Sampled-data feedback system for Example 10–8.

Solution. Let us first establish a formula connecting $O(z, \eta)$ and $I(z)$ for the given system:

$$\text{Summing point: } E(s) = I(s) - O(s) \tag{10–169}$$

$$\text{Forward path: } O(s) = G(s)E^*(s) \tag{10–170}$$

Substituting $O(s)$ from Eq. (10–170) into Eq. (10–169), we get

$$E(s) = I(s) - G(s)E^*(s). \tag{10–171}$$

A direct consequence of Eq. (10–171) is

$$E^*(s) = I^*(s) - G^*(s)E^*(s). \tag{10–172}$$

Hence

$$E^*(s) = \frac{I^*(s)}{1 + G^*(s)} \tag{10–173}$$

or, in Z-transform notation,

$$E(z) = \frac{I(z)}{1 + G(z)}. \tag{10–174}$$

Note that Eq. (10–173) is the same as Eq. (10–117) when we set $H(s) = 1$. By virtue of the relation between Eqs. (10–165) and (10–166), we can write Eq. (10–170) as

$$O(z, \eta) = G(z, \eta)E(z). \tag{10–175}$$

Substitution of $E(z)$ from Eq. (10–174) in Eq. (10–175) gives us the desired formula:

$$O(z, \eta) = \frac{G(z, \eta)I(z)}{1 + G(z)}. \tag{10–176}$$

Equation (10–176) reduces to $O(z)$ when we let $\eta = 0$:

$$O(z) = \frac{G(z)I(z)}{1 + G(z)}. \tag{10–177}$$

We must now find $G(z, \eta)$ from the given $G(s)$. Tables are available in the papers by Barker and Jury for such conversions. The derivation may proceed as follows:

$$G(s) = \frac{1}{s(s + 2)} = \frac{1}{2}\left(\frac{1}{s} - \frac{1}{s + 2}\right), \tag{10–178}$$

$$g(t) = \mathcal{L}^{-1}[G(s)] = \tfrac{1}{2}[U(t) - \epsilon^{-2t}]. \tag{10–179}$$

From Table 10–1, we have

$$G(z) = \frac{1}{2}\left[\frac{z}{z - 1} - \frac{z}{z - \epsilon^{-2T}}\right]. \tag{10–180}$$

From the definition of a sampled time function we write the expression for $g^*(t)$ from Eq. (10–179) as

$$g^*(t) = \tfrac{1}{2}\sum_{n=0}^{\infty}\left[1 - \epsilon^{-2nT}\right]\delta(t - nT). \tag{10–181}$$

In view of Eq. (10–161), we have

$$g^*(t, \eta) = \tfrac{1}{2}\sum_{n=0}^{\infty}\left[1 - \epsilon^{-2(n+\eta)T}\right]\delta(t - nT). \tag{10–182}$$

Since the Z transform of $g^*(t)$ is $G(z)$ in Eq. (10–180), the Z transform of $g^*(t, \eta)$ must be

$$G(z, \eta) = Z[g^*(t, \eta)] = Z_\eta[g^*(t)] = \frac{1}{2}\left[\frac{z}{z-1} - \frac{z\epsilon^{-2\eta T}}{z - \epsilon^{-2T}}\right], \quad (10\text{–}183)$$

which correctly reduces to $G(z)$ in Eq. (10–180) when we let $\eta = 0$.

(a) To find $o(nT)$:

$$G(z) = \frac{1}{2}\left[\frac{z}{z-1} - \frac{z}{z - 0.135}\right] = \frac{0.865z}{2(z-1)(z-0.135)}, \quad (10\text{–}184)$$

$$I(z) = Z[U(t)] = \frac{z}{z-1}. \quad (10\text{–}185)$$

Substituting Eqs. (10–184) and (10–185) into Eq. (10–177), we obtain

$$O(z) = \left(\frac{z}{z-1}\right)\left[\frac{0.865z}{2(z-1)(z-0.135) + 0.865z}\right]$$

$$= \frac{0.865z^2}{2(z^3 - 1.702z^2 + 0.837z - 0.135)}$$

$$= \frac{0.433z^{-1}}{1 - 1.702z^{-1} + 0.837z^{-2} - 0.135z^{-3}}$$

$$= 0.433z^{-1} + 0.737z^{-2} + 0.902z^{-3} + \cdots \quad (10\text{–}186)$$

The answer to part (a) can be tabulated as follows:

n	0	1	2	3	...
$t = nT$	0	1 sec	2 sec	3 sec	...
$o(nT)$	0	0.433	0.737	0.902	...

(b) To find $o(nT + 0.2)$:

$$G(z, 0.2) = \frac{1}{2}\left[\frac{z}{z-1} - \frac{0.670z}{z - 0.135}\right] = \frac{z(0.330z + 0.535)}{2(z-1)(z-0.135)}. \quad (10\text{–}187)$$

Equation (10–176) becomes

$$O(z, 0.2) = \left(\frac{z}{z-1}\right)\left[\frac{z(0.330z + 0.535)}{2(z-1)(z-0.135) + 0.865z}\right]$$

$$= \frac{z^2(0.330z + 0.535)}{2(z^3 - 1.702z^2 + 0.837z - 0.135)}$$

$$= \frac{0.165 + 0.268z^{-1}}{1 - 1.702z^{-1} + 0.837z^{-2} - 0.135z^{-3}}$$

$$= 0.165 + 0.549z^{-1} + 0.796z^{-2} + 0.918z^{-3} + \cdots \quad (10\text{--}188)$$

The answer to part (b) can be tabulated as follows:

n	0	1	2	3	$\cdots$
$t = nT + 0.2$	0.2 sec	1.2 sec	2.2 sec	3.2 sec	$\cdots$
$o(nT + 0.2)$	0.165	0.549	0.796	0.918	$\cdots$

10–8 Additional remarks on the method of Z transforms. Before we conclude this chapter, a few additional remarks on the method of Z transforms are in order. Let us first examine the following assumptions, which have been implied throughout our discussion on Z transformation:

1. The samplers operate periodically with a constant sampling period T.

2. The duration of sampling is negligibly small (compared with T and the time constants of the system).

3. The sampling frequency $(1/T)$ is at least twice the highest frequency component contained in the functions sampled.

Assumption 1 is obviously necessary in order to yield manageable mathematics for describing sampled-data systems. It is conceivable that suitable equations could be formulated even when the sampling period is random or varies in a specified manner, but this would make life difficult for no plausible purpose.

Assumption 2 was implied when we defined the sampled (starred) function in Eq. (10–1) at the very beginning of this chapter. The formulas we have developed in this chapter are exact only if the sampler converts a continuous input function into a train of impulses of zero width with strengths (areas) of the individual impulses equal to the values of the input function at the sampling instants. In practice this condition can only be approximated. If the duration of sampling (width of the pulses) is comparable to the time constants of the system, approximation by ideal impulses will no longer be satisfactory and more refined techniques developed especially for sampled-data systems with finite pulse width will have to be used.†

---

† G. Farmanfarma, Analysis of Linear Sampled-Data Systems with Finite Pulse Width: Open Loop, *Communication and Electronics*, No. 28, pp. 808–819; January 1957.

Assumption 3 deserves a little explanation, since its implication has not been obvious so far. We recall the periodic nature of the Laplace transform of a sampled function from Eq. (10–89), which is repeated below:

$$F^*(s) = \frac{1}{T} \sum_{n=-\infty}^{\infty} F(s + jn\omega_0), \qquad (10\text{–}189)$$

where $\omega_0 = 2\pi/T$. The plot of $|F(j\omega)|$ versus ω was defined in Chapter 5 as the relative frequency distribution of the continuous function $f(t)$. Let us assume that Fig. 10–15(a) is such a distribution, where $\omega_f/2\pi$ is the highest frequency component of the original continuous function $f(t)$ and no significant components of higher frequencies exist. After sampling, the relative frequency distribution of $f^*(t)$ (output of the sampler) will look like Fig. 10–15(b) if the sampling frequency ($\omega_0/2\pi = 1/T$) is greater than twice the highest frequency contained in $f(t)$. It is a periodic distribution repeating at intervals of ω_0 along the ω-axis, the amplitude of each component having been reduced by a factor T, to $1/T$. The function of the sampler is then essentially one of modulation in which the train of uniformly spaced unit impulses is amplitude modulated by the input function $f(t)$. Since the region $-\omega_f \leq \omega \leq +\omega_f$ alone [as well as other pairs of regions, $(-n\omega_0 - \omega_f) \leq \omega \leq (-n\omega_0 + \omega_f)$ and $(n\omega_0 - \omega_f) \leq \omega \leq (n\omega_0 + \omega_f)$, $n = 1, 2, 3, \ldots$] contains all the information necessary to reconstruct the original input function, $f(t)$ can be recovered very simply by means of a low-pass filter with a cutoff frequency $\omega_f/2\pi$. However, if the sampling frequency is less than twice the highest frequency contained in $f(t)$, a relative frequency distribution such as the one shown in Fig. 10–15(c) will result for the sampled function. In this case the frequency components around $\omega = \omega_0/2$ are combinations of the original components in $f(t)$ and those of the lower side band of the neighboring region, and the original function cannot be recovered by linear filtering.[†] Note that we assumed at the outset that the input function $f(t)$ has no significant components of a frequency higher than $\omega_f/2\pi$. In other words, our discussion above pertains to a *band-limited* signal.

In most sampled-data feedback systems, the high-frequency components introduced by the sampler should be filtered out before the input signal reaches the plant of the system. From the point of view of time variations, the filtering operation is, in effect, a smoothing operation

---

[†] J. G. Truxal, *Automatic Feedback Control System Synthesis*, Chapter 9, McGraw-Hill Book Company, New York; 1955.

This ties in directly with C. E. Shannon's famous *sampling theorem:* If a function $f(t)$ contains no frequencies higher than W cycles per second, it is completely specified if its values at a series of points spaced at $1/2W$ seconds apart are given. See C. E. Shannon, Communication in the Presence of Noise, *Proceedings of the IRE*, **37**, p. 11; January 1949.

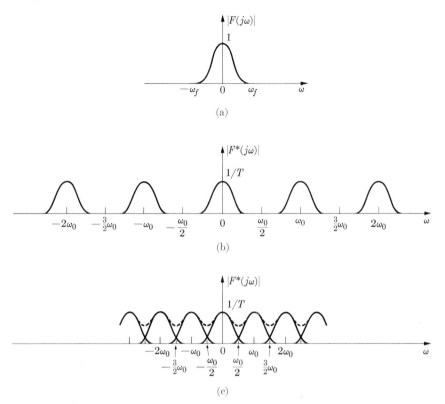

FIG. 10–15. Effect of sampling frequency on the relative frequency distribution of sampled function. (a) Relative frequency distribution of continuous function $f(t)$. (b) Relative frequency distribution of sampled function $f^*(t)$, $\omega_0 > 2\omega_f$. (c) Relative frequency distribution of sampled function $f^*(t)$, $\omega_0 < 2\omega_f$.

which tends to reconstruct the original time function from the impulse train at the output of the sampler. This is accomplished in practice by inserting a *holding circuit* immediately following the sampler, as shown in Fig. 10–16. As the name implies, the main function of the holding circuit is to hold the output of the sampler more closely resembling the original continuous error signal appearing at the input of the sampler. If the holding circuit could function perfectly, it would then reconstruct the error signal faithfully and the sampled-data system would be identical with a continuous system. However, a perfect holding circuit cannot be designed because of the random nature of the input signal, which is not exactly predictable. From the frequency standpoint, an ideal low-pass filter would do the job, but such a filter could be only approximated in practice and the large phase lag (time delay) it introduces would be detri-

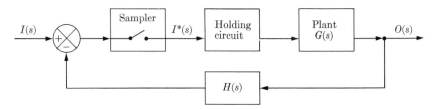

FIG. 10–16. A sampled-data feedback system with a holding circuit.

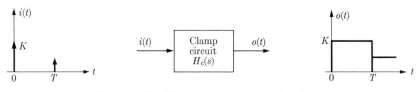

FIG. 10–17. Function of a clamp circuit.

mental to the stability of the system. A particularly simple holding circuit which represents an acceptable compromise between error-signal smoothing and system stability is one whose output at any time stays at the value of the input at the immediately preceding sampling pulse. The output then changes in steps and only at the sampling instants. This type of holding circuit is referred to figuratively as a *clamp circuit* or a *boxcar generator*. Figure 10–17 illustrates its function. The transfer function $H_c(s)$ of a clamp circuit can be found very easily as follows:

$$i(t) = K \, \delta(t) \longrightarrow I(s) = K,$$

$$o(t) = K[U(t) - U(t - T)] \longrightarrow O(s) = K \left(\frac{1}{s} - \frac{1}{s} \, \epsilon^{-Ts} \right).$$

Hence

$$H_c(s) = \frac{O(s)}{I(s)} = \frac{1}{s} \, (1 - \epsilon^{-Ts}). \qquad (10\text{--}190)$$

When a clamp circuit is used as the holding circuit in Fig. 10–16, we have, from Eqs. (10–118) and (10–121),

$$O(s) = \frac{H_c(s)G(s)I^*(s)}{1 + \overline{H_c GH}^*(s)} \qquad (10\text{--}191)$$

and

$$O(z) = \frac{\overline{H_c G}(z)I(z)}{1 + \overline{H_c GH}(z)}. \qquad (10\text{--}192)$$

PROBLEMS

10–1. Find $Z[a^{mt}]$.   **10–2.** Find $Z[t^3]$.   **10–3.** Find $Z[tC^{at}]$.

10–4. Prove that if $Z[f(t)] = F(z)$, then

$$Z[tf(t)] = -Tz \frac{d}{dz} F(z).$$

10–5. Find $Z^{-1}\left[\dfrac{2z^2}{(z+2)^2(z+1)}\right]$.

10–6. Find $Z^{-1}\left[\dfrac{(z-1)^2 - 1}{(z-1)^2 - 3z}\right]$.

10–7. Find $Z^{-1}\left[\dfrac{z}{(z - \epsilon^{-aT})(z - \epsilon^{-bT})}\right]$.

10–8. Find the general solution of the following difference equation by the the method of Z transforms:

$$w(n-1) - 4w(n) + w(n+1) = 0.$$

10–9. Solve the following difference equation by the method of Z transforms, subject to the conditions that both $y(t)$ and $f(t)$ are zero for $t < 0$:

$$y^*(t) - 5y^*(t - T) + 6y^*(t - 2T) = f^*(t).$$

10–10. Twelve 1-ohm resistors are arranged in the form of a spoke wheel, as shown in Fig. 10–18. Find, by solving a difference equation, the current i_{04} when a d-c voltage of 1 volt is applied between points 0 and 1.

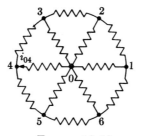

FIGURE 10–18

10–11. Determine the sampled over-all transfer function $H_o(z)$ of a network consisting of a clamp circuit, $H_c(s)$, and a plant, $G_p(s)$, in cascade, as shown in Fig. 10–19.

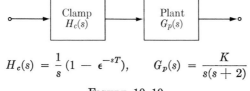

$$H_c(s) = \frac{1}{s}(1 - \epsilon^{-sT}), \qquad G_p(s) = \frac{K}{s(s+2)}$$

FIGURE 10–19

10–12. Repeat problem 10–11, assuming that a sampler is interposed between the clamp and the plant circuits.

Determine the transformed continuous output, $O(s)$, and the transformed sampled output, $O^*(s)$, for each of the sampled-data feedback systems in Figs. 10–20, 10–21, and 10–22. (All samplers operate in synchronism.)

10–13.

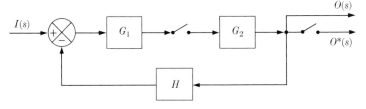

FIGURE 10–20

10–14.

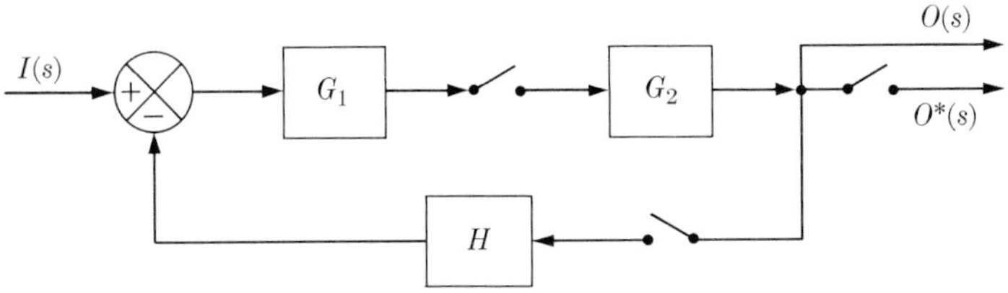

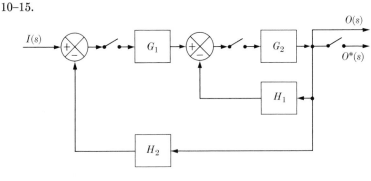

FIGURE 10–21

10–15.

FIGURE 10–22

10–16. Prove that the three conditions in Eqs. (10–150) are necessary and sufficient for both roots of Eq. (10–149) to have an absolute value less than one.

10–17. Examine, both analytically and graphically, the sampled-data feedback system in Fig. 10–23 for stability.

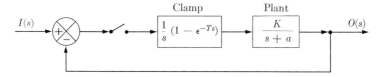

FIGURE 10–23

Find the modified Z transforms, $F(z, \eta)$, for the time functions whose Laplace transforms, $F(s)$, are given in problems 10–18, 10–19, and 10–20.

10–18. $F(s) = \dfrac{1}{s^2}$.

10–19. $F(s) = \dfrac{1}{(s + a)(s + b)}$.

10–20. $F(s) = \dfrac{s}{s^2 + \omega^2}$.

10–21. An exponential voltage $v_i = 10\epsilon^{-t/2}$ volts is applied at the input of an R-C network, as shown in Fig. 10–24. The switch is normally at position a but every second, starting at $t = 0$, it is shifted to position b for a very short time. (a) Determine v_o at $t = 0, 1, 2, 3, 4$ sec. (b) Determine v_o at $t = 0.5$, 1.5, 2.5, 3.5 sec.

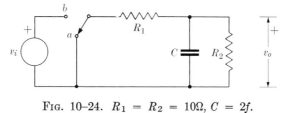

FIG. 10–24. $R_1 = R_2 = 10\Omega$, $C = 2f$.

10–22. A unit step function is applied at the input of the sampled-data feedback system in Fig. 10–23. Assuming $K = 4$, $1/a = 1$ sec and sampling period $T = 1/10$ sec, find the values of the output function (a) at $t = 0, 0.1$, 0.2, 0.3, 0.4 sec, and (b) at $t = 0.04, 0.14, 0.24, 0.34$ sec.

CHAPTER 11

SYSTEMS WITH DISTRIBUTED PARAMETERS

11–1 Introduction. So far we have dealt with only those systems which can be represented by linear ordinary differential or difference equations. These equations involve only one independent variable, which may be time, distance, or some other physical quantity. When the transient response of a system is important, time will always be an independent variable. Such a system can be described by ordinary differential or integro-differential equations only when all its components are lumped. A lumped component is one which can be identified by a single lumped parameter between terminals. Not all physical components are lumped. Some components, such as electrical transmission lines, mechanical torsion bars, heat-insulating slabs, and so on, are distributed components, and distributed parameters are needed to specify their physical properties. In addition to time variation, systems with distributed parameters will also have space variation. More than one independent variable are involved, and partial differential equations are needed for the description of such systems. The recurrent R-C network in Fig. 10–4 at the end of Section 10–3, which is representable by a differential-difference equation, may be looked upon as a transitional situation. With a continuous medium the recurrent structure will no longer be discernible. The discrete variable n (position index) will be changed to a continuous variable x (distance from a reference location) and the differences in the values of the dependent variable at neighboring positions will be converted into partial derivatives with respect to x. A partial differential equation will result for a continuous medium.

A partial differential equation is an equation expressing a relationship between a (dependent) function of several (independent) variables and its partial derivatives with respect to these variables. Understandably, the solution of partial differential equations is inherently more difficult than that of ordinary differential equations. A systematic classification of the various types of partial differential equations and all methods of formal solution is beyond the scope of this book. Briefly, similar to the case for ordinary differential equations, the classical solution of a partial differential equation involves the determination of a complementary function and a particular integral. However, even for linear homogeneous partial differential equations it is usually difficult to obtain a general solution and extremely tedious to particularize it so that the given initial and boundary conditions are satisfied. The usual approach to the problem in most solvable physical situations is to obtain particular solutions of the equation directly by a simple procedure, such as the method of separation of variables, and then combine these solutions in such a way that the given auxiliary conditions are satisfied.*

In this chapter we shall devote our primary attention to transmission lines. The partial differential equations for the voltage and the current along a general transmission line will be derived first. It will be shown that problems involving propagation of waves, vibration of strings and bars, skin-effect phenomena, conduction of heat, diffusion of liquid in a homogeneous medium, and migration of holes in a junction transistor are all analogous to special transmission-line problems and hence are describable by the same types of partial differential equations. Methods of solving linear partial differential equations by Laplace transformation will first be discussed in general terms and then applied to special transmission-line problems. The emphasis will be on basic techniques rather than on an exhaustive coverage of all possible types of problems.

11–2 Transmission lines and analogous sytems. We shall now derive the partial differential equations for the voltage and the current along a general transmission line. Transmission lines are extensively used for the transmission of electric power and signals. Four distributed parameters are required to characterize a transmission line:

$$R = \text{Resistance per unit length,}$$
$$L = \text{Inductance per unit length,}$$
$$G = \text{Conductance per unit length,}$$
$$C = \text{Capacitance per unit length.}$$

We should note here that these parameters are quantities per unit *loop* length of *two* conductors.

Consider the network representation of an elementary length, Δx, of a two-wire line† shown in Fig. 11–1. Two equations based upon Kirchhoff's voltage and current laws may be written.

Voltage equation:

$$v(x + \Delta x, t) + \frac{R}{2} \Delta x[i(x + \Delta x, t) + i(x, t)]$$

$$+ \frac{L}{2} \Delta x \frac{\partial}{\partial t} [i(x + \Delta x, t) + i(x, t)] = v(x, t). \quad (11\text{–}1)$$

---

* A general idea of the usual approach in solving partial differential equations in physics and engineering may be obtained by referring to almost any book on applied or engineering mathematics.

† The voltage and current equations derived for two-wire lines will be applicable without modification to coaxial cables.

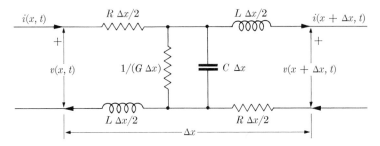

Fig. 11–1. Network representation of an elementary length of a two-wire transmission line.

Transferring $v(x + \Delta x, t)$ to the right side and dividing the entire equation by Δx, we have

$$\frac{1}{2}\left(R + L \frac{\partial}{\partial t}\right)[i(x + \Delta x, t) + i(x, t)] = -\frac{v(x + \Delta x, t) - v(x, t)}{\Delta x}. \quad (11\text{--}2)$$

Now, $i(x + \Delta x, t)$ can be expressed in terms of $i(x, t)$ and Δx by Taylor's expansion:

$$i(x + \Delta x, t) = i(x, t) + \Delta x \frac{\partial i(x, t)}{\partial x} + \cdots \quad (11\text{--}3)$$

Substituting Eq. (11–3) in Eq. (11–2), we have

$$\left(R + L \frac{\partial}{\partial t}\right)\left[i(x, t) + \frac{\Delta x}{2} \frac{\partial i(x, t)}{\partial x} + \cdots\right] = -\frac{v(x + \Delta x, t) - v(x, t)}{\Delta x}. \quad (11\text{--}4)$$

As we let Δx approach zero, all terms multiplied by Δx go to zero and the right side of Eq. (11–4) becomes $-\partial v/\partial x$. We get a first-order linear partial differential equation relating v and i:

$$Ri + L \frac{\partial i}{\partial t} = -\frac{\partial v}{\partial x}, \quad (11\text{--}5)$$

where the arguments (x, t) have been omitted for simplicity. Equation (11–5) correctly tells us that the decrease per unit length (negative rate of change with respect to x) of the voltage along the line is equal to the sum of the voltage drops across the series resistance and the series inductance per unit length.

Current equation:

By an entirely similar procedure we obtain a second partial differential equation relating v and i as follows:

$$Gv + C\,\frac{\partial v}{\partial t} = -\,\frac{\partial i}{\partial x}. \tag{11-6}$$

Equation (11–6) is a statement of the fact that the decrease per unit length (negative rate of change with respect to x) of the current along the line is equal to the sum of the current leaked through the conductance and the current required to charge the capacitance per unit length.

Equations (11–5) and (11–6) form a pair of simultaneous partial differential equations in which v and i are the dependent variables and x and t are the independent variables. We can eliminate i from these two equations by the following procedure:

$$\frac{\partial}{\partial x}\ (11\text{-}5)\text{:}\qquad R\,\frac{\partial i}{\partial x} + L\,\frac{\partial^2 i}{\partial x\,\partial t} = -\,\frac{\partial^2 v}{\partial x^2}, \tag{11-7}$$

$$\frac{\partial}{\partial t}\ (11\text{-}6)\text{:}\qquad G\,\frac{\partial v}{\partial t} + C\,\frac{\partial^2 v}{\partial t^2} = -\,\frac{\partial^2 i}{\partial x\,\partial t}. \tag{11-8}$$

The order of the derivatives with respect to x and t can be reversed because x and t are independent of each other. Substituting Eqs. (11–6) and (11–8) in Eq. (11–7) and rearranging terms, we get a second-order partial differential equation in v alone:

$$\frac{\partial^2 v}{\partial x^2} = LC\,\frac{\partial^2 v}{\partial t^2} + (RC + LG)\,\frac{\partial v}{\partial t} + RGv. \tag{11-9}$$

Similarly, v, instead of i, can be eliminated from Eqs. (11–5) and (11–6). We then have

$$\frac{\partial^2 i}{\partial x^2} = LC\,\frac{\partial^2 i}{\partial t^2} + (RC + LG)\,\frac{\partial i}{\partial t} + RGi. \tag{11-10}$$

We note that Eq. (11–10) in i is exactly the same as Eq. (11–9) in v. Actually, this is not surprising if we observe that the voltage and current equations in (11–5) and (11–6) are duals of each other, with the following analogous quantities:

$$v \leftrightarrow i, \qquad R \leftrightarrow G, \qquad L \leftrightarrow C.$$

When the quantities in Eq. (11–9) are changed to their analogous ones according to the list above, Eq. (11–10) is obtained directly. Equations (11–9) and (11–10) are known as the *telegraphist's equations*.

In solving transmission-line problems there is obviously no need to solve both Eq. (11–9) and Eq. (11–10), inasmuch as they are similar. Moreover, v and i are rigidly related to each other through Eqs. (11–5) and (11–6). Knowing one, the other can be easily found.

There are a number of important special cases for which the telegraphist's equations reduce to simpler forms.

[a] *The lossless line:* $R = 0, \ G = 0$.

$$\frac{\partial^2 v}{\partial x^2} = LC \frac{\partial^2 v}{\partial t^2} = \frac{1}{c^2} \frac{\partial^2 v}{\partial t^2},\tag{11-11}$$

where

$$c = \frac{1}{\sqrt{LC}}\tag{11-12}$$

has the dimension of velocity and is the *velocity of propagation* of waves along the transmission line. We will say more about this quantity later when we come to the solutions in Section 11–4. The current equation is of exactly the same form as the voltage equation. We shall not repeat it here.

Equation (11–11) is a very important type of partial differential equation. It is a one-dimensional *wave equation*. Analogous situations that are describable by one-dimensional wave equations are:

(a–1) Propagation of uniform plane electromagnetic waves: In this case both the electric field intensity E_y and the magnetic field intensity H_z satisfy the wave equation, and the directions of E_y, H_z, and x are mutually perpendicular.[*]

$$\frac{\partial^2 E_y}{\partial x^2} = \frac{1}{c^2} \frac{\partial^2 E_y}{\partial t^2},\tag{11-13}$$

and

$$\frac{\partial^2 H_z}{\partial x^2} = \frac{1}{c^2} \frac{\partial^2 H_z}{\partial t^2},\tag{11-14}$$

where

$$c = \frac{1}{\sqrt{\epsilon\mu}},\tag{11-15}$$

ϵ and μ being the absolute permittivity and the absolute permeability of the medium respectively.

(a–2) Transverse vibration of a stretched flexible string: If $y(x, t)$ denotes the vertical displacement, T is the constant tension, and m is the mass of the string per unit length, then[†]

$$\frac{\partial^2 y}{\partial x^2} = \frac{1}{c^2} \frac{\partial^2 y}{\partial t^2},\tag{11-16}$$

where

$$c = \sqrt{T/m}.\tag{11-17}$$

---

[*] R. F. Harrington, *Introduction to Electromagnetic Engineering*, McGraw-Hill Book Company, Inc., New York, Sections 10–1 and 10–2; 1958.

[†] W. T. Thomson, *Laplace Transformation*, Prentice-Hall, Inc., New York, Chapter 6; 1950.

(a–3) Longitudinal motion of an elastic bar: Let $X(x, t)$ be the longitudinal displacement, E = Young's modulus of elasticity, and ρ = the mass per unit volume of the bar material. We then have*

$$\frac{\partial^2 X}{\partial x^2} = \frac{1}{c^2} \frac{\partial^2 X}{\partial t^2}, \tag{11–18}$$

where

$$c = \sqrt{E/\rho}. \tag{11–19}$$

(a–4) Torsional motion of an elastic shaft: If $\theta(x, t)$ denotes the angular displacement of an elastic cylindrical shaft, E_s is the modulus of elasticity in shear, and ρ is the mass per unit volume, then†

$$\frac{\partial^2 \theta}{\partial x^2} = \frac{1}{c^2} \frac{\partial^2 \theta}{\partial t^2}, \tag{11–20}$$

where

$$c = \sqrt{E_s/\rho}. \tag{11–21}$$

[b] *The distortionless line.* When the distributed parameters satisfy the following condition:

$$\frac{R}{L} = \frac{G}{C}, \tag{11–22}.$$

the transmission line will transmit signals without distortion. The reason for this statement will become clear in Section 11–4. Refer now to the voltage equation (11–9). Let

$$c = \frac{1}{\sqrt{LC}} \tag{11–23}$$

and

$$a = \sqrt{RG}. \tag{11–24}$$

We have, in view of Eq. (11–22),

$$RC + LG = 2RC = 2\sqrt{RGLC} = 2\frac{a}{c}. \tag{11–25}$$

Equation (11–9) becomes

$$\frac{\partial^2 v}{\partial x^2} = \frac{1}{c^2} \frac{\partial^2 v}{\partial t^2} + 2\frac{a}{c}\frac{\partial v}{\partial t} + a^2 v. \tag{11–26}$$

If we introduce a new variable $u(x, t)$ defined by the relation

$$v(x, t) = \epsilon^{-act} u(x, t), \tag{11–27}$$

---

* W. T. Thomson, *Laplace Transformation*, Prentice-Hall, Inc., New York, Chapter 6; 1950.

† R. V. Churchill, *Operational Mathematics*, McGraw-Hill Book Company, Inc., New York, Section 38; 1958.

then substitution of Eq. (11–27) in Eq. (11–26) yields very simply a one-dimensional wave equation in u:

$$\frac{\partial^2 u}{\partial x^2} = \frac{1}{c^2} \frac{\partial^2 u}{\partial t^2}. \tag{11–28}$$

We see that the distribution of the voltage and the current along a distortionless transmission line is the same as that on a lossless line except for an attenuation factor, as shown in Eq. (11–27). More will be said about this in Section 11–4.

[c] *The leakage-free, noninductive cable:* $G = 0$, $L = 0$.

Coaxial cables with good insulation properties can be approximately characterized by transmission-line equations with only two parameters, R and C. On setting both G and L in Eq. (11–9) to zero, we obtain

$$\frac{\partial^2 v}{\partial x^2} = RC \frac{\partial v}{\partial t}. \tag{11–29}$$

This type of partial differential equation is known as a *diffusion equation.* Analogous situations* that are describable by one-dimensional diffusion equations are:

(c–1) Distribution of alternating current in a homogeneous plane conducting medium: In a homogeneous conducting medium of infinite extent and a plane surface the current density $J(x, t)$ satisfies the diffusion equation

$$\frac{\partial^2 J}{\partial x^2} = \mu\sigma \frac{\partial J}{\partial t}, \tag{11–30}$$

where μ and σ are the absolute permeability and conductivity of the medium respectively. This is the well-known one-dimensional *skin-effect* equation.

(c–2) Conduction of heat through a slab of uniform thickness: If $w(x, t)$ denotes the temperature in the slab, k is the thermal conductivity of the material, h is the specific heat, and ρ is the density, then

$$\frac{\partial^2 w}{\partial x^2} = \frac{h\rho}{k} \frac{\partial w}{\partial t}. \tag{11–31}$$

The reciprocal of the coefficient, $k/h\rho$, is called the *thermal diffusivity* of the material.

(c–3) Diffusion of liquid in a homogeneous medium: When a slab of liquid-permeated porous material is dried by evaporation, the concentra-

---

* L. A. Pipes, *Applied Mathematics for Engineers and Physicists,* McGraw-Hill Book Company, Inc., New York, Chapter XVIII; 1946.

tion of the liquid, $U(x, t)$, satisfies the diffusion equation

$$\frac{\partial^2 U}{\partial x^2} = \frac{1}{K} \frac{\partial U}{\partial t},$$

   (11–32)

where K is the *diffusivity constant* of the liquid in length squared per unit time. The same equation governs the concentration of the holes in a plane parallel p-n-p junction transistor, assuming infinite lifetime for the holes.*

11–3 Methods of solution by Laplace transformation. There are two general approaches that we can take in solving linear partial differential equations with constant coefficients by Laplace transformation. We can either convert a partial differential equation with independent variables x and t into an ordinary differential equation by transformation with respect to t and then solve it by classical methods, or we can use repeated Laplace transformation with respect to both t and x, converting the equation into an algebraic one, and then solve the problem entirely by transform methods. We shall discuss and compare these two approaches separately in this section.

A. *Conversion into an ordinary differential equation.* In order to fix ideas, let us use the wave equation as an example. Equation (11–11) is repeated below:

$$\frac{\partial^2 v(x, t)}{\partial x^2} = \frac{1}{c^2} \frac{\partial^2 v(x, t)}{\partial t^2}.$$

   (11–33)

Since this equation contains a second-order derivative with respect to x as well as a second-order derivative with respect to t, a total of four known conditions are required for the determination of all four integration constants. In general, we refer to known conditions with respect to t as *initial conditions* because they are usually values given at $t = 0$. Known conditions at specific values of x are called *boundary conditions*. The following are assumed to be given:

Initial conditions,	$t = 0$: $v(x, 0),$	(11–34a)
	$t = 0$: $v'(x, 0),$	(11–34b)
Boundary conditions,	$x = 0$: $v(0, t),$	(11–34c)
	$x = x_1$: $v(x_1, t),$	(11–34d)

where

$$v'(x, 0) = \left[\frac{\partial}{\partial t} v(x, t) \right]_{t=0}.$$

   (11–35)

---

* A. W. Lo et al., *Transistor Electronics*, Prentice-Hall, Inc., New York, Section 8.2; 1955.

Taking the Laplace transformation of Eq. (11–33) with respect to t and letting

$$\mathcal{L}_t[v(x, t)] = \int_0^\infty v(x, t)\epsilon^{-st}\, dt = V(x, s), \qquad (11\text{–}36)$$

we obtain

$$\frac{d^2 V(x, s)}{dx^2} = \frac{1}{c^2}\, [s^2 V(x, s) - sv(x, 0) - v'(x, 0)]$$

or

$$\frac{d^2 V(x, s)}{dx^2} - \frac{s^2}{c^2}\, V(x, s) = -\frac{1}{c^2}\, [sv(x, 0) + v'(x, 0)]. \qquad (11\text{–}37)$$

We see that Laplace transformation has reduced the number of independent variables by one and converted the given partial differential equation to an ordinary differential equation. In Eq. (11–37) s is considered as a parameter and is not a function of either t or x.

Equation (11–37) can be solved by classical methods quite easily. Let us consider the complementary function and the particular integral separately. The complementary function, $V_c(x, s)$, is the solution of the following homogeneous equation:

$$\frac{d^2 V_c(x, s)}{dx^2} - \frac{s^2}{c^2}\, V_c(x, s) = 0. \qquad (11\text{–}38)$$

We have

$$V_c(x, s) = A\epsilon^{-sx/c} + B\epsilon^{sx/c}. \qquad (11\text{–}39)$$

The particular-integral part of the solution depends upon the given initial conditions. In most cases, both $v(x, 0)$ and $v'(x, 0)$ will be constants; consequently the particular integral, $V_p(x, s)$, will also be a constant. If

$$v(x, 0) = K_1, \qquad (11\text{–}40)$$

$$v'(x, 0) = K_2, \qquad (11\text{–}41)$$

then

$$V_p(x, s) = K(s). \qquad (11\text{–}42)$$

Substitution of this $V_p(x, s)$ in Eq. (11–37) yields directly

$$V_p(x, s) = K(s) = \frac{1}{s^2}\, (K_1 s + K_2). \qquad (11\text{–}43)$$

Combining Eqs. (11–39) and (11–43), we have

$$V(x, s) = V_c(x, s) + V_p(x, s)$$
$$= A\epsilon^{-sx/c} + B\epsilon^{sx/c} + \frac{1}{s^2}\, (K_1 s + K_2), \qquad (11\text{–}44)$$

which is the general solution for Eq. (11–37).

The two arbitrary constants, A and B, in the solution can be determined from the Laplace transforms of the two given boundary conditions (11–34c) and (11–34d):

At $x = 0$,

$$V(0, s) = \mathcal{L}_t[v(0, t)] = A + B + \frac{1}{s^2}(K_1 s + K_2). \tag{11–45}$$

At $x = x_1$,

$$V(x_1, s) = \mathcal{L}_t[v(x_1, t)] = A\epsilon^{-sx_1/c} + B\epsilon^{sx_1/c} + \frac{1}{s^2}(K_1 s + K_2). \tag{11–46}$$

With A and B determined, the desired solution, $v(x, t)$, can be obtained by taking the inverse Laplace transformation of $V(x, s)$ given in Eq. (11–44):

$$v(x, t) = \mathcal{L}_t^{-1}[V(x, s)]. \tag{11–47}$$

The steps involved in the above method of solution can be summarized as follows:

1. Take the $\mathcal{L}_t$ of the given partial differential equation, inserting the given initial conditions. Let $\mathcal{L}_t[v(x, t)] = V(x, s)$.

2. Solve the ordinary differential equation obtained in step 1 for $V(x, s)$ by classical methods. $V(x, s)$ will have a complementary function and a particular integral, $V(x, s) = V_c(x, s) + V_p(x, s)$.

3. Determine the arbitrary constants in the solution for $V(x, s)$ from the $\mathcal{L}_t$ of the given boundary conditions.

4. Obtain the desired solution by taking the $\mathcal{L}_t^{-1}$ of $V(x, s)$. $v(x, t) = \mathcal{L}_t^{-1}[V(x, s)]$.

B. *Conversion into an algebraic equation by repeated Laplace transformation.* This approach deviates from the previous one in that a second Laplace transformation (with respect to x) is applied to the ordinary differential equation (11–37) in order to solve for $V(x, s)$. Let

$$\mathcal{L}_x[V(x, s)] = \int_0^\infty V(x, s)\epsilon^{-px}\, dx = \overline{V}(p, s). \tag{11–48}$$

Then Laplace transformation of Eq. (11–37) with respect to x yields

$$p^2 \overline{V}(p, s) - p V(0, s) - [V(0, s)]' - \frac{s^2}{c^2}\overline{V}(p, s)$$

$$= -\frac{1}{c^2}[s V(p, 0) + V'(p, 0)]$$

or

$$\overline{V}(p, s) = \frac{p V(0, s) + [V(0, s)]' - (1/c^2)[s V(p, 0) + V'(p, 0)]}{p^2 - (s^2/c^2)}, \tag{11–49}$$

where

$$V(0, s) = \mathcal{L}_t[v(0, t)] = \int_0^\infty v(0, t)\epsilon^{-st}\,dt, \qquad (11\text{--}50)$$

$$[V(0, s)]' = \left[\frac{d}{dx} V(x, s)\right]_{x=0}, \qquad (11\text{--}51)$$

$$V(p, 0) = \mathcal{L}_x[v(x, 0)] = \int_0^\infty v(x, 0)\epsilon^{-px}\,dx, \qquad (11\text{--}52)$$

$$V'(p, 0) = \mathcal{L}_x[v'(x, 0)] = \int_0^\infty v'(x, 0)\epsilon^{-px}\,dx. \qquad (11\text{--}53)$$

Equation (11–49) is now algebraic and contains no arbitrary constants. The desired solution, $v(x, t)$, can be obtained by repeated inverse Laplace transformation of $\overline{V}(p, s)$, first with respect to x to give $V(x, s)$, then with respect to t. Thus

$$V(x, s) = \mathcal{L}_x^{-1}[\overline{V}(p, s)], \qquad (11\text{--}54)$$

$$v(x, t) = \mathcal{L}_t^{-1}[V(x, s)]. \qquad (11\text{--}55)$$

We observe that $[V(0, s)]'$ in (11–51) is usually not given; that is, until $V(x, s)$ is solved from Eq. (11–37) we do not know the value of its derivative at $x = 0$. This dilemma can be resolved by finding $V(x, s)$ from the inverse transformation of Eq. (11–49), leaving $[V(0, s)]'$ as an undetermined constant. From condition (11–34d), we have

$$V(x_1, s) = \mathcal{L}_t[v(x_1, t)] = \int_0^\infty v(x_1, t)\epsilon^{-st}\,dt. \qquad (11\text{--}56)$$

By putting $x = x_1$ in $V(x, s)$ obtained from the inverse transformation of Eq. (11–49) and equating it to $V(x_1, s)$ in (11–56), $[V(0, s)]'$ can be determined.

The steps involved in the method of solution by repeated Laplace transformation can be summarized as follows:

1. Take the $\mathcal{L}_t$ of the given partial differential equation, inserting the given initial conditions. Let $\mathcal{L}_t[v(x, t)] = V(x, s)$.

2. Take the $\mathcal{L}_x$ of the ordinary differential equation obtained in step 1, inserting the transforms of the given boundary conditions. Then let $\mathcal{L}_x[V(x, s)] = \overline{V}(p, s)$. $\overline{V}(p, s)$ will be expressible as a ratio of two polynomials in p.

3. Find $V(x, s)$ by taking the $\mathcal{L}_x^{-1}$ of $\overline{V}(p, s)$.

4. Obtain the desired solution $v(x, t)$ by taking the $\mathcal{L}_t^{-1}$ of $V(x, s)$.

In comparing the two methods of solution discussed above, we see that the second method, using repeated Laplace transformation, is the more involved one. Not only is the symbolism more complicated and confusing, it is also necessary to transform all of the given initial and boundary conditions, some with respect to t and others with respect to x. Because of the nature of most practical problems, the quantity $[V(0, s)]'$ is usually not known and has to be determined in a laborious way. The need for performing inverse Laplace transformation twice is also a disadvantage. On the other hand, the first method, which converts a given partial differential equation into an ordinary differential equation and then solves the latter by classical methods, is straightforward and easy to use. For most engineering problems, the solution of the ordinary differential equation (11–37) can be written by inspection; seldom is the work involved in taking a second transformation worth while. We shall use the first method to solve problems in the rest of this chapter.

11–4 The infinite transmission line. Before solving any specific transmission-line problems, let us find the Laplace transformation of the general telegraphist's equations (11–9) and (11–10). Once this is done, the transformed equations can be specialized to suit any given transmission-line situation or any of the analogous problems described in Section 11–2. Let

$$\mathcal{L}[v(x, t)] = V(x, s); \qquad (11–57)^*$$

then

$$\mathcal{L}\left[\frac{\partial^2 v}{\partial x^2}\right] = \frac{\partial^2}{\partial x^2}\{\mathcal{L}[v(x, t)]\} = \frac{d^2}{dx^2} V(x, s), \qquad (11–58)$$

$$\mathcal{L}\left[\frac{\partial v}{\partial t}\right] = sV(x, s) - v(x, 0), \qquad (11–59)$$

$$\mathcal{L}\left[\frac{\partial^2 v}{\partial t^2}\right] = s^2 V(x, s) - sv(x, 0) - v'(x, 0). \qquad (11–60)$$

Taking the Laplace transformation of Eq. (11–9) and making use of Eqs. (11–57) through (11–60), we obtain the following second-order ordinary differential equation in $V(x, s)$:

$$\frac{d^2 V}{dx^2} - (Ls + R)(Cs + G)V = -L[Gv(x, 0) + Cv'(x, 0)]$$
$$- C(Ls + R)v(x, 0), \qquad (11–61)$$

where

$$v'(x, 0) = \left[\frac{\partial v(x, t)}{\partial t}\right]_{t=0}. \qquad (11–62)$$

---

* The operation $\mathcal{L}$ without a subscript will be taken to mean Laplace transformation with respect to the independent variable t.

Quite often the value of $v'(x, 0)$ is not immediately obvious, and it is more convenient to write the terms in brackets on the right side of Eq. (11–61) in another form. Setting $t = 0$ in Eq. (11–6), we have

$$Gv(x, 0) + Cv'(x, 0) = -\frac{di(x, 0)}{dx}. \qquad (11\text{–}63)$$

Hence Eq. (11–61) can be rewritten as

$$\frac{d^2 V}{dx^2} - \gamma^2 V = L\frac{di(x, 0)}{dx} - C(Ls + R)v(x, 0), \qquad (11\text{–}64)$$

where

$$\gamma = \sqrt{(Ls + R)(Cs + G)}. \qquad (11\text{–}65)$$

Equation (11–64) is important because it can serve as the starting point for all transmission-line problems; there is no need to go back as far as Eq. (11–9).

By a similar procedure, if we let

$$\mathcal{L}[i(x, t)] = I(x, s), \qquad (11\text{–}66)$$

a second-order ordinary differential equation in $I(x, s)$ is obtained from the Laplace transformation of Eq. (11–10):

$$\frac{d^2 I}{dx^2} - \gamma^2 I = C\frac{dv(x, 0)}{dx} - L(Cs + G)i(x, 0), \qquad (11\text{–}67)$$

where γ is the same as that given in Eq. (11–65). In fact, Eqs. (11–64) and (11–67) are duals of each other, and it is easy to see that γ is the dual of itself.

We shall now find the step response of an infinite transmission line. The problem may be formulated as follows.

EXAMPLE 11–1. Given an initially relaxed, infinitely long transmission line (Fig. 11–2). A constant voltage of unit magnitude is applied to the sending end ($x = 0$) of the line at $t = 0$. Determine the instantaneous voltage and current at any distance x from the sending end for (a) a lossless line, (b) a distortionless line, and (c) a leakage-free and noninductive cable.

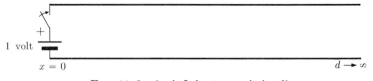

FIG. 11–2. An infinite transmission line.

Solution. For an initially relaxed line, we have the following initial conditions:

$$v(x, 0) = 0, \tag{11–68}$$

$$\frac{di(x, 0)}{dx} = 0. \tag{11–69}$$

The transformed voltage equation for $V(x, s)$ in (11–64) reduces to a homogeneous second-order differential equation:

$$\frac{d^2 V}{dx^2} - \gamma^2 V = 0. \tag{11–70}$$

The general solution of Eq. (11–70) is simply

$$V(x, s) = A\epsilon^{-\gamma x} + B\epsilon^{\gamma x}. \tag{11–71}$$

In order to determine the arbitrary constants A and B, we make use of the boundary conditions:

At $x = 0$, $v(0, t) = U(t)$, which transforms to give

$$V(0, s) = \frac{1}{s}. \tag{11–72}$$

At $x = d \to \infty$, $v(\infty, t)$ must remain finite. Hence

$$V(\infty, s) \text{ is finite.} \tag{11–73}$$

Condition (11–73) requires

$$B = 0 \tag{11–74}$$

and condition (11–72) determines A:

$$A = \frac{1}{s}. \tag{11–75}$$

The solution of Eq. (11–70) that satisfies the boundary conditions of the given problem is therefore

$$V(x, s) = \frac{1}{s} \epsilon^{-\gamma x}, \tag{11–76}$$

where γ is given in Eq. (11–65) for the general transmission line.

(a) *For a lossless line* ($R = 0$ and $G = 0$).

$$\gamma = s\sqrt{LC} = \frac{s}{c}, \tag{11–77}$$

with

$$c = \frac{1}{\sqrt{LC}}. \tag{11–78}$$

Equation (11–76) becomes

$$V(x, s) = \frac{1}{s}\,\epsilon^{-sx/c}. \tag{11–79}$$

Inverse Laplace transformation of Eq. (11–79) yields the step response:

$$v(x, t) = \mathcal{L}^{-1}[V(x, s)] = U\left(t - \frac{x}{c}\right). \tag{11–80}$$

An interpretation of the above expression for $v(x, t)$ may be as follows. From the nature of the unit step function, at any given time, say $t = t_1$, the voltage along the line is unity for $x/c < t_1$ or $x < ct_1$, and zero for $x > ct_1$. On the other hand, at any given distance, say $x = x_1$, from the sending end, the voltage is zero for $t < x_1/c$ and unity for $t > x_1/c$. These two situations are depicted in Fig. 11–3. The unit voltage of the source propagates along the line with a velocity c. In Fig. 11–3(a), when $t = t_1$ the voltage has traveled a distance $x = ct_1$; in Fig. 11–3(b), at $x = x_1$ there is no voltage before the time required (x_1/c) for the voltage to travel the distance from the sending end, and the voltage there will stay at unity for $t > x_1/c$. There is then a *traveling wave* on the line.

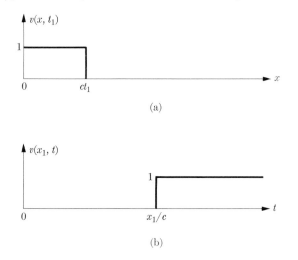

(a)

(b)

FIG. 11–3.   Interpretation of step response of an infinite lossless line. (a) Voltage distribution along the line at $t = t_1$. (b) Time variation of voltage at $x = x_1$.

The instantaneous current, $i(x, t)$, along the line can be similarly solved starting from a specialized form of the transformed current equation (11–67). However, it is simpler to make use of Eq. (11–5), which transforms to the following:

$$(R + Ls)I(x, s) = -\frac{dV(x, s)}{dx} + Li(x, 0). \qquad (11\text{–}81)$$

In our case here, $R = 0$ and $i(x, 0) = 0$. We have

$$I(x, s) = -\frac{1}{Ls}\frac{dV(x, s)}{dx}. \qquad (11\text{–}82)$$

Substituting $V(x, s)$ from Eq. (11–79) in Eq. (11–82), we obtain

$$I(x, s) = \sqrt{\frac{C}{L}}\frac{1}{s}\epsilon^{-sx/c} \qquad (11\text{–}83)$$

and

$$i(x, t) = \mathcal{L}^{-1}[I(x, s)] = \sqrt{\frac{C}{L}}\,U\left(t - \frac{x}{c}\right). \qquad (11\text{–}84)$$

We see that the expression for the current in Eq. (11–84) is the same as that for the voltage in Eq. (11–80) except for the coefficient. In fact, the ratio

$$\frac{v(x, t)}{i(x, t)} = \frac{V(x, s)}{I(x, s)} = Z_0 = \sqrt{\frac{L}{C}} \qquad (11\text{–}85)$$

is a constant and is called the *characteristic impedance* (in this case, a pure resistance) of the lossless line.

(b) *For a distortionless line* $(RC = LG)$. Substitution of the condition (11–22) into the general expression for γ in Eq. (11–65) gives

$$\gamma = \sqrt{\frac{C}{L}}\,(Ls + R). \qquad (11\text{–}86)$$

Consequently, Eq. (11–76) becomes

$$V(x, s) = \frac{1}{s}\epsilon^{-x\sqrt{C/L}(Ls+R)} = \frac{1}{s}\epsilon^{-xR\sqrt{C/L}}\epsilon^{-sx/c}. \qquad (11\text{–}87)$$

Inverse Laplace transformation of Eq. (11–87) yields the step voltage response of an infinite distortionless line:

$$v(x, t) = \epsilon^{-xR\sqrt{C/L}}U\left(t - \frac{x}{c}\right). \qquad (11\text{–}88)$$

The current can be found from Eq. (11–81) by noting that $i(x, 0) = 0$:

$$I(x, s) = -\frac{1}{Ls + R}\frac{dV(x, s)}{dx} = \sqrt{\frac{C}{L}}\frac{1}{s}\epsilon^{-x\sqrt{C/L}(Ls+R)}. \qquad (11\text{–}89)$$

Hence

$$i(x, t) = \mathcal{L}^{-1}[I(x, s)] = \sqrt{\frac{C}{L}}\, \epsilon^{-xR\sqrt{C/L}} U\left(t - \frac{x}{c}\right). \qquad (11\text{-}90)$$

The expressions for $v(x, t)$ and $i(x, t)$ in an infinite distortionless line are thus similar to those in an infinite lossless line except for an attenuation factor $\exp(-xR\sqrt{C/L})$ which varies with x. In other words, there is an attenuated traveling wave along the distortionless line. The characteristic impedance Z_0 of a distortionless line is also a constant and is equal to

$$\frac{V(x, s)}{I(x, s)} = Z_0 = \sqrt{\frac{L}{C}}. \qquad (11\text{-}91)$$

If, instead of the unit step voltage, a general voltage function $f(t)$ is applied to an infinite line at $x = 0$ and $t = 0$, then

$$V(0, t) = \mathcal{L}[f(t)] = F(s) \qquad (11\text{-}92)$$

and the solution of Eq. (11–70) becomes, instead of that given in Eq. (11–76),

$$V(x, s) = F(s)\epsilon^{-\gamma x}. \qquad (11\text{-}93)$$

For a distortionless line, γ takes the value given in Eq. (11–86):

$$V(x, s) = F(s)\epsilon^{-xR\sqrt{C/L}}\epsilon^{-sx/c}. \qquad (11\text{-}94)$$

Inverse transformation of Eq. (11–94) gives directly

$$v(x, t) = \epsilon^{-xR\sqrt{C/L}}f\left(t - \frac{x}{c}\right) U\left(t - \frac{x}{c}\right). \qquad (11\text{-}95)$$

We observe from the above expression that at a given distance x from the sending end the response to an arbitrary applied voltage function $f(t)$ is attenuated by the factor $\exp(-xR\sqrt{C/L})$ and delayed in time by x/c, but *the original waveform is preserved*. This is why we call a transmission line which satisfies the condition $RC = LG$ *distortionless*. The ideal lossless line is, of course, a special case of the distortionless line. The response of a lossless line to an arbitrary voltage function $f(t)$ can be obtained by setting $R = 0$ in Eq. (11–95):

$$v(x, t) = f\left(t - \frac{x}{c}\right) U\left(t - \frac{x}{c}\right). \qquad (11\text{-}96)$$

(c) *For a leakage-free, noninductive cable* ($G = 0$ and $L = 0$).

$$\gamma = \sqrt{RCs}. \qquad (11\text{-}97)$$

The solution given in Eq. (11–76) becomes

$$V(x, s) = \frac{1}{s} \epsilon^{-x\sqrt{RCs}}. \tag{11–98}$$

Equation (11–98) is in a form for which we hitherto have not known the inverse Laplace transform. The formal evaluation by means of the inversion integral (6–16) is an involved process and requires a thorough knowledge of the theory of functions of a complex variable.* We digress here to establish by means of direct transformation a fundamental Laplace transform pair in which the transformed function has a $\sqrt{s}$ factor in an exponent.†

Consider

$$f(t) = \frac{\epsilon^{-a^2/4t}}{\sqrt{\pi t^3}}. \tag{11–99}$$

In order to find its Laplace transform,

$$F(s) = \int_0^\infty \frac{\epsilon^{-a^2/4t}}{\sqrt{\pi t^3}} \epsilon^{-st}\, dt, \tag{11–100}$$

we let

$$\frac{a^2}{4t} = \tau^2, \qquad \frac{dt}{\sqrt{t^3}} = -\frac{4}{a}\, d\tau.$$

Equation (11–100) then becomes

$$F(s) = -\frac{4}{a\sqrt{\pi}} \int_\infty^0 \epsilon^{-[\tau^2 + (sa^2/4\tau^2)]}\, d\tau$$

$$= \frac{4}{a\sqrt{\pi}} \epsilon^{-a\sqrt{s}} \int_0^\infty \epsilon^{-[\tau - (a\sqrt{s}/2\tau)]^2}\, d\tau$$

$$= \frac{4}{a\sqrt{\pi}} \epsilon^{-a\sqrt{s}} \int_0^\infty \epsilon^{-[\tau - (b/\tau)]^2}\, d\tau, \tag{11–101}$$

where $b = a\sqrt{s}/2$. Now let

$$\tau = \frac{b}{\lambda}, \qquad d\tau = -\frac{b}{\lambda^2}\, d\lambda.$$

---

* E. Weber, *Linear Transient Analysis*, Volume II, Chapter 7, John Wiley and Sons, Inc., New York; 1956.

† R. V. Churchill, *Operational Mathematics*, McGraw-Hill Book Company, Inc., New York, Section 23; 1958.

The definite integral in Eq. (11–101) can be put in a different form:

$$\int_0^\infty \epsilon^{-[\tau-(b/\tau)]^2} \, d\tau = -\int_\infty^0 \frac{b}{\lambda^2} \, \epsilon^{-[(b/\lambda)-\lambda]^2} \, d\lambda$$

$$= \int_0^\infty \frac{b}{\lambda^2} \, \epsilon^{-[\lambda-(b/\lambda)]^2} \, d\lambda. \qquad (11-102)$$

Since λ in the definite integral on the right side of Eq. (11–102) is a dummy variable, we can replace it with τ and obtain

$$\int_0^\infty \epsilon^{-[\tau-(b/\tau)]^2} \, d\tau = \int_0^\infty \frac{b}{\tau^2} \, \epsilon^{-[\tau-(b/\tau)]^2} \, d\tau. \qquad (11-103)$$

We can also add the integral on the left to both sides of Eq. (11–103) and get

$$2\int_0^\infty \epsilon^{-[\tau-(b/\tau)]^2} \, d\tau = \int_0^\infty \left(1 + \frac{b}{\tau^2}\right) \epsilon^{-[\tau-(b/\tau)]^2} \, d\tau. \qquad (11-104)$$

If we substitute

$$x = \tau - \frac{b}{\tau}, \qquad dx = \left(1 + \frac{b}{\tau^2}\right) d\tau,$$

we finally obtain from Eq. (11–104)

$$2\int_0^\infty \epsilon^{-[\tau-(b/\tau)]^2} \, d\tau = \int_{-\infty}^\infty \epsilon^{-x^2} \, dx = \sqrt{\pi} \qquad (11-105)$$

and, from Eq. (11–101),

$$F(s) = \mathcal{L}\left[\frac{\epsilon^{-a^2/4t}}{\sqrt{\pi t^3}}\right] = \frac{2}{a} \epsilon^{-a\sqrt{s}} \qquad (11-106)$$

or

$$\mathcal{L}^{-1}[\epsilon^{-a\sqrt{s}}] = \frac{a\epsilon^{-a^2/4t}}{2\sqrt{\pi t^3}}. \qquad (11-107)$$

In view of the integration theorem in Eq. (6–62):

$$\mathcal{L}^{-1}\left[\frac{F(s)}{s}\right] = \int_0^t f(t) \, dt, \qquad (11-108)$$

we also have

$$\mathcal{L}^{-1}\left[\frac{1}{s} \epsilon^{-a\sqrt{s}}\right] = \frac{a}{2\sqrt{\pi}} \int_0^t \epsilon^{-a^2/4t} \frac{dt}{\sqrt{t^3}}$$

$$= -\frac{2}{\sqrt{\pi}} \int_{\infty}^{a/2\sqrt{t}} \epsilon^{-\tau^2} d\tau$$

$$= \frac{2}{\sqrt{\pi}} \left[\int_0^{\infty} \epsilon^{-\tau^2} d\tau - \int_0^{a/2\sqrt{t}} \epsilon^{-\tau^2} d\tau \right].$$

Making use of relations (8–104) and (8–105), we have

$$\mathcal{L}^{-1} \left[\frac{1}{s} \epsilon^{-a\sqrt{s}} \right] = 1 - \text{erf} \left(\frac{a}{2\sqrt{t}} \right) = \text{erfc} \left(\frac{a}{2\sqrt{t}} \right). \quad (11\text{–}109)$$

The error function erf (x) was defined in Section 8–6; its graph was shown in Fig. 8–13. The *complementary error function* erfc (x) is defined as follows:

$$\text{erfc} (x) = 1 - \text{erf} (x) = \frac{2}{\sqrt{\pi}} \int_x^{\infty} \epsilon^{-u^2} du. \quad (11\text{–}110)$$

Returning to Eq. (11–98), we find the step voltage response of an infinite leakage-free and noninductive cable to be

$$v(x, t) = \mathcal{L}^{-1} \left[\frac{1}{s} \epsilon^{-x\sqrt{RCs}} \right] = \text{erfc} \left(\frac{x}{2} \sqrt{\frac{RC}{t}} \right). \quad (11\text{–}111)$$

The response $v(x, t)$ in Eq. (11–111) is usually plotted versus the square of the reciprocal of the argument, $4t/x^2RC$; it is shown as the solid curve in Fig. 11–4.

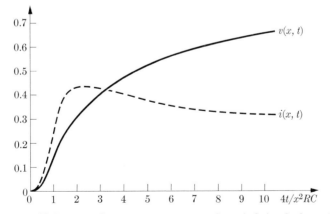

Fig. 11–4. Voltage and current responses of an infinite leakage-free, noninductive cable to a unit step voltage.

To find the current response, we write from Eq. (11–81), with $i(x, 0) = 0$ and $L = 0$:

$$I(x, s) = -\frac{1}{R}\frac{dV(x, s)}{dx} = \sqrt{\frac{C}{Rs}}\,\epsilon^{-x\sqrt{RCs}}. \qquad (11\text{–}112)$$

The inverse transformation of the above expression can be obtained by applying relation (6–65) concerning multiplication by t:

$$\mathcal{L}^{-1}\left[-\frac{d}{ds}F(s)\right] = tf(t) \qquad (11\text{–}113)$$

to the transform pair in (11–107):

$$\mathcal{L}^{-1}\left[-\frac{2}{a}\frac{d}{ds}\,\epsilon^{-a\sqrt{s}}\right] = \mathcal{L}^{-1}\left[\frac{1}{\sqrt{s}}\,\epsilon^{-a\sqrt{s}}\right] = \frac{\epsilon^{-a^2/4t}}{\sqrt{\pi t}}. \qquad (11\text{–}114)$$

Therefore, from Eqs. (11–112) and (11–114) we have

$$i(x, t) = \sqrt{\frac{C}{R}}\,\mathcal{L}^{-1}\left[\frac{1}{\sqrt{s}}\,\epsilon^{-x\sqrt{RCs}}\right] = \sqrt{\frac{C}{\pi Rt}}\,\epsilon^{-x^2 RC/4t}. \qquad (11\text{–}115)$$

The graph for $i(x, t)$ is sketched as the dashed curve in Fig. 11–4 versus the reciprocal of the exponent, $4t/x^2 RC$, for a fixed distance x. We observe from Fig. 11–4 that there is a buildup time for both the voltage and the current. This phenomenon imposes a severe limit on the speed of a signal that can be transmitted over a leakage-free, noninductive cable, especially when the distance is great. It explains the need for inductive loading on transoceanic cables in order that the condition for distortionless transmission, $RC = LG$, is more nearly satisfied.

The characteristic impedance of a leakage-free, noninductive cable is no longer a constant. From Eqs. (11–98) and (11–112),

$$Z_0 = \frac{V(x, s)}{I(x, s)} = \sqrt{\frac{R}{Cs}}, \qquad (11\text{–}116)$$

which is now an irrational function of s. Note that the ratio of $v(x, t)$ to $i(x, t)$ is a function of both x and t; it *does not* define the characteristic impedance.

EXAMPLE 11–2.   Determine the input current of an initially relaxed, infinitely long, general transmission line if a constant voltage of unit magnitude is applied to the sending end $(x = 0)$ at $t = 0$.

Solution.   Figure 11–2 in Example 11–1 still applies. The relation between the transformed current and the transformed voltage can be obtained from Eq. (11–81) by setting $i(x, 0) = 0$.

$$I(x, s) = -\frac{1}{Ls + R} \frac{dV(x, s)}{dx}. \qquad (11\text{–}117)$$

For an infinite line with an applied unit step voltage, the transformed voltage is, from Eq. (11–76),

$$V(x, s) = \frac{1}{s} \epsilon^{-\gamma x}. \qquad (11\text{–}118)$$

Substitution of Eq. (11–118) into (11–117) gives

$$I(x, s) = \frac{\gamma}{s(Ls + R)} \epsilon^{-\gamma x}. \qquad (11\text{–}119)$$

Since we are interested only in the current at the input or the sending end, we put $x = 0$ in Eq. (11–119) and obtain

$$I(0, s) = \frac{\gamma}{s(Ls + R)} = \frac{1}{s} \sqrt{\frac{Cs + G}{Ls + R}} \qquad (11\text{–}120)$$

for a general transmission line. We note that the characteristic impedance of a general transmission line is

$$Z_0 = \frac{V(x, s)}{I(x, s)} = \sqrt{\frac{Ls + R}{Cs + G}}. \qquad (11\text{–}121)$$

In order to determine the required input current, we must find the inverse transform of $I(0, s)$ in Eq. (11–120):

$$I(0, s) = \sqrt{\frac{C}{L}} \frac{s + G/C}{s\sqrt{(s + R/L)(s + G/C)}}$$

$$= \sqrt{\frac{C}{L}} \frac{s + G/C}{s\sqrt{s^2 + [(R/L) + (G/C)]s + (RG/LC)}}$$

$$= \sqrt{\frac{C}{L}} \frac{s + G/C}{s\sqrt{(s + a)^2 - b^2}}$$

$$= \sqrt{\frac{C}{L}} \left[\frac{1}{\sqrt{(s + a)^2 - b^2}} + \frac{G}{Cs\sqrt{(s + a)^2 - b^2}} \right], \qquad (11\text{–}122)$$

where

$$a = \frac{1}{2}\left(\frac{R}{L} + \frac{G}{C}\right) \qquad (11\text{–}123)$$

and

$$b = \frac{1}{2}\left(\frac{R}{L} - \frac{G}{C}\right). \qquad (11\text{–}124)$$

From Eq. (8–121), we deduce that

$$\mathcal{L}^{-1}\left[\frac{1}{\sqrt{(s+a)^2 - b^2}}\right] = \epsilon^{-at}I_0(bt), \qquad (11\text{--}125)$$

where I_0 is the modified Bessel function of the first kind of the zeroth order.

The integration theorem in Eq. (6–62) tells us that

$$\mathcal{L}^{-1}\left[\frac{G}{Cs\sqrt{(s+a)^2 - b^2}}\right] = \frac{G}{C}\int_0^t \epsilon^{-at}I_0(bt)\,dt. \qquad (11\text{--}126)$$

Combining the results from (11–125) and (11–126) in Eq. (11–122), we obtain the expression for the required input current:

$$i(0, t) = \mathcal{L}^{-1}[I(0, s)] = \sqrt{\frac{C}{L}}\left[\epsilon^{-at}I_0(bt) + \frac{G}{C}\int_0^t \epsilon^{-at}I_0(bt)\,dt\right],$$
$$(11\text{--}127)$$

where the values of a and b are as given in Eqs. (11–123) and (11–124).

11–5 Finite transmission lines. A direct consequence of the assumption that a given transmission line is infinitely long is that the term $\epsilon^{\gamma x}$ cannot exist in the expression for either the transformed voltage or the transformed current, since both voltage and current must remain finite. Thus, from the preceding section, the solution to Eq. (11–70) for an initially relaxed infinite line is simply

$$V(x, s) = A\epsilon^{-\gamma x}, \qquad (11\text{--}128)$$

since B in Eq. (11–71) must vanish. However, this is no longer true for a finite line and both terms must be retained:

$$V(x, s) = A\epsilon^{-\gamma x} + B\epsilon^{\gamma x}. \qquad (11\text{--}129)$$

The solutions for finite transmission line problems are therefore inherently more laborious. In this section we shall discuss how the boundary conditions at both ends of a finite line can be applied to determine the arbitrary constants and how inverse Laplace transformation of transcendental functions can be effected by expansion into series, resulting in traveling-wave solutions.

EXAMPLE 11–3. A d-c voltage E is applied to an initially relaxed lossless transmission line of length d, which is shorted at the other end. Determine (a) $v(x, t)$, and (b) $i(x, t)$.

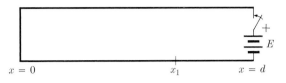

Fig. 11–5. A shorted finite lossless line.

Solution. Figure 11–5 illustrates the given situation, in which the d-c voltage E is assumed to be applied at $x = d$. For an initially relaxed line, the equation for the transformed voltage is, from Eq. (11–70),

$$\frac{d^2 V}{dx^2} - \gamma^2 V = 0, \qquad 0 \leq x \leq d, \tag{11–130}$$

which has a solution

$$V(x, s) = A\epsilon^{-\gamma x} + B\epsilon^{\gamma x}. \tag{11–131}$$

The given boundary conditions are:

At $x = 0$, $v(0, t) = 0$, hence

$$V(0, s) = 0. \tag{11–132}$$

At $x = d$, $v(d, t) = E$, which transforms to give

$$V(d, s) = \frac{E}{s}. \tag{11–133}$$

Substitution of conditions (11–132) and (11–133) into Eq. (11–131) yields two simultaneous equations:

$$0 = A + B, \tag{11–134}$$

$$\frac{E}{s} = A\epsilon^{-\gamma d} + B\epsilon^{\gamma d}, \tag{11–135}$$

from which A and B can be solved:

$$A = -B = \frac{E}{s(\epsilon^{-\gamma d} - \epsilon^{\gamma d})}. \tag{11–136}$$

Therefore

$$V(x, s) = \frac{E}{s} \left[\frac{\epsilon^{-\gamma x} - \epsilon^{\gamma x}}{\epsilon^{-\gamma d} - \epsilon^{\gamma d}} \right] = \frac{E}{s} \frac{\sinh \gamma x}{\sinh \gamma d}. \tag{11–137}$$

Note that the above solution applies to a shorted general line; we have not yet made use of the fact that the given line is lossless.

(a) To find $v(x, t)$, we expand $V(x, s)$ into a series of exponential terms with negative exponents:

$$V(x, s) = \frac{E}{s} \frac{\epsilon^{\gamma x}(1 - \epsilon^{-2\gamma x})}{\epsilon^{\gamma d}(1 - \epsilon^{-2\gamma d})}$$

$$= \frac{E}{s} \epsilon^{-\gamma(d-x)}(1 - \epsilon^{-2\gamma x})(1 - \epsilon^{-2\gamma d})^{-1}$$

$$= \frac{E}{s} [\epsilon^{-\gamma(d-x)} - \epsilon^{-\gamma(d+x)}](1 + \epsilon^{-2\gamma d} + \epsilon^{-4\gamma d} + \cdots),$$

or

$$V(x, s) = \frac{E}{s} \{[\epsilon^{-\gamma(d-x)} - \epsilon^{-\gamma(d+x)}] + [\epsilon^{-\gamma(3d-x)} - \epsilon^{-\gamma(3d+x)}]$$

$$+ [\epsilon^{-\gamma(5d-x)} - \epsilon^{-\gamma(5d+x)}] + \cdots\}. \tag{11-138}$$

For the lossless line, $R = 0$, $G = 0$, and $\gamma = s/c$, with $c = 1/\sqrt{LC}$, Eq. (11–138) becomes

$$V(x, s) = \frac{E}{s} \{[\epsilon^{-s(d-x)/c} - \epsilon^{-s(d+x)/c}]$$

$$+ [\epsilon^{-s(3d-x)/c} - \epsilon^{-s(3d+x)/c}]$$

$$+ [\epsilon^{-s(5d-x)/c} - \epsilon^{-s(5d+x)/c}] + \cdots\}. \tag{11-139}$$

Inverse transformation of Eq. (11–139) yields the desired answer:

$$v(x, t) = E \left\{ \left[U\left(t - \frac{d-x}{c}\right) - U\left(t - \frac{d+x}{c}\right) \right] \right.$$

$$+ \left[U\left(t - \frac{3d-x}{c}\right) - U\left(t - \frac{3d+x}{c}\right) \right]$$

$$\left. + \left[U\left(t - \frac{5d-x}{c}\right) - U\left(t - \frac{5d+x}{c}\right) \right] + \cdots \right\}. \tag{11-140}$$

The above solution is plotted versus t in Fig. 11–6 for an arbitrary distance x_1 from the shorted end. Figure 11–6 can be explained by the concept of traveling waves. When the voltage E is applied to the line at

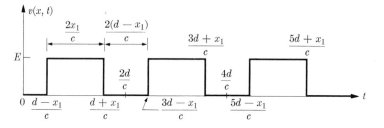

FIG. 11–6. Voltage variation at $x = x_1$ on the line shown in Fig. 11–5.

$t = 0$, nothing happens at $x = x_1$ until the wave with amplitude E traveling with a velocity c has reached that point, at $t = (d - x_1)/c$. The wave continues its journey toward the left until it discovers a short circuit at $x = 0$, where the voltage must be zero. The zero voltage at $x = 0$ can be considered as the sum of the incident wave with an amplitude E and a reflected wave with an amplitude $-E$. A *voltage reflection coefficient*, Γ, at a terminating or load impedance Z_L on a transmission line with characteristic impedance Z_0 is defined in the following way:

$$\Gamma = \frac{Z_L - Z_0}{Z_L + Z_0}. \tag{11-141}$$

This gives the ratio between the reflected voltage and the incident voltage. For a shorted end, $Z_L = 0$ and $\Gamma = -1$, the reflected voltage is of equal magnitude but of opposite phase to the incident voltage.

At $t = (d - x_1)/c + 2x_1/c = (d + x)/c$, the reflected wave reaches the point $x = x_1$, cancelling the original voltage there. The reflected wave travels to $x = d$ at $t = 2d/c$, where another reflection with a reversal of phase takes place. This second reflected wave travels toward the left again, reaching $x = x_1$ at $t = 2d/c + (d - x_1)/c = (3d - x_1)/c$ and creating a voltage E there. The voltage variation shown in Fig. 11-6 can therefore be visualized as the result of traveling waves and multiple reflections at the ends.

We note from Eq. (11-141) that $\Gamma = 0$ if $Z_L = Z_0$; hence there is no reflection from a termination if the terminating impedance is equal to the characteristic impedance of the line. *Insofar as the input end is concerned, a line terminated in its characteristic impedance behaves as if it were infinitely long.* Consequently, *the effect of an infinite transmission line in a circuit is the same as that of a lumped impedance equal to the characteristic impedance of the line.*

(b) To find $i(x, t)$, we use Eq. (11-82) for the lossless line:

$$I(x, s) = -\frac{1}{Ls} \frac{dV(x, s)}{dx}$$

$$= \frac{E\gamma}{Ls^2} \left[\frac{\epsilon^{-\gamma x} + \epsilon^{\gamma x}}{\epsilon^{-\gamma d} - \epsilon^{\gamma d}} \right]$$

$$= -\frac{E\gamma}{Ls^2} \frac{\cosh \gamma x}{\sinh \gamma d}, \tag{11-142}$$

where substitution of $V(x, s)$ from Eq. (11-137) has been made. Since $\gamma = s/c = s\sqrt{LC}$, we have

$$I(x, s) = \frac{E}{R_0 s} \left[\frac{\epsilon^{-sx/c} + \epsilon^{sx/c}}{\epsilon^{-sd/c} - \epsilon^{sd/c}} \right], \tag{11-143}$$

where R_0 is the characteristic resistance $\sqrt{L/C}$ of the lossless line. Following a procedure similar to that for $V(x, s)$, we expand $I(x, s)$ into a series of exponential terms with negative exponents:

$$I(x, s) = -\frac{E}{R_0 s}\left[\epsilon^{-s(d-x)/c} + \epsilon^{-s(d+x)/c}\right](1 - \epsilon^{-2sd/c})^{-1}$$

$$= -\frac{E}{R_0 s}\left\{\left[\epsilon^{-s(d-x)/c} + \epsilon^{-s(d+x)/c}\right]\right.$$

$$\left. + \left[\epsilon^{-s(3d-x)/c} + \epsilon^{-s(3d+x)/c}\right] + \cdots\right\}. \qquad (11\text{-}144)$$

Inverse transformation of the above expression gives us the desired current function:

$$i(x, t) = -\frac{E}{R_0}\left\{\left[U\left(t - \frac{d-x}{c}\right) + U\left(t - \frac{d+x}{c}\right)\right]\right.$$

$$\left. + \left[U\left(t - \frac{3d-x}{c}\right) + U\left(t - \frac{3d+x}{c}\right)\right] + \cdots\right\}. \qquad (11\text{-}145)$$

The presence of the minus sign on the right side of Eq. (11–145) is due to the fact that the current in the upper wire of the transmission line flows in the direction of decreasing x, contrary to the reference current direction shown in Fig. 11–1. Figure 11–7 is a graph of $|i(x_1, t)|$ versus t at an arbitrary point $x = x_1$ from the shorted end. It can be similarly explained by traveling waves, the major difference being that the current reflection coefficient is the negative of the voltage reflection coefficient defined in (11–141). At the shorted end, both the amplitude and the phase of the reflected current wave will be equal to those of the incident current wave. The current therefore builds up on successive reflections, increasing eventually to infinity, as we would expect for the case of a shorted lossless line.

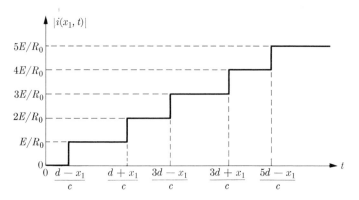

FIG. 11–7. Current variation at $x = x_1$ on the line shown in Fig. 11–5.

Up to this point we have discussed only those problems in which the transmission line is initially relaxed. In the following example we shall consider a case in which the initial conditions are not quiescent.

EXAMPLE 11–4. A lossless transmission line of length d is initially charged uniformly to a voltage E. With one end open, a load resistance R_L is connected to the other end at $t = 0$, as shown in Fig. 11–8. Determine (a) $i(x, t)$, and (b) $v(x, t)$.

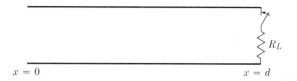

$x = 0$ $x = d$

FIG. 11–8. An initially charged finite lossless line.

Solution. Let us first examine the given initial and boundary conditions.

Initial conditions:

$$t = 0, \quad v(x, 0) = E, \tag{11–146}$$

$$t = 0, \quad i(x, 0) = 0. \tag{11–147}$$

Boundary conditions:

$$x = 0, \quad i(0, t) = 0, \tag{11–148a}$$

$$I(0, s) = 0. \tag{11–148b}$$

$$x = d, \quad v(d, t) = R_L i(d, t), \tag{11–149a}$$

$$V(d, s) = R_L I(d, s). \tag{11–149b}$$

We note that although we do not know either $v(d, t)$ or $i(d, t)$ at $x = d$, relation (11–149a) at the load resistance should always be satisfied. Here we have a mixed boundary condition involving both the voltage and the current, and it is an indication that we should expect to consider the transformed voltage and current equations simultaneously.

Inserting the initial conditions (11–146) and (11–147) in Eq. (11–64), we have the transformed voltage equation for the lossless line:

$$\frac{d^2 V}{dx^2} - \gamma^2 V = -sLCE, \tag{11–150}$$

where $\gamma = s\sqrt{LC} = s/c$. The solution of the above second-order ordinary differential equation consists of a complementary function and a particular integral which is independent of x.

$$V(x, s) = A\epsilon^{-\gamma x} + B\epsilon^{\gamma x} + \frac{E}{s}. \tag{11-151}$$

We must stop here momentarily because we cannot proceed to determine the constants A and B from the given boundary conditions in Eqs. (11–148) and (11–149).

Let us now turn to the transformed current equation. From Eq. (11–67) we obtain for our problem a homogeneous second-order ordinary differential equation for the transformed current:

$$\frac{d^2 I}{dx^2} - \gamma^2 I = 0, \tag{11-152}$$

which has a solution

$$I(x, s) = A'\epsilon^{-\gamma x} + B'\epsilon^{\gamma x}. \tag{11-153}$$

The constants A' and B' in the above solution are definitely related to the constants A and B in the solution for $V(x, s)$ in (11–151). The relation can be found from the Laplace transformation of either Eq. (11–5) or Eq. (11–6). For a lossless line, Eq. (11–6) simplifies to

$$C \frac{\partial v}{\partial t} = - \frac{\partial i}{\partial x}. \tag{11-154}$$

Laplace transformation of Eq. (11–154), together with the initial condition in (11–146), yields

$$C(sV - E) = - \frac{dI}{dx},$$

or

$$V = - \frac{1}{Cs} \frac{dI}{dx} + \frac{E}{s}. \tag{11-155}$$

The expressions for V and I in Eqs. (11–151) and (11–153) respectively can now be substituted in Eq. (11–155):

$$A\epsilon^{-\gamma x} + B\epsilon^{\gamma x} + \frac{E}{s} = - \frac{1}{Cs} (-\gamma A'\epsilon^{-\gamma x} + \gamma B'\epsilon^{\gamma x}) + \frac{E}{s},$$

or

$$A\epsilon^{-\gamma x} + B\epsilon^{\gamma x} = \frac{\gamma}{Cs} (A'\epsilon^{-\gamma x} - B'\epsilon^{\gamma x}). \tag{11-156}$$

If Eq. (11–156) is to hold for all values of x, the coefficients for like exponential terms on both sides of the equation must be equal. Thus,

$$A = \frac{\gamma}{Cs} A' = \sqrt{\frac{L}{C}} A' = R_0 A' \tag{11-157}$$

and

$$B = - \frac{\gamma}{Cs} B' = - \sqrt{\frac{L}{C}} B' = -R_0 B', \tag{11-158}$$

where R_0 has been written for the characteristic resistance $\sqrt{L/C}$ of the lossless line.

Applying the transformed boundary condition (11–148b) to Eq. (11–153), we have

$$I(0, s) = A' + B' = 0,$$

or

$$B' = -A'. \tag{11–159}$$

Solutions for $V(x, s)$ and $I(x, s)$ in Eqs. (11–151) and (11–153) can now be written as follows:

$$V(x, s) = R_0 A'(\epsilon^{-\gamma x} + \epsilon^{\gamma x}) + \frac{E}{s}, \tag{11–160}$$

$$I(x, s) = A'(\epsilon^{-\gamma x} - \epsilon^{\gamma x}). \tag{11–161}$$

Finally, we substitute the above two expressions into the transformed boundary condition (11–149b) by setting $x = d$, and solve for A' to obtain

$$A' = -\frac{E}{s(R_L + R_0)\{\epsilon^{\gamma d} - [(R_L - R_0)/(R_L + R_0)]\epsilon^{-\gamma d}\}}$$

$$= -\frac{E}{s(R_L + R_0)[\epsilon^{\gamma d} - \Gamma\epsilon^{-\gamma d}]} = -B', \tag{11–162}$$

where Γ is the voltage reflection coefficient at the load resistance R_L:

$$\Gamma = \frac{R_L - R_0}{R_L + R_0}. \tag{11–163}$$

(a) *To find the current.* From Eqs. (11–161) and (11–162) we have

$$I(x, s) = \frac{E}{s(R_L + R_0)}\left[\frac{\epsilon^{\gamma x} - \epsilon^{-\gamma x}}{\epsilon^{\gamma d} - \Gamma\epsilon^{-\gamma d}}\right]. \tag{11–164}$$

To determine $i(x, t)$, we expand $I(x, s)$ into a series of exponential terms with negative exponents:

$$I(x, s)$$

$$= \frac{E}{s(R_L + R_0)}\left[\epsilon^{-\gamma(d-x)} - \epsilon^{-\gamma(d+x)}\right](1 - \Gamma\epsilon^{-2\gamma d})^{-1}$$

$$= \frac{E}{s(R_L + R_0)}\left[\epsilon^{-\gamma(d-x)} - \epsilon^{-\gamma(d+x)}\right](1 + \Gamma\epsilon^{-2\gamma d} + \Gamma^2\epsilon^{-4\gamma d} + \cdots)$$

$$= \frac{E}{s(R_L + R_0)}\left\{\left[\epsilon^{-\gamma(d-x)} - \epsilon^{-\gamma(d+x)}\right] + \Gamma\left[\epsilon^{-\gamma(3d-x)} - \epsilon^{-\gamma(3d+x)}\right]\right.$$

$$\left. + \Gamma^2\left[\epsilon^{-\gamma(5d-x)} - \epsilon^{-\gamma(5d+x)}\right] + \cdots\right\}. \tag{11–165}$$

Remembering that $\gamma = s/c$ for a lossless line, we perform the inverse transformation of $I(x, s)$ to obtain the current:

$$i(x, t) = \frac{E}{R_L + R_0} \left\{ \left[U\left(t - \frac{d - x}{c}\right) - U\left(t - \frac{d + x}{c}\right) \right] \right.$$

$$+ \Gamma \left[U\left(t - \frac{3d - x}{c}\right) - U\left(t - \frac{3d + x}{c}\right) \right]$$

$$\left. + \Gamma^2 \left[U\left(t - \frac{5d - x}{c}\right) - U\left(t - \frac{5d + x}{c}\right) \right] + \cdots \right\}. \quad (11\text{--}166)$$

The traveling-wave interpretation of the solution in (11–166) is similar to that in Example 11–3 except that in the present Example the voltage reflection coefficient at the open end $(x = 0)$ is $+1$ and that at the load end $(x = d)$ is Γ given by Eq. (11–163). Hence every time the wave has traveled twice the length of the line (which takes $2d/c$ in time) the terms must be multiplied by Γ.

Of particular interest is the special case when the load resistance is equal to the characteristic resistance,

$$R_L = R_0 = \sqrt{L/C};$$

then

$$\Gamma = 0$$

and only the first two terms in Eq. (11–166) remain:

$$i(x, t) = \frac{E}{2R_0} \left[U\left(t - \frac{d - x}{c}\right) - U\left(t - \frac{d + x}{c}\right) \right]. \quad (11\text{--}167)$$

Although we expect that the current will be in the form of a rectangular pulse because of the difference between two step functions shifted by unequal amounts, it is hard to visualize what happens in the entire line, since the current is a function of both distance and time. The two-dimensional plot in Fig. 11–9 will provide aid in this visualization. A third dimension can be imagined as being the constant magnitude $E/2R_0$, perpendicular to the page.

In constructing Fig. 11–9, we write Eq. (11–167) as

$$i(x, t) = \frac{E}{2R_0} (U_1 - U_2),$$

where

$$U_1 = U\left(t - \frac{d - x}{c}\right), \qquad U_2 = U\left(t - \frac{d + x}{c}\right).$$

The graph is divided into three regions, (a), (b), and (c). Region (a) is excluded because in that region both U_1 and U_2 are zero. Region (c) is

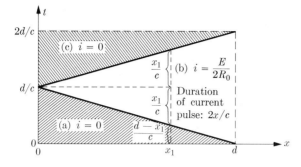

FIG. 11–9. Graph for visualizing current distribution on the line in Fig. 11–8.

also excluded because both U_1 and U_2 will be unity there, resulting in a zero difference. At any distance x_1 from the open end, the vertical extent of region (b) determines the duration of the current pulse. Thus Fig. 11–9 gives us a clear picture of the variation of the pulse width as a function of the distance. In particular, at $x = d$ a current pulse of duration $2d/c$ will exist in the load resistance. We have a pulse-forming line.

(b) *To find the voltage.* From Eqs. (11–160) and (11–162) we have

$$V(x, s) = -\frac{R_0 E}{s(R_L + R_0)}\left[\frac{\epsilon^{\gamma x} + \epsilon^{-\gamma x}}{\epsilon^{\gamma d} - \Gamma\epsilon^{-\gamma d}}\right] + \frac{E}{s}. \qquad (11\text{–}168)$$

Again we expand into a series of exponential terms with negative exponents:

$$V(x, s) = -\frac{R_0 E}{s(R_L + R_0)}\left\{\left[\epsilon^{-\gamma(d-x)} + \epsilon^{-\gamma(d+x)}\right]\right.$$
$$+ \Gamma\left[\epsilon^{-\gamma(3d-x)} + \epsilon^{-\gamma(3d+x)}\right]$$
$$\left. + \Gamma^2\left[\epsilon^{-\gamma(5d-x)} + \epsilon^{-\gamma(5d+x)}\right] + \cdots\right\} + \frac{E}{s}. \qquad (11\text{–}169)$$

Inverse transformation of the above expression gives us the voltage:

$$v(x, t) = -\frac{R_0 E}{R_L + R_0}\left\{\left[U\left(t - \frac{d-x}{c}\right) + U\left(t - \frac{d+x}{c}\right)\right]\right.$$
$$+ \Gamma\left[U\left(t - \frac{3d-x}{c}\right) + U\left(t - \frac{3d+x}{c}\right)\right]$$
$$\left. + \Gamma^2\left[U\left(t - \frac{5d-x}{c}\right) + U\left(t - \frac{5d+x}{c}\right)\right] + \cdots\right\}$$
$$+ EU(t). \qquad (11\text{–}170)$$

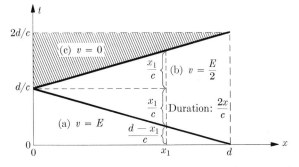

FIG. 11–10. Graph for visualizing voltage distribution on the line in FIG. 11–8.

For the special case $R_L = R_0$ and $\Gamma = 0$, Eq. (11–170) reduces to the following:

$$v(x, t) = E \left[U(t) - \frac{1}{2} U \left(t - \frac{d - x}{c} \right) - \frac{1}{2} U \left(t - \frac{d + x}{c} \right) \right]. \quad (11\text{–}171)$$

We can construct a graph similar to that in Fig. 11–9 by writing

$$v(x, t) = E[U - \tfrac{1}{2}U_1 - \tfrac{1}{2}U_2].$$

In region (a) of Fig. 11–10, $U = 1$, $U_1 = U_2 = 0$, hence $v = E$. In region (b), $U = 1$, $U_1 = 1$, $U_2 = 0$, resulting in a voltage pulse of amplitude $E/2$, its duration being determined by the vertical extent of this region. In region (c), $U = U_1 = U_2 = 1$ and the voltage is zero.

Across the load resistance $R_L = R_0$ at $x = d$, Eq. (11–171) becomes

$$v(d, t) = \frac{E}{2} \left[U(t) - U \left(t - \frac{2d}{c} \right) \right], \quad (11\text{–}172)$$

which is a pulse of amplitude $E/2$, starting at $t = 0$ and having a duration $2d/c$.

PROBLEMS

11–1. Find the solution of the following partial differential equation:

$$\frac{\partial^2 y}{\partial \theta^2} = 4 \frac{\partial^2 y}{\partial x^2}$$

such that $y(x, 0) = 0$, $y(x, \pi) = 0$, $y(0, \theta) = 2 \sin 2\theta$, and $y_x(0, \theta) = 0$, where y_x denotes $\partial y / \partial x$.

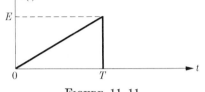

FIGURE 11–11

11–2. A triangular voltage pulse $e(t)$, as shown in Fig. 11–11, is applied to the sending end of an infinitely long, leakage-free, noninductive cable. (a) Determine $i(x, t)$. (b) Obtain the expression for the input current and plot it versus time.

11–3. A rectangular voltage pulse of amplitude E and duration T is applied to an infinitely long lossless transmission line through a lumped, series capacitance C_0. Determine the input voltage $v(0, t)$ and the input current $i(0, t)$.

11–4. Repeat problem 11–3 for an infinitely long, leakage-free, noninductive cable.

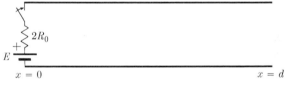

FIGURE 11–12

11–5. A d-c voltage E is applied to the sending end ($x = 0$) of a lossless transmission line of length d through a resistance $2R_0$ at $t = 0$, where R_0 is the characteristic resistance of the lossless line (Fig. 11–12). The line is open at the far end ($x = d$). (a) Find the expression for the voltage at the sending end, $v(0, t)$. (b) Plot $v(0, t)$ versus t.

11–6. A uniform bar clamped at one end, as shown in Fig. 11–13, has the following physical characteristics: length d, cross-sectional area A, mass den-

FIGURE 11–13

sity ρ, modulus of elasticity E. If a horizontal force $f(t) = F \cos \omega t$ is applied at the free end at $t = 0$, determine the instantaneous longitudinal displacement of the bar, $\chi(x, t)$.

11-7. A uniform bar of length d is made of homogeneous material with thermal diffusivity K. The bar is insulated on its sides by material impervious to heat and has an initial temperature of 50°C. At $t = 0$, the temperature of one end of the bar is suddenly and permanently lowered tò 0°C, and the other end is raised to 100°C. Determine the distribution of the temperature in the bar for $t > 0$.

11-8. A sinusoidal voltage $e = E \sin \omega t$ is applied to the input end of an initially relaxed distortionless transmission line of length d. The far end of the line is terminated with a lumped resistance R_d. Determine $v(x, t)$ and $i(x, t)$.

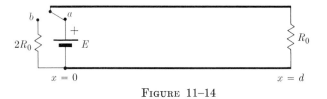

FIGURE 11–14

11-9. The circuit containing a lossless line of characteristic resistance R_0 and length d shown in Fig. 11–14 is initially under steady-state conditions. At $t = 0$, the switch at the sending end is closed to position b. Determine $v(d, t)$ and $i(d, t)$.

APPENDIX A

NUMERICAL SOLUTION OF ALGEBRAIC EQUATIONS

A-1 Introduction. In the analysis of linear systems we are invariably faced with the necessity of determining the roots of an algebraic equation or of finding the zeros of a polynomial. The classical solution of a differential equation requires that the roots of its characteristic equation be found in order to obtain the complementary function. When the method of Laplace transforms is used, the transformed solution is, in general, expressed as a ratio of two polynomials, and it is necessary to determine the zeros of the polynomial in the denominator before the inverse transformation can be effected.

Formulas exist for the solution of general quadratic, cubic, and quartic algebraic equations with literal coefficients. However, ordinary operations of algebra cannot give the exact solutions of algebraic equations of a degree higher than four; we must resort to numerical methods. Even for cubic and quartic equations the general formulas are so complicated and the process so involved that it is difficult for the user to be conscious of what he is doing in each step; one is then left with an unsatisfying feeling even though he may be able to grind out the answers by following cookbook formulas.

Many methods for the numerical solution of algebraic equations are available.* The general approach is to start with a crude trial solution and refine it by an iterative process. A sequence of successive approximations is generated which converges to the true solution if the method is successful. The various methods differ in their iterative procedure, thus resulting in different rates of convergence. In order to converge, some methods demand a trial solution fairly close to the true value, while others may converge with a crude trial solution for some problems but diverge for other problems even when the starting trial solution is close to the true value. Some methods are simple and possess good convergence properties, but cannot be used to determine complex roots. There is unfortunately no one method which can be said categorically to be superior to all others with respect to simplicity of iterative procedure, rapidity of convergence, and versatility for the numerical solution of high-degree algebraic equations.

---

* A very good summary of the different numerical methods can be found in F. B. Hildebrand, *Introduction to Numerical Analysis*, pp. 443–476, McGraw-Hill Book Co., New York; 1956.

In this appendix we choose to present Graeffe's root-squaring method. This method has a number of definite advantages over other methods. First, it yields *all* the roots of a given equation at the *same* time without the necessity of guessing at the probable values of the roots. Second, it can be used to determine repeated roots as well as complex roots. Third, there is no question of the existence of ultimate convergence of the iterative procedure. Very large numbers will be dealt with when Graeffe's method is used, and care must be exercised at all times in the iterative procedure, since this method cannot lead to correct answers if a mistake occurs at any stage. (An error in some other iterative methods may mean only a reduction in the rate of convergence; they may be said to possess self-correcting properties.)

Before we develop Graeffe's method in detail, it is helpful to make a few general observations about the properties of algebraic equations. This will be done in the next section.

A-2 General properties of algebraic equations. The following properties are common to all algebraic equations having *real* coefficients:

1. Every algebraic equation of degree n:

$$f(x) = a_n x^n + a_{n-1} x^{n-1} + a_{n-2} x^{n-2} + \cdots + a_1 x + a_0 = 0$$

has exactly n roots. (A root of multiplicity m is counted as m roots.)

2. Complex roots occur in conjugate pairs.

3. An odd-degree equation has at least one real root.

4. If $f(a)$ and $f(b)$ are of opposite sign, where a and b are both real, then there exists at least one real root of $f(x) = 0$ between a and b. The total number of such roots is odd.

5. The number of positive real roots of $f(x) = 0$ is either equal to the number of variations of sign between successive terms in $f(x)$ when arranged in descending powers of x, or less than that number by an even integer. The number of negative real roots of $f(x) = 0$ is either equal to the number of variations of sign between successive terms in $f(-x)$, or less than that number by an even integer (Descartes' rule of signs).

A-3 Graeffe's root-squaring method. General principle. The Graeffe's root-squaring method for the numerical determination of the roots of an algebraic equation consists of forming a sequence of equations such that the roots of each new equation are the *squares* of the roots of the preceding equation in the sequence. That is, if $x_1, x_2, \ldots, x_n$ are the roots of the original equation, the first new equation to be formed will have $x_1^2, x_2^2, \ldots, x_n^2$ as its roots, the second new equation in the sequence will have $x_1^4, x_2^4, \ldots, x_n^4$ as its roots, and so on. Ultimately, an equation will be obtained whose roots are $x_1^m, x_2^m, \ldots, x_n^m$, where $m = 2^p$. When m is sufficiently large, the roots $x_1^m, x_2^m, \ldots, x_n^m$ will be *widely separated* in magnitude and they can be determined by a simple process.

Let the given equation be

$$f(x) = a_n x^n + a_{n-1} x^{n-1} + a_{n-2} x^{n-2} + \cdots + a_1 x + a_0 = 0. \quad \text{(A–1)}$$

If $x_1, x_2, \ldots, x_n$ are the roots of this equation, we can write it in the following factored form:

$$f(x) = a_n(x - x_1)(x - x_2) \cdots (x - x_n) = 0. \quad \text{(A–2)}$$

Substituting $-x$ for x in Eq. (A–2), we have

$$f(-x) = a_n(-x - x_1)(-x - x_2) \cdots (-x - x_n)$$
$$= a_n(-1)^n(x + x_1)(x + x_2) \cdots (x + x_n). \quad \text{(A–3)}$$

Now we form the product

$$(-1)^n f(x)f(-x) = a_n^2(x^2 - x_1^2)(x^2 - x_2^2) \cdots (x^2 - x_n^2) = 0. \quad \text{(A–4)}$$

When we let $x^2 = y$, Eq. (A–4) becomes

$$f_2(y) = a_n^2(y - x_1^2)(y - x_2^2) \cdots (y - x_n^2) = 0. \quad \text{(A–5)}$$

The roots of $f_2(y) = 0$ in Eq. (A–5) are thus $x_1^2, x_2^2, \ldots, x_n^2$, which are the squares of the roots of the original equation $f(x) = 0$.

In view of Eq. (A–1), the product on the left side of Eq. (A–4) can be written as

$$(-1)^n f(x)f(-x)$$
$$= (-1)^n[a_n x^n + a_{n-1} x^{n-1} + a_{n-2} x^{n-2} + \cdots + a_1 x + a_0]$$
$$\times [a_n(-x)^n + a_{n-1}(-x)^{n-1} + a_{n-2}(-x)^{n-2} + \cdots + a_1(-x) + a_0]$$
$$= [a_n x^n + a_{n-1} x^{n-1} + a_{n-2} x^{n-2} + \cdots + a_1 x + a_0]$$
$$\times [a_n x^n - a_{n-1} x^{n-1} + a_{n-2} x^{n-2} - \cdots + (-1)^{n+1} a_1 x + (-1)^n a_0].$$
$$\text{(A–6)}$$

The multiplication in the above equation can be carried out with detached coefficients, as shown in Table A–1. The new coefficients, $b_n, b_{n-1}, b_{n-2}, b_{n-3}, b_{n-4}, \ldots$, in the last row of Table A–1 are the sums of the squared coefficients and the doubled products in the respective columns; they are the coefficients of the new equation, $f_2(y) = 0$, whose

Table A–1

Graeffe's Schedule for Root Squaring

Original coefficients:	a_n a_n	a_{n-1} $-a_{n-1}$	a_{n-2} a_{n-2}	a_{n-3} $-a_{n-3}$	a_{n-4} a_{n-4}	... ...
Squared coefficients:	$+a_n^2$	$-a_{n-1}^2$	$+a_{n-2}^2$	$-a_{n-3}^2$	$+a_{n-4}^2$	...
Doubled products		$+2a_na_{n-2}$	$-2a_{n-1}a_{n-3}$ $+2a_na_{n-4}$	$+2a_{n-2}a_{n-4}$ $-2a_{n-1}a_{n-5}$ $+2a_na_{n-6}$	$-2a_{n-3}a_{n-5}$ $+2a_{n-2}a_{n-6}$ $-2a_{n-1}a_{n-7}$ $+2a_na_{n-8}$	... ...
New coefficients:	b_n	b_{n-1}	b_{n-2}	b_{n-3}	b_{n-4}	...

roots are the squares of those of the original equation. Examination of Table A–1 reveals the following formula for the b coefficients:

$$b_{n-k} = \sum_{i=n-2k}^{n} (-1)^{n-i} a_i a_{2(n-k)-i}, \tag{A–7}$$

or

$$b_{n-k} = (-1)^k [a_{n-k}^2 - 2a_{(n-k)+1}a_{(n-k)-1} + 2a_{(n-k)+2}a_{(n-k)-2} - \cdots + (-1)^k 2a_n a_{n-2k}]. \tag{A–8}$$

Note that the last term in Eq. (A–8) for b_k,

$$(-1)^{2k} 2a_n a_{n-2k} = 2a_{(n-k)+k}a_{(n-k)-k},$$

always carries a positive sign, irrespective of whether n is even or odd. Note further that the sum of the two subscripts for the a coefficients in each of the terms in b_{n-k} is always equal to $2(n - k)$, and that there are altogether $(k + 1)$ different combinations of a coefficients in the expression for b_{n-k}.

In practice, the second row of coefficients in Table A–1 is usually omitted and Graeffe's iteration table is constructed with the following rows:

First row: Coefficients (including their signs) of the original equation arranged in descending powers of x. A zero is written for the coefficients of missing powers of x.

Second row: Squares of the individual coefficients in the first row.

Third row: Twice the products of the two coefficients adjacent to the one directly above in the first row.

Fourth row: Twice the products of the next two equally removed coefficients in the first row. This process is continued and new rows are formed until the last element in each column is $2a_n a_{n-2k}$.

Two steps are needed after all the squared coefficients and the doubled products of coefficients have been tabulated in order to obtain the coefficients of the new equation whose roots are the squares of those for the original equation.

1. The signs of *alternate* elements, starting from the second element in the second row and all subsequent rows, are changed.

2. The new coefficients are obtained by adding the squared coefficients and the doubled products in the corresponding columns *after* the sign changes have been made. These new coefficients then become the first row in the subsequent iteration.

The above procedure is repeated and coefficients for new equations are found until the roots of the resulting equation after ν squarings are widely separated in magnitude. Assume that the roots of the original equation are such that

$$|x_1| > |x_2| > |x_3| > \cdots > |x_n|. \qquad \text{(A–9)}$$

Then the root-squaring procedure will be stopped when the following condition is satisfied within the desired accuracy:

$$|x_1^m| \gg |x_2^m| \gg |x_3^m| \gg \cdots \gg |x_n^m|, \qquad \text{(A–10)}$$

where

$$m = 2^\nu. \qquad \text{(A–11)}$$

Let us write the final equation, whose roots are now $x_1^m, x_2^m, x_3^m, \ldots, x_n^m$, as follows:

$$f_m(y) = A_n y^n + A_{n-1} y^{n-1} + A_{n-2} y^{n-2}$$

$$+ \cdots + A_1 y + A_0 = 0. \qquad \text{(A–12)}$$

Equation (A–12) can be broken down into very simple subequations from which the original roots $x_1, x_2, \ldots, x_n$ can be found. In the next section we shall discuss the methods of formulating these subequations as well as the criteria for judging when condition (A–10) is satisfied.

A–4 Graeffe's root-squaring method. Applications. We shall treat three different situations separately.

I. *All roots real and distinct.* The relations between the roots and the coefficients of Eq. (A–12) can be obtained by noting that

$$y^n + \frac{A_{n-1}}{A_n} y^{n-1} + \frac{A_{n-2}}{A_n} y^{n-2} + \cdots + \frac{A_1}{A_n} y + \frac{A_0}{A_n}$$

$$= (y - x_1^m)(y - x_2^m)(y - x_3^m) \cdots (y - x_n^m). \quad \text{(A-13)}$$

Hence

$$\frac{A_{n-1}}{A_n} = -(\text{sum of all the roots})$$

$$= -(x_1^m + x_2^m + x_3^m + \cdots + x_n^m), \quad \text{(A-14)}$$

$$\frac{A_{n-2}}{A_n} = +(\text{sum of products of the roots taken two at a time})$$

$$= +(x_1^m x_2^m + x_1^m x_3^m + \cdots + x_2^m x_3^m + \cdots), \quad \text{(A-15)}$$

$$\frac{A_{n-3}}{A_n} = -(\text{sum of products of the roots taken three at a time})$$

$$= -(x_1^m x_2^m x_3^m + x_1^m x_2^m x_4^m + \cdots), \quad \text{(A-16)}$$

$$\vdots$$

$$\frac{A_0}{A_n} = (-1)^n(\text{product of all the roots})$$

$$= (-1)^n x_1^m x_2^m x_3^m \cdots x_n^m. \quad \text{(A-17)}$$

Under the hypothesis that condition (A–10) holds, the first terms in Eqs. (A–14) through (A–16) will be very much larger than the succeeding terms, and the following approximate, simple subequations consisting only of the first terms are obtained:

$$\frac{A_{n-1}}{A_n} = -x_1^m, \quad \text{(A-18)}$$

$$\frac{A_{n-2}}{A_n} = +x_1^m x_2^m, \quad \text{(A-19)}$$

$$\frac{A_{n-3}}{A_n} = -x_1^m x_2^m x_3^m, \quad \text{(A-20)}$$

$$\vdots$$

Dividing each of the above equations after Eq. (A–18) by the respective preceding one, we have

$$\frac{A_{n-2}}{A_{n-1}} = -x_2^m, \quad \text{(A-21)}$$

$$\frac{A_{n-3}}{A_{n-2}} = -x_3^m, \quad \text{(A-22)}$$

$$\vdots$$

The *magnitudes* of the roots of the original equation can then be obtained from Eqs. (A–18), (A–21), (A–22), etc., by taking the logarithm:

$$\log |x_1| = \frac{1}{m} \log \left(- \frac{A_{n-1}}{A_n} \right), \qquad \text{(A-23)}$$

$$\log |x_2| = \frac{1}{m} \log \left(- \frac{A_{n-2}}{A_{n-1}} \right), \qquad \text{(A-24)}$$

$$\log |x_3| = \frac{1}{m} \log \left(- \frac{A_{n-3}}{A_{n-2}} \right), \qquad \text{(A-25)}$$
$$\vdots$$

Graeffe's method does *not* give the signs of the roots, which must be determined either by applying Descartes' rule of signs or by actual substitution of the roots into the original equation.

In carrying out Graeffe's iteration for obtaining new coefficients, the process can be stopped when another doubling of m produces new coefficients which are essentially equal to the squares of the corresponding coefficients of the equation already obtained. The reasoning behind this statement can be seen quite easily by recalling from Table A–1 the way in which new coefficients are formed in conjunction with the approximate subequations (A–18) to (A–20).

EXAMPLE A–1. Determine the roots of the following equation:

$$x^3 - 3x^2 + 3 = 0. \qquad \text{(A-26)}$$

Solution. Graeffe's iteration for the given equation is carried out in Table A–2 to slide-rule accuracy in accordance with the schedule shown in Table A–1. It is seen that we could stop at $m = 16$, since further iteration yields new coefficients which are essentially the squares of the corresponding coefficients of the equation for $m = 16$. From Eqs. (A–23), (A–24), and (A–25), we obtain

$$\log |x_1| = \frac{1}{16} \log (2.85 \times 10^6) = 0.404$$

or

$$|x_1| = 2.53,$$

$$\log |x_2| = \frac{1}{16} \log \left(\frac{3.35 \times 10^8}{2.85 \times 10^6} \right) = 0.129$$

or

$$|x_2| = 1.35,$$

$$\log |x_3| = \frac{1}{16} \log \left(\frac{4.31 \times 10^7}{3.35 \times 10^8} \right) = -0.0557$$

or

$$|x_3| = 0.880.$$

TABLE A-2

GRAEFFE'S ITERATION FOR EXAMPLE A-1

	x^3	x^2	x^1	x^0
$m = 1$	1	-3	0	$+3$
Squared coefficients	1	-9	0	-9
Doubled products		0	$+18$	
$m = 2$	1	-9	$+18$	-9
Squared coefficients	1	-81	$+3.24 \times 10^2$	-81
Doubled products		$+36$	-1.62×10^2	
$m = 4$	1	-45	$+1.62 \times 10^2$	-81
Squared coefficients	1	-2.025×10^3	$+2.62 \times 10^4$	-6.56×10^3
Doubled products		$+0.324 \times 10^3$	-0.73×10^4	
$m = 8$	1	-1.701×10^3	$+1.89 \times 10^4$	-6.56×10^3
Squared coefficients	1	-2.89×10^6	$+3.57 \times 10^8$	-4.31×10^7
Doubled products		$+0.04 \times 10^6$	-0.22×10^8	
$m = 16$	1	-2.85×10^6	$+3.35 \times 10^8$	-4.31×10^7
Squared coefficients	1	-8.12×10^{12}	$+1.12 \times 10^{17}$	-1.86×10^{15}
Doubled products		negligible	negligible	

Substitution of the numerical values of the individual roots into the given equation will reveal quite simply that the first two roots are positive and the last root is negative. This is in agreement with the predictions of Descartes' rule of signs. Thus

$$x_1 = +2.53, \qquad x_2 = +1.35, \qquad x_3 = -0.88.$$

The fact that $x_1 + x_2 + x_3 = 3$ (negative of the coefficient of the x^2 term) provides another check for the answers.

II. *All roots real, some numerically equal.* Let us assume that two of the roots of the given equation (A-1), say x_2 and x_3, are numerically equal:

$$|x_1| > |x_2| = |x_3| > |x_4| > \cdots > |x_n|. \qquad (A-27)$$

Graeffe's iteration will be carried out in the usual way, resulting in Eq. (A-12), whose roots are the mth power of those for the original equation.

Similar to Eqs. (A–18), (A–19), (A–20), etc., we will now have, in view of Eqs. (A–14) through (A–17),

$$\frac{A_{n-1}}{A_n} = -x_1^m, \tag{A-28}$$

$$\frac{A_{n-2}}{A_n} = +x_1^m(x_2^m + x_3^m) = +2x_1^m x_2^m, \tag{A-29}$$

$$\frac{A_{n-3}}{A_n} = -x_1^m x_2^{2m}, \tag{A-30}$$

$$\vdots$$

Equation (A–28) yields $|x_1|$ directly:

$$\log |x_1| = \frac{1}{m} \log \left(-\frac{A_{n-1}}{A_n} \right), \tag{A-31}$$

which is the same as Eq. (A–23). $|x_2| = |x_3|$ can be obtained from the quotient of (A–30) and (A–28):

$$\frac{A_{n-3}}{A_{n-1}} = x_2^{2m}.$$

Hence

$$\log |x_2| = \log |x_3| = \frac{1}{2m} \log \left(\frac{A_{n-3}}{A_{n-1}} \right). \tag{A-32}$$

Other distinct roots are obtained in the normal manner.

If we carry the root-squaring process a step beyond Eq. (A–12), whose roots are the mth power of those of the original equation and are already widely separated ($|x_1^m| \gg |x_2^m| = |x_3^m| \gg |x_4^m| \gg \cdots$), and call the new coefficients A_n', A_{n-1}', A_{n-2}', $\ldots$, we will have

$$A_n' = A_n^2, \tag{A-33}$$

$$A_{n-1}' = -A_{n-1}^2 + 2A_n A_{n-2} = A_n^2 \left[-\left(\frac{A_{n-1}}{A_n} \right)^2 + 2 \frac{A_{n-2}}{A_n} \right]$$

$$= A_n^2[-x_1^{2m} + 4x_1^m x_2^m] \cong A_n^2[-x_1^{2m}] = -A_{n-1}^2, \tag{A-34}$$

$$A_{n-2}' = A_{n-2}^2 - 2A_{n-1}A_{n-3} + 2A_n A_{n-4}$$

$$= A_n^2 \left[\left(\frac{A_{n-2}}{A_n} \right)^2 - 2\left(\frac{A_{n-1}}{A_n} \right)\left(\frac{A_{n-3}}{A_n} \right) + 2 \frac{A_{n-4}}{A_n} \right]$$

$$= A_n^2[4x_1^{2m} x_2^{2m} - 2x_1^{2m} x_2^{2m} + 2x_1^m x_2^{2m} x_4^m]$$

$$\cong A_n^2[2x_1^{2m} x_2^{2m}] = \tfrac{1}{2}A_{n-2}^2, \tag{A-35}$$

$$\vdots$$

Note that A'_{n-2} in Eq. (A–35) does *not* numerically approach the square of the preceding coefficient, but is only *half as large*. The appearance of this type of peculiarity indicates the presence of roots with equal numerical values. *In general, if the coefficient of y^{n-k} in the Graeffe's iteration approaches one-half of the square of the preceding coefficient, then the kth and the $(k+1)$th roots are numerically equal. The ratio of the coefficients, $b_{n-(k+1)}/b_{n-(k-1)}$, gives the $(2m)$th power of these roots.*

EXAMPLE A–2. Determine the roots of the following equation:

$$2x^3 + 7x^2 - 12x - 42 = 0. \tag{A–36}$$

Solution. With a view to simplifying the Graeffe's iteration procedure, we first divide the given equation by 2 and obtain

$$x^3 + 3.5x^2 - 6x - 21 = 0. \tag{A–37}$$

The iteration is carried out in Table A–3 to slide-rule accuracy.

TABLE A–3

GRAEFFE'S ITERATION FOR EXAMPLE A–2

	x^3	x^2	x^1	x^0
$m = 1$	1	$+3.5$	-6	-21
Squared coefficients Double products	1	-1.225×10 -1.200×10	$+0.36 \times 10^2$ $+1.47 \times 10^2$	-4.41×10^2
$m = 2$	1	-2.425×10	$+1.83 \times 10^2$	-4.41×10^2
Squared coefficients Doubled products	1	-5.88×10^2 $+3.66 \times 10^2$	$+3.35 \times 10^4$ -2.14×10^4	-1.945×10^5
$m = 4$	1	-2.22×10^2	$+1.21 \times 10^4$	-1.945×10^5
Squared coefficients Doubled products	1	-4.93×10^4 $+2.42 \times 10^4$	$+1.464 \times 10^8$ -0.864×10^8	-3.78×10^{10}
$m = 8$	1	-2.51×10^4	$+0.600 \times 10^8$	-3.78×10^{10}
Squared coefficients Doubled products	1	-6.30×10^8 $+1.20 \times 10^8$	$+3.60 \times 10^{15}$ -1.89×10^{15}	-1.429×10^{21}
$m = 16$	1	-5.10×10^8	$+1.71 \times 10^{15}$	-1.429×10^{21}
Squared coefficients Doubled products	1	-2.60×10^{17} $+0.034 \times 10^{17}$	$+2.92 \times 10^{30}$ -1.46×10^{30}	-2.04×10^{42}
$m = 32$	1	-2.57×10^{17}	$+1.46 \times 10^{30}$	-2.04×10^{42}

We can stop the iteration at $m = 32$, when the coefficient for $y^2 (y = x^m)$ is essentially equal to the square of the corresponding coefficient at $m = 16$. The fact that the coefficient for y^1 at $m = 32$ is approximately one-half of the square of that at $m = 16$ indicates that $|x_2| = |x_3|$. From Eqs. (A–31) and (A–32), we have

$$\log |x_1| = \frac{1}{32} \log (2.57 \times 10^{17}) = 0.544$$

or

$$|x_1| = 3.50,$$

$$\log |x_2| = \log |x_3| = \frac{1}{64} \log \left(\frac{2.04 \times 10^{42}}{2.57 \times 10^{17}}\right) = 0.389$$

or

$$|x_2| = |x_3| = 2.45.$$

We note by applying Descartes' rule of signs that there are one positive and two negative roots. Substitution of $|x_1| = 3.50$ into the given equation reveals that x_1 must be negative in order to satisfy the equation. Hence the roots of the given equation are

$$x_1 = -3.50, \qquad x_2 = +2.45, \qquad x_3 = -2.45.$$

Note that $-(x_1 + x_2 + x_3) = 3.5$ and $-x_1 x_2 x_3 = -21$. These results check with the coefficients of x^2 and x^0 in Eq. (A–37).

III. *Complex roots.* Graeffe's root-squaring method, with suitable modifications in its formulas, can be used to determine complex roots after the iteration process (carried out in the usual manner) reveals the existence of these roots. Complex roots are determined at the same time as the real roots, and it is not necessary to know the number of each kind of root ahead of time. Let us assume that the roots of the given equation satisfy relation (A–27) and that x_2 and x_3 are now a pair of complex conjugate roots:

$$x_2 = r\epsilon^{j\theta} = r(\cos \theta + j \sin \theta) = u + jv, \qquad (A\text{–}38)$$

$$x_3 = r\epsilon^{-j\theta} = r(\cos \theta - j \sin \theta) = u - jv. \qquad (A\text{–}39)$$

Graeffe's root-squaring procedure is applied to the given equation. After ν squarings, an equation of the form of Eq. (A–12) or (A–13) is obtained whose roots are the mth power ($m = 2^\nu$) of those for the original equation and are widely separated. Owing to the presence of the complex conjugate roots x_2 and x_3, Eqs. (A–14) through (A–17) now become approximately

$$\frac{A_{n-1}}{A_n} = -x_1^m, \tag{A-40}$$

$$\frac{A_{n-2}}{A_n} = +x_1^m r^m(\epsilon^{jm\theta} + \epsilon^{-jm\theta}) = +2x_1^m r^m \cos m\theta, \tag{A-41}$$

$$\frac{A_{n-3}}{A_n} = -x_1^m r^{2m}, \tag{A-42}$$

$$\vdots$$

We note from Eq. (A–41) that A_{n-2} contains a cosine factor. Instead of becoming and remaining positive as the root-squaring process is repeated, this coefficient will *fluctuate in sign*. This peculiar behavior identifies the presence of complex roots. There will be as many coefficients behaving in this manner as there are pairs of complex conjugate roots. *In general, if the coefficient of y^{n-k} in the Graeffe's iteration continually fluctuates in sign, then the kth and the $(k + 1)$th roots are conjugate complex roots.*

The numerical values of real roots are found in the usual manner. For instance, from Eq. (A–40) we have

$$\log |x_1| = \frac{1}{m} \log \left(- \frac{A_{n-1}}{A_n} \right). \tag{A-43}$$

The 2mth power of the *magnitude* of the complex roots is equal to the ratio of the coefficient immediately following the irregular one to that immediately preceding the irregular one. From Eqs. (A–42) and (A–40), we obtain

$$\log r = \frac{1}{2m} \log \left(\frac{A_{n-3}}{A_{n-1}} \right). \tag{A-44}$$

In order to determine the complex roots completely, we go back to the original equation (A-1) and note that

$$\frac{a_{n-1}}{a_n} = -(x_1 + x_2 + x_3 + x_4 + \cdots) = -(x_1 + 2u + x_4 + \cdots). \tag{A-45}$$

The real part, u, of x_2 and x_3 can be determined from Eq. (A–45) after all real roots have been found. The imaginary part is then

$$v = \sqrt{r^2 - u^2}. \tag{A-46}$$

When there are more than one pair of conjugate complex roots, all the real roots can be found by determining the mth root of the quotients of adjacent pairs of regularly behaved coefficients as before. The magnitudes of the various complex roots can also be found as before by taking the $(2m)$th root of the ratios of the coefficients immediately after and immediately before each irregularly behaved one. However, modifications* are necessary in order to determine the real parts of the complex roots. In general, it is better to keep the basic concept of Graeffe's root-squaring method in mind so that each special case can be individually handled, rather than to formulate and remember elaborate rules for all possible cases.

EXAMPLE A–3. Determine the roots of the following equation:

$$2x^4 + 9x^3 + 9x^2 + 6x + 2 = 0. \tag{A–47}$$

Solution. We first divide the given equation by 2, the coefficient of the highest power in x. This will simplify the iteration process.

$$x^4 + 4.5x^3 + 4.5x^2 + 3x + 1 = 0. \tag{A–48}$$

Graeffe's iteration for Eq. (A–48) is carried out in Table A–4 to slide-rule accuracy.

We can stop the iteration at $m = 64$, when the coefficients for y^3 and y^1 $(y = x^m)$ are essentially equal to the square of the corresponding coefficients at $m = 32$. The coefficient for y^2 behaves irregularly and continually fluctuates in sign. This indicates the presence of a pair of conjugate complex roots, x_2 and x_3. Let us first determine all the real roots:

$$\log |x_1| = \frac{1}{64} \log (1.353 \times 10^{34}) = 0.5333$$

or

$$|x_1| = 3.414,$$

$$\log |x_4| = \frac{1}{64} \log \left(\frac{1}{7.30 \times 10^{14}} \right) = -0.2322$$

or

$$|x_4| = 0.586.$$

Since there are no variations of sign in the given equation, x_1 and x_4 must both be negative. Hence

$$x_1 = -3.414, \qquad x_4 = -0.586.$$

---

* R. E. Doherty and E. G. Keller, *Mathematics of Modern Engineering,* Volume I, pp. 98–128, John Wiley and Sons; 1936.

TABLE A–4

GRAEFFE'S ITERATION FOR EXAMPLE A–3

	x^4	x^3	x^2	x^1	x^0
$m = 1$	1	$+4.5$	$+4.5$	$+3$	$+1$
Squared coefficients	1	-20.25	$+20.25$	-9	$+1$
Doubled products		$+9.00$	-27.00	$+9$	
			$+2.00$		
$m = 2$	1	-11.25	-4.75	0	$+1$
Squared coefficients	1	-1.266×10^2	$+2.26 \times 10$	0	$+1$
Doubled products		-0.095×10^2	0	-9.5	
			$+0.20 \times 10$		
$m = 4$	1	-1.361×10^2	$+2.46 \times 10$	-9.5	$+1$
Squared coefficients	1	-1.852×10^4	$+0.61 \times 10^3$	-9.025×10	$+1$
Doubled products		$+0.005 \times 10^4$	-2.59×10^3	$+4.92 \times 10$	
			negligible		
$m = 8$	1	-1.847×10^4	-1.98×10^3	-4.105×10	$+1$
Squared coefficients	1	-3.41×10^8	$+3.92 \times 10^6$	-1.685×10^3	$+1$
Doubled products		negligible	-1.516×10^6	-3.96×10^3	
			negligible		
$m = 16$	1	-3.41×10^8	$+2.404 \times 10^6$	-5.645×10^3	$+1$
Squared coefficients	1	-1.163×10^{17}	$+5.78 \times 10^{12}$	-3.19×10^7	$+1$
Doubled products		negligible	-3.85×10^{12}	$+0.48 \times 10^7$	
			negligible		
$m = 32$	1	-1.163×10^{17}	$+1.93 \times 10^{12}$	-2.71×10^7	$+1$
Squared coefficients	1	-1.353×10^{34}	$+3.72 \times 10^{24}$	-7.34×10^{14}	$+1$
Doubled products		negligible	-6.30×10^{24}	$+0.04 \times 10^{14}$	
			negligible		
$m = 64$	1	-1.353×10^{34}	-2.58×10^{24}	-7.30×10^{14}	$+1$

The absolute value of the complex roots can be obtained from Eq. (A–44):

$$\log r = \frac{1}{128} \log \left(\frac{7.30 \times 10^{14}}{1.353 \times 10^{34}}\right) = -0.1505,$$

or

$$r = 0.707.$$

From the coefficient of x^3 in the given equation (A–47) or Eq. (A–48) we have, in view of Eq. (A–45),

$$4.5 = -(x_1 + 2u + x_4),$$

or

$$u = -\tfrac{1}{2}(4.5 + x_1 + x_4) = -\tfrac{1}{2}(4.5 - 3.414 - 0.586) = -0.250,$$

which is the real part of the complex roots. The imaginary part is then

$$v = \sqrt{r^2 - u^2} = \sqrt{0.707^2 - 0.250^2} = 0.6615.$$

Hence the four roots of the given equation (A–47) are:

$$x_1 = -3.414, \qquad x_2 = -0.250 + j0.6615,$$

$$x_3 = -0.250 - j0.6615, \qquad x_4 = -0.586.$$

APPENDIX B

TABLES OF LAPLACE TRANSFORMS

Two tables of Laplace transforms are included in this appendix for easy reference. Table B–1 is a tabulation of operation-transform pairs, and Table B–2 is a collection of function-transform pairs. All of the transform pairs either have been derived in the text, or can be derived with the techniques that have been presented in the text.

TABLE B–1

OPERATION-TRANSFORM PAIRS

No.	Operation	$f(t)$	$F(s) = \mathcal{L}[f(t)]$
1	Definition of $\mathcal{L}$	$f(t)$	$\int_0^\infty f(t)\epsilon^{-st}\,dt$
2	Shift in t	$f(t - t_0)U(t - t_0)$	$\epsilon^{-t_0 s}F(s)$
3	First derivative	$\dfrac{d}{dt}f(t)$	$sF(s) - f(0+)$
4	Second derivative	$\dfrac{d^2}{dt^2}f(t)$	$s^2F(s) - sf(0+)$ $- f'(0+)$
5	nth derivative	$\dfrac{d^n}{dt^n}f(t)$	$s^nF(s) - s^{n-1}f(0+)$ $- s^{n-2}f'(0+)$ $- \cdots - f^{(n-1)}(0+)$
6	Indefinite integral	$\int f(t)\,dt$	$\dfrac{1}{s}[F(s) + f^{-1}(0+)]$
7	Definite integral	$\int_0^t f(t)\,dt$	$\dfrac{1}{s}F(s)$
8	Multiplication by t	$t f(t)$	$-\dfrac{d}{ds}F(s)$
9	Division by t	$\dfrac{1}{t}f(t)$	$\int_s^\infty F(s)\,ds$
10	Scale change	$f(at)$	$\dfrac{1}{a}F\left(\dfrac{s}{a}\right)$
11	Shift in s	$\epsilon^{-at}f(t)$	$F(s + a)$
12	Initial value	$\lim\limits_{t\to\infty+}f(t)$	$\lim\limits_{s\to\infty} sF(s)$
13	Final value	$\lim\limits_{t\to 0}f(t)$	$\lim\limits_{s\to 0} sF(s)$
14	Convolution	$f_1(t)*f_2(t)$ $= \int_0^t f_1(\tau)f_2(t - \tau)\,d\tau$	$F_1(s)F_2(s)$

TABLE B–2

FUNCTION-TRANSFORM PAIRS

No.	$F(s)$	$f(t), t \geq 0$
1	1	$\delta(t)$
2	$\dfrac{1}{s}$	$U(t)$
3	$\dfrac{1}{s^n} \; (n = 1, 2, 3, \ldots)$	$\dfrac{t^{n-1}}{(n-1)!}$
4	$\dfrac{1}{s+a}$	ϵ^{-at}
5	$\dfrac{1}{(s+a)(s+b)}$	$\dfrac{1}{b-a}\,(\epsilon^{-at} - \epsilon^{-bt})$
6	$\dfrac{s}{(s+a)(s+b)}$	$\dfrac{1}{a-b}\,(a\epsilon^{-at} - b\epsilon^{-bt})$
7	$\dfrac{1}{(s+a)(s+b)(s+c)}$	$\dfrac{\epsilon^{-at}}{(b-a)(c-a)} + \dfrac{\epsilon^{-bt}}{(a-b)(c-b)}$ $+ \dfrac{\epsilon^{-ct}}{(a-c)(b-c)}$
8	$\dfrac{s}{(s+a)(s+b)(s+c)}$	$-\dfrac{a\epsilon^{-at}}{(b-a)(c-a)} - \dfrac{b\epsilon^{-bt}}{(a-b)(c-b)}$ $-\dfrac{c\epsilon^{-ct}}{(a-c)(b-c)}$
9	$\dfrac{s^2}{(s+a)(s+b)(s+c)}$	$\dfrac{a^2\epsilon^{-at}}{(b-a)(c-a)} + \dfrac{b^2\epsilon^{-bt}}{(a-b)(c-b)}$ $+ \dfrac{c^2\epsilon^{-ct}}{(a-c)(b-c)}$
10	$\dfrac{1}{(s+a)^2}$	$t\epsilon^{-at}$
11	$\dfrac{s}{(s+a)^2}$	$(1-at)\epsilon^{-at}$
12	$\dfrac{1}{(s+a)(s+b)^2}$	$\dfrac{1}{(a-b)^2}\,\epsilon^{-at} + \dfrac{(a-b)t-1}{(a-b)^2}\,\epsilon^{-bt}$
13	$\dfrac{s}{(s+a)(s+b)^2}$	$-\dfrac{a}{(a-b)^2}\,\epsilon^{-at} - \dfrac{b(a-b)t-a}{(a-b)^2}\,\epsilon^{-bt}$
14	$\dfrac{s^2}{(s+a)(s+b)^2}$	$\dfrac{a^2}{(a-b)^2}\,\epsilon^{-at}$ $+ \dfrac{b^2(a-b)t + b^2 - 2ab}{(a-b)^2}\,\epsilon^{-bt}$

TABLE B–2 (*Continued*)

FUNCTION-TRANSFORM PAIRS

No.	$F(s)$	$f(t),\ t \geq 0$
15	$\dfrac{1}{(s+a)^n}$	$\dfrac{1}{(n-1)!}\, t^{n-1} \epsilon^{-at}$
16	$\dfrac{1}{s^2 + a^2}$	$\dfrac{1}{a}\sin at$
17	$\dfrac{s}{s^2 + a^2}$	$\cos at$
18	$\dfrac{1}{s^2 - a^2}$	$\dfrac{1}{a}\sinh at$
19	$\dfrac{s}{s^2 - a^2}$	$\cosh at$
20	$\dfrac{1}{s(s^2 + a^2)}$	$\dfrac{1}{a^2}\,(1 - \cos at)$
21	$\dfrac{1}{s^2(s^2 + a^2)}$	$\dfrac{1}{a^3}\,(at - \sin at)$
22	$\dfrac{1}{(s+a)(s^2+b^2)}$	$\dfrac{1}{a^2 + b^2}$ $\left[\epsilon^{-at} + \dfrac{1}{b}\sqrt{a^2 + b^2}\,\sin(bt - \theta)\right],$ $\theta = \tan^{-1}\left(\dfrac{b}{a}\right)$
23	$\dfrac{s}{(s+a)(s^2+b^2)}$	$-\dfrac{a}{a^2 + b^2}$ $\left[\epsilon^{-at} - \dfrac{1}{a}\sqrt{a^2 + b^2}\,\sin(bt + \theta)\right],$ $\theta = \tan^{-1}\left(\dfrac{a}{b}\right)$
24	$\dfrac{s^2}{(s+a)(s^2+b^2)}$	$\dfrac{a^2}{a^2 + b^2}$ $\left[\epsilon^{-at} - \dfrac{b}{a^2}\sqrt{a^2 + b^2}\,\sin(bt - \theta)\right],$ $\theta = \tan^{-1}\left(\dfrac{b}{a}\right)$
25	$\dfrac{1}{s[(s+a)^2 + b^2]}$	$\dfrac{1}{a^2 + b^2}$ $\left[1 - \dfrac{1}{b}\sqrt{a^2 + b^2}\,\epsilon^{-at}\sin(bt + \theta)\right],$ $\theta = \tan^{-1}\left(\dfrac{b}{a}\right)$

TABLE B–2 (Continued)

FUNCTION-TRANSFORM PAIRS

No.	$F(s)$	$f(t),\ t \geq 0$
26	$\dfrac{1}{s^2[(s+a)^2+b^2]}$	$\dfrac{1}{a^2+b^2}$ $\left[t - \dfrac{2a}{a^2+b^2} + \dfrac{1}{b}\,\epsilon^{-at}\sin(bt+\theta)\right],$ $\theta = 2\tan^{-1}\left(\dfrac{b}{a}\right)$
27	$\dfrac{s}{(s^2+a^2)(s^2+b^2)}$	$\dfrac{1}{b^2-a^2}\,(\cos at - \cos bt)$
28	$\dfrac{1}{(s^2+a^2)^2}$	$\dfrac{1}{2a^3}\,(\sin at - at\cos at)$
29	$\dfrac{s}{(s^2+a^2)^2}$	$\dfrac{t}{2a}\sin at$
30	$\dfrac{s^2}{(s^2+a^2)^2}$	$\dfrac{1}{2a}\,(\sin at + at\cos at)$
31	$\dfrac{s^2-a^2}{(s^2+a^2)^2}$	$t\cos at$
32	$\dfrac{1}{s(s^2+a^2)^2}$	$\dfrac{1}{a^4}\,(1-\cos at) - \dfrac{1}{2a^3}\,t\sin at$
33	$\dfrac{4a^3}{s^4+4a^4}$	$\sin at\cosh at - \cos at\sinh at$
34	$\dfrac{s}{s^4+4a^4}$	$\dfrac{1}{2a^2}\sin at\sinh at$
35	$\dfrac{1}{s^4-a^4}$	$\dfrac{1}{2a^3}\,(\sinh at - \sin at)$
36	$\dfrac{s}{s^4-a^4}$	$\dfrac{1}{2a^2}\,(\cosh at - \cos at)$
37	$\dfrac{s^2}{s^4-a^4}$	$\dfrac{1}{2a}\,(\sinh at + \sin at)$
38	$\dfrac{s^3}{s^4-a^4}$	$\dfrac{1}{2}\,(\cosh at + \cos at)$
39	$\dfrac{1}{\sqrt{s}}$	$\dfrac{1}{\sqrt{\pi t}}$
40	$\dfrac{1}{(s-a)\sqrt{s}}$	$\dfrac{1}{\sqrt{a}}\,\epsilon^{at}\,\mathrm{erf}\,(\sqrt{at})$

TABLE B–2 (*Continued*)

FUNCTION-TRANSFORM PAIRS

No.	$F(s)$	$f(t)$, $t \geq 0$
41	$\dfrac{1}{s\sqrt{s+a}}$	$\dfrac{1}{\sqrt{a}}\,\text{erf}\,(\sqrt{at})$
42	$\dfrac{1}{\sqrt{s}+\sqrt{a}}$	$\dfrac{1}{\sqrt{\pi t}} - \sqrt{a}\epsilon^{at}\,\text{erfc}\,(\sqrt{at})$
43	$\dfrac{1}{\sqrt{s}(\sqrt{s}+\sqrt{a})}$	$\epsilon^{at}\,\text{erfc}\,(\sqrt{at})$
44	$\dfrac{1}{\sqrt{s^2+a^2}}$	$J_0(at)$
45	$\dfrac{1}{\sqrt{s^2-a^2}}$	$I_0(at)$
46	$\dfrac{1}{\sqrt{s}\sqrt{s+a}}$	$\epsilon^{-at/2}I_0(at/2)$
47	$\dfrac{1}{s+\sqrt{s^2+a^2}}$	$\dfrac{1}{at}J_1(at)$
48	$\dfrac{1}{(s+\sqrt{s^2+a^2})\sqrt{s^2+a^2}}$	$\dfrac{1}{a}J_1(at)$
49	$\epsilon^{-a\sqrt{s}}$	$\dfrac{a}{2\sqrt{\pi t^3}}\,\epsilon^{-a^2/4t}$
50	$\dfrac{1}{s}\epsilon^{-a\sqrt{s}}$	$\text{erfc}\left(\dfrac{a}{2\sqrt{t}}\right)$
51	$\dfrac{1}{\sqrt{s}}\epsilon^{-a\sqrt{s}}$	$\dfrac{1}{\sqrt{\pi t}}\,\epsilon^{-a^2/4t}$

ANSWERS TO PROBLEMS

CHAPTER 1

1-4. (a) $w = c_1 \cos 2t + c_2 \sin 2t$
   (b) $a_0 = 4, \quad a_1 = 0$

1-6. $d^2e/dt^2 + 377^2e = 0$

1-7. (a) $\dfrac{d^2y}{dx^2} + 2\dfrac{dy}{dx} + y = 4 \cos x$
   (b) $c_1 = c_2 = \frac{1}{2}$

1-8. (a), (d), and (e) are linearly independent; (b), (c), and (f) are linearly dependent.

CHAPTER 2

2-1. $i = c\epsilon^{-Rt/L} + \dfrac{E}{\sqrt{R^2 + \omega^2L^2}} \sin [\omega t - \tan^{-1}(\omega L/R)]$

2-2. $y = (c + \epsilon^x)/x^2$

2-3. $w = \dfrac{1}{2t^2}\left(c - \dfrac{1}{\beta} \cos \beta t\right)$

2-4. $\alpha = c\epsilon^{-\sin \theta} + 1$

2-5. $y = (x + 1)[c + \ln (x + 1)] - 2$

2-6. $x = \epsilon^y(c + 3\epsilon^y)$ or $y = \ln (-c \pm \sqrt{c^2 + 12x}) - 1.79$

2-7. $\dfrac{dz}{dx} + (1 - n)a(x)z = (1 - n)b(x)$

2-8. $y = \pm 1/x\sqrt{c + 6x}$

2-9. $w = (ct^{-3/2} + \sqrt{3}t/5)^2$

2-11. $y = c_1\epsilon^{5x/2} + c_2\epsilon^{-2x} - (78 - 70x + 49x^2)\epsilon^{-x}/7^3$

2-12. $y = c_1 \sin x + (c_2 - x/2) \cos x$

2-13. $y = c_1 + c_2\epsilon^{2x} + c_3\epsilon^{-2x} - 2x$

2-14. $y = c_1\epsilon^{2x} + (c_2 \cos \sqrt{3}x + c_3 \sin \sqrt{3}x)\epsilon^{-x} - 3x/8 - \epsilon^x/7$

2-15. $y = (c_1 + c_2x + x^2/8)\epsilon^{-x/2} + x - 4$

2-16. $i = (c_1 \cos \sqrt{2}t/3 + c_2 \sin \sqrt{2}t/3)\epsilon^{-2t/3}$
   $+ \dfrac{125}{137} (15 \sin t/5 - 7 \cos t/5)\epsilon^{-t/5}$

2-17. $w = c_1\epsilon^{-t} + c_2\epsilon^{-2t} + c_3\epsilon^{-3t} + \dfrac{1}{6}\left(\dfrac{85}{18} - \dfrac{11}{3}t + t^2\right)$

2-18. $w = c_1 + (c_2 + c_3x)\epsilon^{-x} + (3x - \frac{1}{2} \sin x)$

2-19. (b) $y = (c_1 \cos \log x + c_2 \sin \log x)/x + \frac{1}{2}(\log x - 1)$

2-20. $y = (c_1 + c_2 x - \ln x)\epsilon^{-2x}$

2-21. $w = (c_0 + c_1 t + c_2 t^2)\epsilon^{-t} + 24(t - 6)\epsilon^{-t/2}$

2-22. $y = c_1 + c_2 x + c_3 \epsilon^{\sqrt{2}x} + c_4 \epsilon^{-\sqrt{2}x}$

$z = c_1 + c_2 x - c_3 \epsilon^{\sqrt{2}x} - c_4 \epsilon^{-\sqrt{2}x}$

2-23. $u = c_1 \epsilon^{\sqrt{2}x} + c_2 \epsilon^{-\sqrt{2}x} + c_3 \cos \sqrt{2}x + c_4 \sin \sqrt{2}x$

$v = c_1 \epsilon^{\sqrt{2}x} + c_2 \epsilon^{-\sqrt{2}x} - c_3 \cos \sqrt{2}x - c_4 \sin \sqrt{2}x$

2-24. $w = c_1 \epsilon^{2x} + c_2 \epsilon^{-2x} - 2$

$y = -\dfrac{c_1}{2}\epsilon^{2x} + \dfrac{c_2}{2}\epsilon^{-2x} + c_3 - 3 + x$

$z = \frac{3}{2}c_1 \epsilon^{2x} - \frac{3}{2}c_2 \epsilon^{-2x} + c_3 - x$

2-25. $i_1 = c_0 + (c_1 + c_2 t)\epsilon^{-t/2CR} + \dfrac{E}{4R}(\sin \omega t - 4\omega C R \cos \omega t)$

$i_2 = (4C Rc_2 - c_1 - c_2 t)\epsilon^{-t/2CR} + \dfrac{E}{4R}\sin \omega t$

CHAPTER 3

3-10. $i_1 = \dfrac{L_2}{L_1 L_2 - M^2}\displaystyle\int_0^t v_1\, dt - \dfrac{M}{L_1 L_2 - M^2}\displaystyle\int_0^t v_2\, dt + i_1(0)$

$i_2 = -\dfrac{M}{L_1 L_2 - M^2}\displaystyle\int_0^t v_1\, dt + \dfrac{L_1}{L_1 L_2 - M^2}\displaystyle\int_0^t v_2\, dt + i_2(0)$

3-11. (a) $i_m = V\sqrt{\dfrac{C}{L}}\,\epsilon^{-\alpha t_m}; \quad t_m = \dfrac{1}{b}\tanh^{-1}\left(\dfrac{b}{\alpha}\right)$

where $\alpha = \dfrac{R}{2L}$ and $b = \sqrt{\left(\dfrac{R}{2L}\right)^2 - \dfrac{1}{LC}}$

(b) $i_m = \dfrac{2V}{\epsilon R}; \quad t_m = \dfrac{2L}{R}$

(c) $i_m = V\sqrt{\dfrac{C}{L}}\,\epsilon^{-\alpha t_m}; \quad t_m = \dfrac{1}{\beta}\tan^{-1}\left(\dfrac{\beta}{\alpha}\right)$

where $\alpha = \dfrac{R}{2L}$ and $\beta = \sqrt{\dfrac{1}{LC} - \left(\dfrac{R}{2L}\right)^2}$

3-12. $i_c = V_g\sqrt{\dfrac{R_2^2 + \omega^2(L_2 + M)^2}{a^2 + b^2}}$

$$\cos\left[\omega t + \alpha + \tan^{-1}\dfrac{\omega a(L_2 + M) - b R_2}{a R_2 + \omega b(L_2 + M)}\right]$$

where $a = R_1R_2 + R_2R_3 + R_1R_3 - \omega^2(L_1L_2 - M^2)$

$$- \frac{1}{C}(L_1 + L_2 + 2M) \quad \text{and}$$

$$b = \omega[R_1L_2 + R_2L_1 + R_3(L_1 + L_2 + 2M)]$$

$$- \frac{1}{\omega C}(R_1 + R_2)$$

3-13. (a) $v_L = 10\epsilon^{-20t}$ volts

(b) $v_L = \frac{5}{13}[(\sin \alpha - 5 \cos \alpha)\epsilon^{-20t} + 25 \sin(100t + \alpha)$
$+ 5 \cos(100t + \alpha)]$ volts

(c) $\alpha = 78.7°$

3-14. (a) $v_c = 5(2 - 3\epsilon^{-10t})$ volts

(b) $v_c = 5[(\cos \alpha - 1 - \sin \alpha)\epsilon^{-10t} + \sin(10t + \alpha) - \cos(10t + \alpha)]$
volts

(c) $\alpha = 0°$

3-15. $v_c = -\frac{E}{3}[1 + \frac{1}{2}\epsilon^{-3t/C(3R_2 + 2R_1)}]$ volts (reference + polarity on top plate)

3-16. (a) $v(t) = 150(1 - \epsilon^{-100t})$ volts $(0 < t < 0.011$ sec)

(b) $v(t) = 99.254\epsilon^{-2.01 \times 10^4 t} + 0.746$

(d) Time constant for charging $= 0.01$ sec
Time constant for discharging $= 4.98 \times 10^{-5}$ sec

3-17. (a) $v_c = 100[1 - 0.9\epsilon^{-5000t}(\sin 5000t + \cos 5000t)]$ volts

(b) $v_c = \epsilon^{-5000t}(80.1 \sin 5000t + 81.5 \cos 5000t)$
$+ 71.5 \sin(7000t - 89.2°)$ volts

(c) $(v_c)_{ss} = \dfrac{V_g}{\sqrt{(1 - \omega^2LC)^2 + (\omega CR)^2}} \sin\left(\omega t - \tan^{-1}\dfrac{\omega CR}{1 - \omega^2LC}\right)$

Resonant frequency $= 1130$ cps

3-18. (a) $v(t) = 333.3\epsilon^{-333.3t}$ volts.

(b) Thévenin's equivalent network to the left of terminals 1–1': a d-c voltage source $v_0 = 500$ volts in series with R_2 and uncharged C.

3-19. (a) $v_c = \epsilon^{-750t}(56.8 \sin 660t - 50 \cos 660t) + 50$ volts (reference + polarity on top plate).

(b) Norton's equivalent network to the left of terminals 1–1': a current source $i_0 = 0.5 + 500t$ amperes in parallel with R_1, R_2, L, and C.

3-20. (a) Thévenin's equivalent.

Voltage source: $v_0 = \dfrac{\omega^2 L^2 E}{R_1^2 + \omega^2 L^2}\left(\cos \omega t - \dfrac{R_1}{\omega L}\sin \omega t - \epsilon^{-R_1 t/L}\right)$.

Passive network: parallel combination of R_1 and L.

(b) Norton's equivalent.

Current source: $i_0 = \dfrac{E}{R_1} (\cos \omega t - 1)$.

Passive network: parallel combination of R_1 and L

$$i_{R_2} = \frac{2E}{3R_1[1 + (R_1/3\omega L)^2]} \left(-\epsilon^{-R_1 t/3L} + \cos \omega t - \frac{R_1}{3\omega L} \sin \omega t \right)$$

CHAPTER 4

4-1. (a)

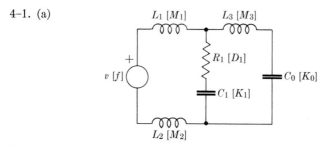

4-2. (b)

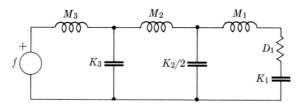

4-3. (a)

4-4. (b)

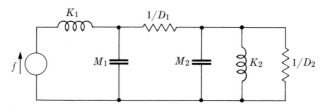

4–5. (a)

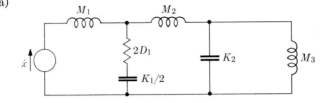

4–6. (b)

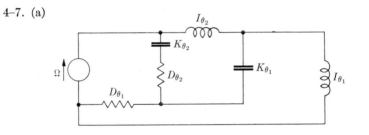

4–7. (a)

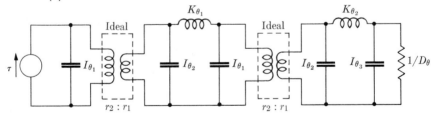

4–8. (b)

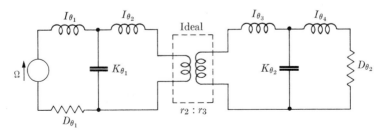

4–9. (a)

4–10. (b)

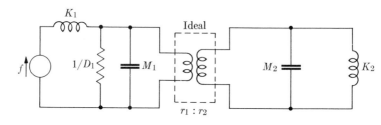

4–11. (a)

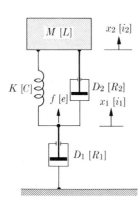

4–12. (b)

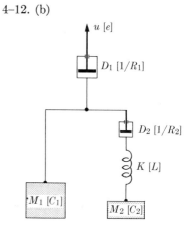

4–14.

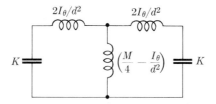

4–15.

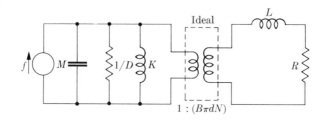

4–16.

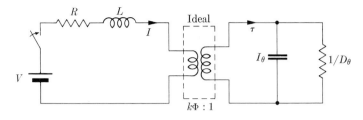

CHAPTER 5

5–1. (a) $f(t) = \dfrac{A}{2} + \dfrac{A}{\pi} \displaystyle\sum_{n=1}^{\infty} \dfrac{1}{n} \sin (2n\pi t/T)$

5–2. (a) $f(x) = \dfrac{A}{\pi} \left[1 + \dfrac{\pi}{2} \cos 2x + 2 \displaystyle\sum_{k=1}^{\infty} \dfrac{(-1)^{k+1}}{(4k^2 - 1)} \cos 4kx \right]$

5–3. (a) $f(\theta) = \dfrac{2A}{\pi} \left[(\theta_0 \csc \theta_0 + \cos \theta_0) \sin \theta + \displaystyle\sum_{n=3,5,7} \dfrac{2}{n^2 - 1} \right.$

$\left. \left(\cot \theta_0 \sin n\theta_0 - \dfrac{1}{n} \cos n\theta_0 \right) \sin n\theta \right]$

5–4. (a) $f(t) = 0.4A \left\{ 1 + \displaystyle\sum_{n=\text{odd}} \dfrac{1}{n\pi} \left[5 \sin (2n\pi t/T) - \dfrac{2}{n\pi} \cos (2n\pi t/T) \right] \right\}$

5–5. (a) $f(t) = \dfrac{4A}{\pi} \displaystyle\sum_{k=1}^{\infty} \dfrac{2k}{(2k)^2 - 1} \sin (4k\pi t/T)$

5–6. (a) $f(x) = \tfrac{1}{2}(1 - \epsilon^{-2}) + \displaystyle\sum_{n=1}^{\infty} \dfrac{4}{4 + n^2\pi^2} [1 + (-1)^{n+1}\epsilon^{-2}] \cos (n\pi x/2)$

5–7. (a) $f(x) = \dfrac{1}{3} + \dfrac{1}{2\pi^2} \displaystyle\sum_{n=1}^{\infty} \dfrac{1}{n^2} [(1 + jn\pi)\epsilon^{j2n\pi x} + (1 - jn\pi)\epsilon^{-j2n\pi x}]$

5–8. (a) $a_{2k+1} = 0, \quad b_{2k} = 0$

$a_{2k} = \dfrac{2}{\pi} \displaystyle\int_{-\pi/2}^{\pi/2} f(\theta) \cos 2k\theta \, d\theta$

$b_{2k+1} = \dfrac{2}{\pi} \displaystyle\int_{-\pi/2}^{\pi/2} f(\theta) \sin (2k + 1)\theta \, d\theta$

(b) $a_{2k} = 0, \quad b_{2k+1} = 0$

$$a_{2k+1} = \frac{2}{\pi} \int_{-\pi/2}^{\pi/2} f(\theta) \cos (2k+1)\theta \, d\theta$$

$$b_{2k} = \frac{2}{\pi} \int_{-\pi/2}^{\pi/2} f(\theta) \sin 2k\theta \, d\theta$$

(c) $a_{4k+2} = 0, \quad b_{4k} = 0$

$$a_{4k+1} = b_{4k+1} = \frac{1}{\pi} \int_{-3\pi/4}^{\pi/4} f(\theta) \, [\cos (4k+1)\theta + \sin (4k+1)\theta] \, d\theta$$

$$a_{4k+3} = -b_{4k+3} = \frac{1}{\pi} \int_{-3\pi/4}^{\pi/4} f(\theta) \, [\cos (4k+3)\theta - \sin (4k+3)\theta] \, d\theta$$

(d) $a_{(n \neq 4k)} = 0, \quad b_{(n \neq 4k)} = 0$

$$a_{4k} = \frac{4}{\pi} \int_0^{\pi/2} f(\theta) \cos 4k\theta \, d\theta$$

$$b_{4k} = \frac{4}{\pi} \int_0^{\pi/2} f(\theta) \sin 4k\theta \, d\theta$$

5-10. (b) $\frac{1}{3}A^2 T; \quad \sum_{k=1}^{\infty} \frac{1}{(2k-1)^4} = \frac{\pi^4}{96}$

5-11. $g(\omega) = \dfrac{A}{\omega^2 T} \, [(1 - \cos \omega T) + j (\sin \omega T - \omega T)]$

5-12. $g(\omega) = \dfrac{4\pi A/T}{(2\pi/T)^2 - \omega^2} \sin \left(\dfrac{\omega T}{2}\right) \epsilon^{j(\pi - \omega T)/2}$

5-13. $g(\omega) = \dfrac{E_m}{4\pi} \left[-\dfrac{\omega}{\omega^2 - (99\omega_a)^2} + \dfrac{\omega}{\omega^2 - (101\omega_a)^2} + j \, \dfrac{400\omega_a}{\omega^2 - (100\omega_a)^2} \right]$

$$\sin \left(\frac{\omega T_a}{2}\right)$$

5-14. $g(\omega) = \dfrac{A}{j\omega} \, (1 - \epsilon^{-j\omega t_0})(1 + \epsilon^{-j2\omega t_0})$

5-15. $g(\omega) = -\dfrac{A}{\omega^2 t_0} \, (1 - \epsilon^{-j\omega t_0})[1 - \epsilon^{-j\omega(T - t_0)}]$

5-17. (a) $\omega_0/[(j\omega + \alpha)^2 + \omega_0^2]$

5-18. $\mathcal{F}[f_2(t)] = g_1(-\omega) \epsilon^{-j\omega t_0}$

5-19. $u_1(t) = \dfrac{10}{\pi} \sum_{n=1}^{\infty} \dfrac{(-1)^{n+1}}{n|Z_n|} \sin (n\pi t - \psi_n)$

$$|Z_n| = \frac{n}{(1 + n^2)^2} \, [(4 - 13n^2 + n^4)^2 + 4n^6]^{1/2}$$

$$\psi_n = \tan^{-1} (4 - 13n^2 + n^4)/2n^3$$

5–20. (a) 40.5 volts, (b) 1280 volts

5–21. $e_o(t) = \dfrac{AK}{\pi} \{Si[\omega_c(t - t_d)] - Si[\omega_c(t - t_d - t_0)]\}$;

$Si(x) = \displaystyle\int_0^x \dfrac{\sin u}{u}\, du$, called the sine integral of x, is a tabulated function.

CHAPTER 6

6–1. (a) $2/(s + m)^3$

(b) $(6 - 2s^2 + s^3)/s^4$

(c) $6\epsilon^{-5s}(1 + 5s + \frac{25}{2}s^2)/s^3$

(d) $\epsilon^{-T(s+\alpha)}[T(s + \alpha) + 1]/(s + \alpha)^2$

(e) $2\epsilon^{-s}(1 + 3s + \frac{9}{2}s^2)/s^3$

(f) $[(s^2 - \omega^2)\cos\theta - 2s\omega \sin\theta]/(s^2 + \omega^2)^2$

(g) $(2\omega \cos 2\omega t_0 - s \sin 2\omega t_0)/(s^2 + 4\omega^2)$

(h) $\epsilon^{-2t_0 s}(2\omega \cos 2\omega t_0 + s \sin 2\omega t_0)/(s^2 + 4\omega^2)$

(i) $3\{[(s + 4)^2 - 4]\cos 2 + 4(s + 4)\sin 2\}/[(s + 4)^2 + 4]^2$

(j) $(s + 5)/(s^2 + 10s + 16)$

6–2. (b) $bF(bs + 1)$

(c) $bF(bs + b^2)$

6–3. (b) (i) $1/(s + 2)$

(ii) $\epsilon^{-(s+2)}/(s + 2)$

(iii) $\epsilon^2/(s + 2)$

(iv) $\epsilon^{-s}/(s + 2)$

6–4. $F(s) = E(1 - \epsilon^{-t_0 s})[1 - \epsilon^{-(T-t_0)s}]/t_0 s^2$

6–5. $F(s) = E(\epsilon^{-t_1 s} + \epsilon^{-t_2 s})(1 - \epsilon^{-Ts})/s$

6–6. $F(s) = \dfrac{E}{4s}\left(\dfrac{1 - \epsilon^{-Ts}}{1 - \epsilon^{-Ts/4}} - 4\epsilon^{-Ts}\right)$

6–7. $F(s) = E(1 - \epsilon^{-s/f_a})$

$$\left[\dfrac{10^4 \pi f_a}{s^2 + (10^4 \pi f_a)^2} + \dfrac{0.4s}{s^2 + (9998\pi f_a)^2} - \dfrac{0.4s}{s^2 + (10002\pi f_a)^2}\right]$$

6–8. $F(s) = \dfrac{2E}{Ts^2}\tanh(Ts/4)$

6–9. $F(s) = \dfrac{E}{s}\tanh(Ts/4)$

6-10. $F(s) = \dfrac{2\pi E\epsilon^{-t_0 s}(1 - \epsilon^{-\tau s})}{\tau[s^2 + (2\pi/\tau)^2](1 - \epsilon^{-Ts})}$

6-11. $F(s) = \dfrac{A}{Ts^2}\left[\dfrac{Ts - 0.4(1 - \epsilon^{-Ts/2})}{1 + \epsilon^{-Ts/2}}\right]$

6-12. $\mathcal{L}\left[\displaystyle\iint \cdots \int f(t)\,(dt)^n\right] = \dfrac{F(s)}{s^n} + \displaystyle\sum_{k=1}^{n} \dfrac{f^{(-k)}(0+)}{s^{n-k+1}}$

6-13. (b) $(s^4 + 18s^2 + 648)/s^2(s^2 + 36)^2$

6-14. (b) $\ln(1 + \alpha/s)$

6-15. (a) $\dfrac{1}{s}\tan^{-1}\left(\dfrac{1}{s}\right)$,   (b) $\dfrac{1}{s}\tan^{-1}\left(\dfrac{a}{s}\right)$

CHAPTER 7

7-1. $f(t) = \sinh t + 2\,(\cosh t - 1)$

7-2. $f(t) = \tfrac{1}{4}[1 - \cos(t - 1)]U(t - 1)$

7-3. $f(t) = \epsilon^{-2(t-1/2)}\left[\cos\sqrt{5}(t - \tfrac{1}{2}) + \dfrac{1}{\sqrt{5}}\sin\sqrt{5}(t - \tfrac{1}{2})\right]U(t - \tfrac{1}{2})$

7-4. $f(t) = \tfrac{5}{4} - \tfrac{1}{4}\left(\cos\sqrt{3}t + \dfrac{9}{\sqrt{3}}\sin\sqrt{3}t\right)\epsilon^{-t}$

7-5. $f(t) = \tfrac{1}{3}(\cos t - \cos 2t)$

7-6. $f(t) = \tfrac{1}{40}[(t^2 + 8t + \tfrac{15}{2}) + (7t - \tfrac{15}{2})\epsilon^{2t}]$

7-7. $f(t) = \tfrac{3}{4}\left[(t - \tfrac{1}{3}) - \dfrac{3}{2\sqrt{2}}\sin\sqrt{2}(t - \tfrac{1}{3}) + \tfrac{1}{2}(t - \tfrac{1}{3})\cos\sqrt{2}(t - \tfrac{1}{3})\right]$
$\times U(t - \tfrac{1}{3})$

7-8. $f(t) = \tfrac{1}{4}[2t - 3(1 - \epsilon^{-2t/3})] - \tfrac{1}{4}\{2(t - 4) - 3[1 - \epsilon^{-2(t-4)/3}]\}$
$\times U(t - 4)$

7-9. $f(t) = (2\cos 3t - 1)/t$

7-10. $v_L = 2.5\epsilon^{-37.5t}$ volts

7-11. (a) $v_c = 10\,(\cos\alpha - \sin\alpha - \tfrac{1}{2})\epsilon^{-10t}$
$+ 10\,[\sin(10t + \alpha) - \cos(10t + \alpha)]$

    (b) $\alpha = 2n\pi$ $(n = 0, 1, 2, \ldots)$

7-12. $i_2 = 0.89\,[(\cosh 123t - 1.32\sinh 123t)\epsilon^{-161t}$
$- \cos 100t + 0.022\sin 100t]$ amp

7-13. (a) $v_o(t) = 40.8(-\epsilon^{-357t} + \cos 10^4 t + 28 \sin 10^4 t)$ volts

(b) $v_o(t) = 1140[\epsilon^{-357t}U(t) - \epsilon^{-357(t-0.001)}U(t - 0.001)]$ volts

7-14. $v_c(t) = \dfrac{E}{1 + 4\pi^2} \{\sin \pi t - 2\pi \cos \pi t + 2\pi \epsilon^{-t/2} - [\sin \pi(t - 4)$

$- 2\pi \cos \pi(t - 4) + 2\pi \epsilon^{-(t-4)/2}]U(t - 4)\} + \tfrac{1}{2}\epsilon^{-t/2}$ volts

7-15. $u_1(t) = \dfrac{1000}{3}\{[\tfrac{3}{50} t - 1 + (\cosh 0.436t + 1.01 \sinh 0.436t)\epsilon^{-t/2}]U(t)$

$- 2\{\tfrac{3}{50}(t - \tfrac{1}{2}) - 1$

$+ [\cosh 0.436(t - \tfrac{1}{2}) + 1.01 \sinh 0.436(t - \tfrac{1}{2})]\epsilon^{-(t/2-1/4)}\}U(t - \tfrac{1}{2})$

$+ \{\tfrac{3}{50}(t - 1) - 1$

$+ [\cosh 0.436(t - 1) + 1.01 \sinh 0.436(t - 1)]\epsilon^{-(t-1)/2}\}U(t - 1)\}$

ft/sec

7-16. $\theta_B(t) = \dfrac{\theta_0}{a} [\cos \sqrt{c - b^2/4}\,t + (b/2\sqrt{c - b^2/4}) \sin \sqrt{c - b^2/4}\,t]\epsilon^{-bt/2}$

where $a = r_B/r_A$, $b = (D_{\theta B} + a^2 D_{\theta A})/(I_{\theta B} + a^2 I_{\theta A})$ and

$c = 1/K_{\theta B}(I_{\theta B} + a^2 I_{\theta A})$

7-17. $y = -(x + \sinh x)$, $z = -\cosh x$

7-18. $v = \tfrac{1}{2}(\epsilon^x - \cos x + 5 \sin x) - x$, $w = \tfrac{1}{2}(\epsilon^x - 5 \cos x - \sin x) + 1$

7-19. Initial values: 1, 0,0.

Final values: $\tfrac{5}{4}$; Final-Value Theorem cannot be applied to the function in Problems 7-5 and 7-6 because $sF(s)$ becomes infinite (is not analytic) at $s = \pm j1$ and at $s = \pm j2$ for the function in Problem 7-5, and at $s = 0$ and $s = +2$ for the function in Problem 7-6.

7-20. $w = -2x$

7-21. $[v_R(t)]_{tr} = E\epsilon^{-Rt/L}\left[\dfrac{1 - \epsilon^{Rt/L}}{\epsilon^{RT/L} - 1}\right]$

$[v_R(t)]_{ss} = v_1(t) - \dfrac{E}{\epsilon^{RT/L} - 1} \{\epsilon^{-R(t-nT)/L} - \epsilon^{-R[t-\tau-(n-1)T]/L}\},$

$(n - 1)T \leq t \leq (n - 1)T + \tau$

$[v_R(t)]_{ss} = v_1(t) - \dfrac{E}{\epsilon^{RT/L} - 1} \{\epsilon^{-R(t-nT)/L} - \epsilon^{-R(t-\tau-nT)/L}\},$

$(n - 1)T + \tau \leq t \leq nT.$

7-22. $v_c(t) = \tfrac{5}{4}(3.87\epsilon^{-2t} - 20.1\epsilon^{-10t})$, $0.3 < t < 0.4$ sec

7-23. $[v_o(t)]_{ss} = \dfrac{250\pi E}{T[300^2 + (\pi/T)^2]}$

$\times \left\{\dfrac{2\epsilon^{-300t}}{1 - \epsilon^{-300T}} - \cos\left(\dfrac{\pi}{T} t\right) + \dfrac{300T}{\pi}\sin\left(\dfrac{\pi}{T} t\right)\right\}, \quad 0 \leq t \leq T.$

CHAPTER 8

8-1. (a) $i_1 = \dfrac{V}{R_1}\left[1 - \dfrac{L_2}{L_1 + L_2}\,\epsilon^{-R_1 t/(L_1+L_2)}\right], \quad t > 0.$

(b) $v_{L_1} = \dfrac{L_1 L_2 V}{(L_1 + L_2)R_1}\left[-\delta(t) + \dfrac{R_1}{L_1 + L_2}\,\epsilon^{-R_1 t/(L_1+L_2)}\,U(t)\right].$

8-2. $v_o(t) = E\{[1 - \tfrac{2}{3}\epsilon^{-t/3C_1 R}]U(t) - [1 - \tfrac{2}{3}\epsilon^{-(t-T)/3C_1 R}]U(t - T)\}$

8-3. (a) $t^2/2,$ (b) $\delta(t) + (t - 2)\epsilon^{-t},$ (c) $\tfrac{1}{4}(1 - \cos 2t + 2 \sin 2t),$

(d) $(\cos \omega_1 t - \cos \omega_2 t)/(\omega_2^2 - \omega_1^2),$

(e) $tU(t) - 2(t - a)U(t - a) + (t - 2a)U(t - 2a).$

8-4. $f(t) = -\tfrac{1}{2}\epsilon^{-t}(1 - 4\epsilon^{-t} + 3\epsilon^{-2t}).$

8-5. $f(x) = \tfrac{1}{2}\cos\sqrt{3/2}\,\beta x.$

8-6. (a) $h(t) = \tfrac{1}{4}[\delta(t) - \tfrac{3}{8}\epsilon^{-3t/8}]$ (b) $i_2(t) = \dfrac{E}{116}\,\epsilon^{-4t}(32 - 3\epsilon^{29t/8})$

(c) $w_u(t) = \tfrac{1}{4}\epsilon^{-3t/8}$

8-7. (a) $w_u(t) = \tfrac{1}{2}\epsilon^{-20t}$

(b) $v_L(t) = \dfrac{E}{40T}\{[(1 + 20T)\epsilon^{-20t} - 1]U(t) + [1 - \epsilon^{-20(t-T)}]U(t - T)\}$

(c) $h(t) = \tfrac{1}{2}[\delta(t) - 20\epsilon^{-20t}]$

8-8. (a) $w_u(t) = \epsilon^{-D_2 t/M}$

$f_M(t) = \dfrac{F\omega D_2/M}{(D_2/M)^2 + \omega^2}$

$\times\left\{\left[\cos\omega(t - t_1) + \dfrac{\omega M}{D_2}\sin\omega(t - t_1) - \epsilon^{-D_2(t-t_1)/M}\right]\right.$

$\times U(t - t_1)$

$-\left[\cos\omega(t - t_1) + \dfrac{\omega M}{D_2}\sin\omega(t - t_1) + \epsilon^{-D_2(t-t_1-\pi/\omega)/M}\right]$

$\left.\times U\left(t - t_1 - \dfrac{\pi}{\omega}\right)\right\}$

(b) $K = 1/\omega^2 M$

8-10. $h_o(t) = \tfrac{1}{4}\left[\delta(t) + \tfrac{2}{3}\left(1 + \dfrac{t}{6}\right)\epsilon^{-t/3}U(t)\right]$

8-11. $w_{uo}(t) = 4(1 - 2t)\epsilon^{-2t}U(t)$

8-12. (a) $\dfrac{4}{3\sqrt{\pi}}\,t^{3/2}$ (b) $\tfrac{1}{2}[\epsilon^{-(t-2)}\,\text{erf}\,2\sqrt{t - 2}]U(t - 2)$

(c) $\epsilon^{-t/2} I_0(t/2)$

(d) $\epsilon^{at}(1 - \text{erf}\sqrt{at}) = \epsilon^{at} \text{erfc}\sqrt{at}$

(e) $\dfrac{1}{\sqrt{\pi t}} - \epsilon^t \text{erfc}\sqrt{t}$

(f) $\epsilon^{-2t} J_0(3t)$

(g) $-\dfrac{1}{a} \dfrac{dJ_0(at)}{dt} = J_1(at)$

(h) $\dfrac{a}{2} \epsilon^{-at/2}[I_1(at/2) - I_0(at/2)]$

CHAPTER 9

9–1.

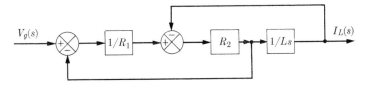

9–2.

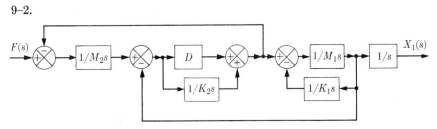

9–3.

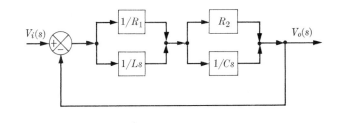

9–4.

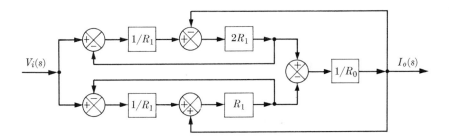

9–6.

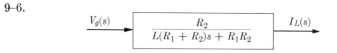

$$V_g(s) \longrightarrow \boxed{\dfrac{R_2}{L(R_1 + R_2)s + R_1 R_2}} \longrightarrow I_L(s)$$

9–7.

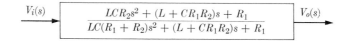

$$F(s) \longrightarrow \boxed{\dfrac{K_1(K_2 D s + 1)}{M_1 M_2 K_1 K_2 s^4 + K_1 K_2 D(M_1 + M_2)s^3 + (M_1 K_1 + M_2 K_1 + M_2 K_2)s^2 + K_2 D s + 1}} \longrightarrow X_1(s)$$

9–8.

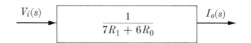

$$V_i(s) \longrightarrow \boxed{\dfrac{LCR_2 s^2 + (L + CR_1 R_2)s + R_1}{LC(R_1 + R_2)s^2 + (L + CR_1 R_2)s + R_1}} \longrightarrow V_o(s)$$

9–9.

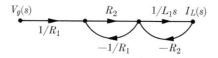

$$V_i(s) \longrightarrow \boxed{\dfrac{1}{7R_1 + 6R_0}} \longrightarrow I_o(s)$$

9–10.

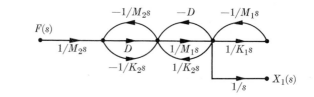

9–11.

INDEX

INDEX

Active elements, 43, 99
Algebraic equations, 383
  complex roots of, 23
  general properties of, 281, 384
  numerical solution of, 383
Algebraic function
  irrational, 247
  rational, 185
Analogous systems, 97, 267
  for transmission line problems, 349
Analogy
  direct, 103
  force-current, 106
  force-voltage, 103
  impedance-type, 103
  inverse, 107
  mobility-type, 107
Analysis
  by Fourier methods, 146 ff.
  by Laplace transformation, 196 ff.
  by Z transformation, 320 ff.
  loop, 52
  node, 58
Antiderivative, 46
Arbitrary constants, 7, 9
  determination of, 11, 68
Auxiliary equation, *see* Characteristic
  equation

Bandwidth, 81, 145
Barker, R. H., 336
Barnes, J. L., 285
Beranek, L. L., 111
Bernoulli's equation, 41
Bessel's function of the first kind, 252
  modified, 253, 370
Biot, M. A., 313
Block-diagram
  representation, 258
  transformations, 262, 264
Borel's theorem, 233, 323
Boundary conditions, 355

Boxcar generator, 344
Branch, 53
  directed, 268
Branch transmittance, 268

Capacitance, 47
  with initial charge, equivalent circuit for, 87
Carslaw, H. S., 122
Cauchy's differential equation, 41
Characteristic equation, 21, 25
  with distinct roots, 24
  with multiple roots, 33
  with zero roots, 30
Characteristic function, 183, 185
Characteristic impedance, 373
  of distortionless line, 364
  of general transmission line, 369
  of leakage-free noninductive cable, 368
  of lossless line, 363
Churchill, R. V., 353, 365
Clamp circuit, 344
Coefficient of coupling, 51
Complementary error function, 367
Complementary function, 8, 67, 185, 207, 313
  for complex conjugate roots, 24
  for distinct roots of characteristic equation, 24, 25
  for multiple (repeated) roots, 33
Compliance of spring
  for rotational motion, 100
  for translational motion, 99
Conditions
  boundary, 355
  Dirichlet, 122, 143
  initial, 9, 49, 355
Conductance, 64, 108
Conduction of heat through uniform slab, 354
Conductivity, 354

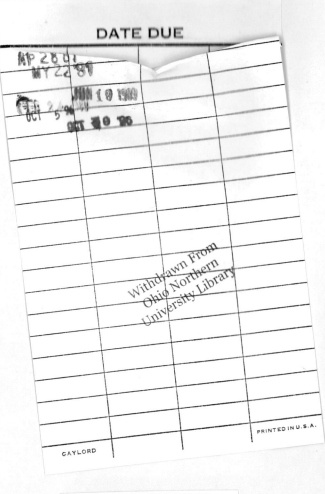